Geo-Information Technology and Its Applications

Geo-Information Technology and Its Applications

Editors

Weicheng Wu
Yalan Liu
Mingxing Hu

MDPI • Basel • Beijing • Wuhan • Barcelona • Belgrade • Manchester • Tokyo • Cluj • Tianjin

Editors
Weicheng Wu
East China University of Technology
China

Yalan Liu
Chinese Academy of Sciences
China

Mingxing Hu
Si Pailou Campus of Southeast University
China

Editorial Office
MDPI
St. Alban-Anlage 66
4052 Basel, Switzerland

This is a reprint of articles from the Special Issue published online in the open access journal *ISPRS International Journal of Geo-Information* (ISSN 2220-9964) (available at: https://www.mdpi.com/journal/ijgi/special_issues/%28ICGITA2019%29).

For citation purposes, cite each article independently as indicated on the article page online and as indicated below:

LastName, A.A.; LastName, B.B.; LastName, C.C. Article Title. *Journal Name* **Year**, *Volume Number*, Page Range.

ISBN 978-3-0365-5995-7 (Hbk)
ISBN 978-3-0365-5996-4 (PDF)

Contents

About the Editors . . . . **vii**

Preface to "Geo-Information Technology and Its Applications" . . . . **ix**

Weicheng Wu, Yalan Liu and Mingxing Hu
Editorial on Special Issue "Geo-Information Technology and Its Applications"
Reprinted from: *ISPRS Int. J. Geo-Inf.* **2022**, *11*, 347, doi:10.3390/ijgi11060347 . . . . **1**

Bing Han, Mingxing Hu and Jialing Wang
Site Selection for Pre-Hospital Emergency Stations Based on the Actual Spatiotemporal Demand: A Case Study of Nanjing City, China
Reprinted from: *ISPRS Int. J. Geo-Inf.* **2020**, *9*, 559, doi:10.3390/ijgi9100559 . . . . **7**

Bing Han, Mingxing Hu, Jiemin Zheng and Tan Tang
Site Selection of Fire Stations in Large Cities Based on Actual Spatiotemporal Demands: A Case Study of Nanjing City
Reprinted from: *ISPRS Int. J. Geo-Inf.* **2021**, *10*, 542, doi:10.3390/ijgi10080542 . . . . **27**

Zilai Zheng, Takehiro Morimoto and Yuji Murayama
A GIS-Based Bivariate Logistic Regression Model for the Site-Suitability Analysis of Parcel-Pickup Lockers: A Case Study of Guangzhou, China
Reprinted from: *ISPRS Int. J. Geo-Inf.* **2021**, *10*, 648, doi:10.3390/ijgi10100648 . . . . **47**

Bo Liu, Yu Shi, Da-Jun Li, Yan-Dong Wang, Gabriela Fernandez and Ming-Hsiang Tsou
An Economic Development Evaluation Based on the OpenStreetMap Road Network Density: The Case Study of 85 Cities in China
Reprinted from: *ISPRS Int. J. Geo-Inf.* **2020**, *9*, 517, doi:10.3390/ijgi9090517 . . . . **69**

Liujia Chen, Xiaojing Yao, Yalan Liu, Yujiao Zhu, Wei Chen, Xizhi Zhao and Tianhe Chi
Measuring Impacts of Urban Environmental Elements on Housing Prices Based on Multisource Data—A Case Study of Shanghai, China
Reprinted from: *ISPRS Int. J. Geo-Inf.* **2020**, *9*, 106, doi:10.3390/ijgi9020106 . . . . **95**

Xuan Zhao, Jianjun Liu, Hongke Hao and Yanzheng Yang
Quantifying the Spatial Heterogeneity and Driving Factors of Aboveground Forest Biomass in the Urban Area of Xi'an, China
Reprinted from: *ISPRS Int. J. Geo-Inf.* **2020**, *9*, 744, doi:10.3390/ijgi9120744 . . . . **119**

Rui Alexandre Castanho, José Manuel Naranjo Gomez, Ana Vulevic and Gualter Couto
The Land-Use Change Dynamics Based on the CORINE Data in the Period 1990–2018 in the European Archipelagos of the Macaronesia Region: Azores, Canary Islands, and Madeira
Reprinted from: *ISPRS Int. J. Geo-Inf.* **2021**, *10*, 342, doi:10.3390/ijgi10050342 . . . . **135**

Mingyue Wang, Jun'e Fu, Zhitao Wu and Zhiguo Pang
Spatiotemporal Variation of NDVI in the Vegetation Growing Season in the Source Region of the Yellow River, China
Reprinted from: *ISPRS Int. J. Geo-Inf.* **2020**, *9*, 282, doi:10.3390/ijgi9040282 . . . . **177**

Jian Yin, Yuanhong Qiu and Bin Zhang
Identification of Poverty Areas by Remote Sensing and Machine Learning: A Case Study in Guizhou, Southwest China
Reprinted from: *ISPRS Int. J. Geo-Inf.* **2021**, *10*, 11, doi:10.3390/ijgi10010011 . . . . **195**

Yang Zhang, Weicheng Wu, Yaozu Qin, Ziyu Lin, Guiliang Zhang, Renxiang Chen, Yong Song, Tao Lang, Xiaoting Zhou, Wenchao Huangfu, Penghui Ou, Lifeng Xie, Xiaolan Huang, Shanling Peng and Chongjian Shao
Mapping Landslide Hazard Risk Using Random Forest Algorithm in Guixi, Jiangxi, China
Reprinted from: *ISPRS Int. J. Geo-Inf.* **2020**, *9*, 695, doi:10.3390/ijgi9110695 . . . . **215**

Ting Chen, Haiqing He, Dajun Li, Puyang An and Zhenyang Hui
Damage Signature Generation of Revetment Surface along Urban Rivers Using UAV-Based Mapping
Reprinted from: *ISPRS Int. J. Geo-Inf.* **2020**, *9*, 283, doi:10.3390/ijgi9040283 . . . . **231**

Bo Liu, Xuechao Liu, Dajun Li, Yu Shi, Gabriela Fernandez and Yandong Wang
A Vector Line Simplification Algorithm Based on the Douglas–Peucker Algorithm, Monotonic Chains and Dichotomy
Reprinted from: *ISPRS Int. J. Geo-Inf.* **2020**, *9*, 251, doi:10.3390/ijgi9040251 . . . . **253**

Hongxiang Guo, Guojin He, Wei Jiang, Ranyu Yin, Lei Yan and Wanchun Leng
A Multi-Scale Water Extraction Convolutional Neural Network (MWEN) Method for GaoFen-1 Remote Sensing Images
Reprinted from: *ISPRS Int. J. Geo-Inf.* **2020**, *9*, 189, doi:10.3390/ijgi9040189 . . . . **267**

Keqing Li, Changyong Liang, Wenxing Lu, Chu Li, Shuping Zhao and Binyou Wang
Forecasting of Short-Term Daily Tourist Flow Based on Seasonal Clustering Method and PSO-LSSVM
Reprinted from: *ISPRS Int. J. Geo-Inf.* **2020**, *9*, 676, doi:10.3390/ijgi9110676 . . . . **285**

About the Editors

Weicheng Wu

Weicheng Wu, with a PhD in geography from the University of Paris I (Pantheon-Sorbonne), is a full-time Professor and Team Leader at the Key Lab of Digital Land and Resources, and a PhD supervisor at the Faculty of Earth Sciences of the East China University of Technology (ECUT). Dr Wu was formerly a remote sensing specialist with the University of Sassari in Italy and the International Center for Agricultural Research in the Dry Areas (ICARDA). Before beginning his present work, he had led and/or participated in more than 20 international cooperation projects focused in Central and Western Asia and Northern Africa (CWANA) and European Mediterranean regions, funded by different international consortia such as ESA, EU, IFAD, USAID, AusAID and CGIAR, etc. Since joining ECUT in 2018, he has been managing several projects on the assessment of the sustainable utilization of land resources and of the impacts of natural disasters on food security and society with geo-information technology and artificial intelligence techniques in a project supported by ECUT and the Jiangxi Government. He was the Chair of the International Conference on Geo-information Technology and its Applications (ICGITA 2019) and is Associate Editor of the *International Journal of Remote Sensing* and Guest Editor of *Remote Sensing* and *ISPRS International Journal of Geo-Information*. His research interests include analysis of human–nature interactions, carbon sequestration/emission, natural disaster risk zoning and food security. He has published more than 100 scientific papers, and in 2018, Dr Wu was granted the title "Ten Leading Chinese Talents on Science and Technology in Europe" and selected by the "Talent Program of Jiangxi Government".

Yalan Liu

Yalan Liu earned PhD in cartography and remote sensing from the University of the Chinese Academy of Sciences in 2004. She is a full-time Professor at the Aerospace Information Research Institute of the Chinese Academy of Sciences (AIR/CAS), and a part-time Professor in the Regional Centre for Space Science and Technology Education in Asia and the Pacific (China) (Affiliated to the United Nations). She is also the Director of the Lab of Spatial Information Integration Technology at AIR/CAS, and has specialized in remote sensing and geographic information systems (GIS) including technology, theory and applications since 1996. Her research interests include intelligent information extraction, integration of spatial information and its application to environment monitoring, disaster mitigation, smart cities and decision support for sustainable development. She has been leading three projects funded by the National Natural Science Foundation, the National Key Research and Development Program, and the China High-resolution Earth Observation System Program, and has participated in two projects supported by the National Science and Technology Support Program and two international cooperation projects. She has served as a Guest Editor and reviewer of several international journals such as *ISPRS International Journal of Geo-Information* and *Remote Sensing*, and has published more than 100 scientific papers. Dr Liu received several prizes of Scientific and Technological Progress Awards (STPA), including one First Class Prize of STPA of the Chinese Academy of Sciences, one Third Class Prize of STPA of the Beijing Government, and one Second Class Prize of the State STPA of China.

Mingxing Hu

Mingxing Hu received his PhD in remote sensing from the China University of Mining and Technology, and is full-time Professor and PhD supervisor in the Department of Urban Planning,

School of Architecture, Southeast University of China. Dr Hu has been engaged in research work, mainly focuses on the application of spatial information technology in urban planning and management since 2001. During his academic career, Dr Hu has presided over one and participated in four national projects funded by the National Natural Science Foundation of China. In the meantime, Dr Hu has published over 60 academic papers, 3 monographs, and received 6 national and provincial awards including one Second Class Prize of The State Scientific and Technological Progress Award (SSTPA). Dr Hu has pioneered the application of the GIS technique in the protection planning of historical and cultural cities and districts in China, and established a new method for current situation survey and planning formulation to improve the scientificity and technicality of protection planning. He served as Guest Editor of the *ISPRS International Journal of Geo-Information*.

Preface to "Geo-Information Technology and Its Applications"

This Special Issue (SI) aims to demonstrate the state of the art in development and application of the geo-information technology in various fields. In particular, we focus on smart city and urban planning, land resource investigation and management, geo-disaster risk assessment and the optimal touristic flow management. This work was motivated by the high-quality presentations in the International Conference of Geo-information Technology and its Applications (ICGITA 2019), and we expect to present the updated outcomes to the international geo-informatic community. This book will have utility as a reference for graduates and scientists in environmental science, urban planning, geoscience and big data mining. The guest editors are grateful to the authors for their contributions of high quality research, and, specifically, to the MDPI team for their effective assistance and cooperation, without whom none of this would have been possible.

Weicheng Wu, Yalan Liu, and Mingxing Hu
Editors

International Journal of *Geo-Information*

MDPI

Editorial

Editorial on Special Issue "Geo-Information Technology and Its Applications"

Weicheng Wu [1,*], Yalan Liu [2] and Mingxing Hu [3]

1 Key Laboratory of Digital Land, Resources and Faculty of Earth Sciences, East China University of Technology, Nanchang 330013, China
2 Aerospace Information Research Institute, Chinese Academy of Sciences, Beijing 100101, China; liuyl@aircas.ac.cn
3 School of Architecture, Southeast University, 2 Sipailou, Nanjing 210096, China; 101009930@seu.edu.cn
* Correspondence: wuwch@ecut.edu.cn

Abstract: Geo-information technology plays a critical role in urban planning and management, land resource quantification, natural disaster risk and damage assessment, smart city development, land cover change modeling and touristic flow management. In particular, the development of big data mining and machine learning techniques (including deep learning) in recent years has expanded the potential applications of geo-information technology and promoted innovation in approaches to mining in different fields. In this context, the International Conference on Geo-Information Technology and its Applications (ICGITA 2019) was held in Nanchang, Jiangxi, China, 11–13 October 2019, co-organized by the Key Laboratory of Digital Land and Resources, East China University of Technology, the Institute of Remote Sensing and Digital Earth (RADI) of the Chinese Academy of Sciences (CAS), which was renamed in 2017 the Aerospace Information Research Institute (AIR), CAS, and the Institute of Space and Earth Information Science of the Chinese University of Hong Kong. The outstanding papers presented at this event and some other original articles were collected and published in this Special Issue "Geo-Information Technology and Its Applications" in the *International Journal of Geo-Information*. This Special Issue consists of 14 high-quality and innovative articles that explore and discuss the typical applications of geo-information technology in the above-mentioned domains.

Citation: Wu, W.; Liu, Y.; Hu, M. Editorial on Special Issue "Geo-Information Technology and Its Applications". *ISPRS Int. J. Geo-Inf.* **2022**, *11*, 347. https://doi.org/10.3390/ijgi11060347

Received: 1 June 2022
Accepted: 6 June 2022
Published: 13 June 2022

1. Urban Planning and Infrastructure Optimization

In the field of urban planning and optimal service management, applications of geo-information technology are manifested in aiding analysis and decision-making, information collection and database building, simulation, and forecasting to promote scientific planning. Six articles on GIS-based spatiotemporal big data and related modeling to assist urban infrastructure site selection, assess urban economic development and housing price, and characterize green spaces are published in this section. The first one is the research by Zheng et al. [1] on the boundary of the suitable area based on point of interest (POI) data to obtain the location of parcel-pickup lockers (PPLs). Their research includes construction of a bivariate logistic regression (LR) model to solve the suitability classification problem and training the dataset to filter the critical factors affecting the site selection. Testing results showed that the best LR model had excellent performance, and an optimal solution was obtained for Guangzhou (GZ), China. The findings of the study can assist planning managers in using the suitable areas as the site-selection ranges for PPLs, reducing the difficulties and time costs of analysis.

Spatiotemporal big data can provide new technical methods and data sources for the planning and layout of urban emergency service facilities. For emergency and fire services, which exhibit random occurrence and extremely time-consuming requirements, the theories and methods for studying the spatial and temporal characteristics of the event occurrence and the scientific layout based on spatiotemporal big data have become emerging research fields in recent years. Taking Nanjing as an example, the second and third articles by Han et al. [2,3] used multi-source big data, including ambulance-carried GPS data, Amap-recorded traffic congestion data, and survey data of emergency rescue facilities to study the layout of pre-hospital emergency points and fire stations. In the case of siting emergency stations, the Location Set Covering model was chosen to integrate first aid demands with traffic states, reducing the negative impacts of the random occurrence of demand in space and traffic delays on the planning of pre-hospital emergency stations, and also improving the accuracy of the emergency location model. Doing so also improves the accuracy of the Emergency Medical Services (EMS) siting model, satisfying both the planning conditions and the actual traffic constraints by randomly simulating how the EMS demand can be reached within the target time. Various required factors are determined based on modeling and analysis by processing and analyzing the current data, agreeing on a target time. Calculation of the in-transit time from a large number of randomly distributed EMS simulation points to a facility point may shed light on the model conditions and find solutions for the locations. The frequent occurrence of fires has brought new challenges to urban fire safety, and the spatial layout of fire stations is crucial to firefighting security. In the siting of fire stations [3], multi-source big data, including the full data of the fire outbreak history in the past five years, were collected for a comprehensive analysis. Based on a set of preprocessing tasks—e.g., analyzing the regularity of fire occurrence, selecting the factors related to fire risks, assigning weights to the indicators using the entropy weighting approach, and assessing the risk sore for each single grid—the spatial distribution probability at points in each cluster was calculated according to the clustering analysis, and the random fire outbreak points were generated via Monte Carlo simulation. The travel time from massive randomly distributed simulated fire points to the candidate facility was calculated to obtain the site location. This type of approach incorporates a spatiotemporal big data perspective and provides ideas for improving the siting model and efficiency of emergency site planning.

The fourth paper, by Liu et al. (2020a) [4], took advantage of the OpenStreetMap (OSM) data to evaluate the economic development of cities in China, and the authors found that there is a significant correlation between the OSM road network density and the municipal gross domestic product (GDP). Hence, OSM road network density can be used as a spatial metric to evaluate and forecast the urban economic development in China and possibly also in other countries.

The role of spatiotemporal big data in urban planning is also reflected in the provision of information and services for policy decisions and the public. To assess the impact of urban environmental elements on housing prices, Chen, L. et al. [5] obtained street-view data and high-resolution remote sensing data of Shanghai and calculated the green view index, sky view index, urban green coverage rate, etc. The study also extracted house price data and used the Shapley Additive Explanation (SHAP) method to explain the impact of housing prices. The results show significant differences in the effects of urban greenery coverage and greenery view index on house prices as home buyers in Shanghai are only willing to pay the premium for greenery view houses when the greenery view index or urban greenery coverage is high. The sky visibility index has a negative effect on house prices, probably due to the fact that high-density and high-rise residential areas tend to have better amenities, and residents are more willing to pay the premium for houses in neighborhoods with more water coverage. This study provides a modeling tool to reveal the decisions by homebuyers and property developers and to provide policy support for urban land sales, property development and urban environmental improvements [5].

To provide useful insights for configuring urban greenspace, the sixth paper, by Zhao et al. [6], employed the geographic detector to investigate the spatial distribution of urban forest biomass and the impacts of four potential geographical factors (GFs) on the aboveground biomass distribution of urban forests in 1480 plots in Xi'an, China. The results indicate that the aboveground biomass and four GFs show obvious heterogeneity regarding their spatial distribution, and the interactions among these four GFs also tend to contribute to the distribution pattern of aboveground biomass. Their research reveals that the approach of using geographical detector is a useful tool in the urban area and is able to demonstrate the driving pattern of aboveground biomass and provide a reference for urban planning and management.

2. Land Cover Change Analysis

The application of geo-information technology is able to effectively address the challenges of monitoring land use change dynamics on a large scale. Islands, as peripheral and ultra-peripheral areas, are often highlighted as areas that are ecologically sensitive to human activities as the latter may provoke biodiversity and habitat loss. Hence, understanding land use dynamics and trends in island areas and super-peripheries is essential to maintaining the regional sustainability. Castanho et al. [7] analyzed the trends and dynamics of land use change in the European Archipelagos of the Macaronesia Region (EAR) from 1990 to 2018 based on CORINE data. The study found significant changes in landscape and the need for measures to be taken to mitigate negative environmental impacts. Another article by Wang et al. [8] in this special section, also based on remote sensing data, observed the spatiotemporal variation in vegetation in the source region of the Yellow River (SRYR) during the vegetation growing season. The paper investigates the changes in SRYR using the normalized vegetation index (NDVI) and its response to climate change during the growing seasons of 1998–2016 in combination with climate data based on trend analysis, the Mann–Kendall trend test, and partial correlation analysis. Finally, an NDVI–climate mathematical model was developed to project NDVI trends from 2020 to 2038, reflecting long-term vegetation trends through the past and future variations in vegetation.

3. Poverty Assessment

The extraction of geographical features from remote sensing data is a novel breakthrough in research focusing on social issues. As an objective social phenomenon, poverty has always accompanied the changes in human society and is a long-term problem that hinders the development of human civilization. Based on the Random Forest machine learning method, Yin et al. [9] extracted 23 spatial features based on nighttime lights and geographical data, and they conducted a poverty assessment for Guizhou Province, China, for the period 2012–2019. The authors found that when poor counties are in close proximity to non-poor countries, it makes it easier to eradicate their poverty. This conclusion provides a reference for the identification of poor counties using remote sensing imagery and for research on the potential management of poverty eradication.

4. Geohazard Prediction and Mapping

Geo-information technology is indispensable for georisk analysis and assessment of natural hazards. Geodisaster risk prediction and early warning may serve for decision-making on risk prevention and is the basis for urban safety and protection of human well-being. The paper by Zhang et al. [10] took Guixi County in eastern Jiangxi Province as an example and conducted a landslide risk prediction and zoning study. A comprehensive dataset of 21 geo-information layers, including lithology, faults, rainfall, altitude, slope, distances to faults, roads and rivers, and thickness of the weathering crust, was prepared after assigning weights to the geo-environmental factors based on their landslide propensity. Landslide locations and non-risk zones (mostly flat areas) were vectorized into polygons and randomly divided into two groups to create a training set (70%) and a validation set (30%). With this training set, landslide risk modeling and prediction assessment were achieved using the Random Forest (RF) algorithm, and the validation showed a high reliability of the risk prediction (91.23%).

5. Unmanned Aerial Vehicle (UAV) Application

In the domain of environmental monitoring, the UAV market is expanding at an extremely high rate, and various new UAV technologies are emerging. The gradual maturation of geo-information technology and faster data acquisition will enable UAV applications in environmental monitoring and mapping with higher timeliness. Chen, T. et al. [11] proposed a low-cost UAV photogrammetry point cloud method that can effectively recognize the signatures of revetment damage. In order to recover the finely detailed surface of a revetment in a quick and accurate manner, an object-based dense matching approach was used to generate pixel-by-pixel dense point clouds for characterizing the signatures of revetment damage. Extraction of the damaged areas with different sizes in the slope intensity image of the revetment was effectuated through damage identification using a self-adaptive and multiscale gradient operator in the slope intensity image of the revetment. A revetment with slope protection along urban rivers was selected to evaluate the performance of damage recognition. The results demonstrated that the method could not only restore the fine features of the embankment surface but also significantly improve the accuracy of the damage recognition.

6. Algorithm Improvement

Studies have also explored algorithmic improvements such as the paper by Liu et al. (2020b) [12], who addressed the problem that the Douglas–Peucker (D–P) algorithm is prone to self-intersection and other errors when compressing more complex curves, and this hinders its application in data compression. A new vector line simplification algorithm based on the D–P algorithm, monotonic chains, and dichotomy was hence proposed. The method first used the D–P algorithm to compress complex curves and divided these curves into a number of monotonic chains; second, it applied the dichotomy approach to quickly and precisely locate and process the intersecting monotonic chains so as to solve the self-intersection problem. The improved D–P algorithm was experimentally verified, showing better results in terms of algorithmic efficiency, compression rate, and algorithmic accuracy when dealing with self-intersection problems in vector data compression. Guo et al. [13] proposed a new convolutional neural network for speed segmentation, i.e., the multi-scale water body extraction network (MWEN), and for the automatic extraction of water bodies from the Gaofen-1 satellite images. Three convolutional neural networks, including the fully convolutional network (FCN), Unet, and Deeplab V3+, for semantic segmentation were used to compare the performance of MWEN for water body extraction and visual comparison and five evaluation metrics were harnessed for this purpose. The results illustrate that the improved algorithm produced better extraction than others and has great potential for mapping different types of water bodies with reduced noise.

7. Touristic Management

Li et al. [14] presented a tourist flow forecasting method based on seasonal clustering. The K-means algorithm and the particle swarm optimization least squares support vector machine (PSO-LSSVM) algorithm, which took seasonal variations into account, were used to forecast the tourist flow in the scenic area. The LSSVM was also applied to compare the performance of the proposed model with that of the existing ones. Experiments based on a dataset comprising the daily tourist data for the Huangshan Mountains during the period between 2014 and 2017 were effectuated. Results showed that seasonal clustering is an effective approach to improve tourist flow prediction and management.

Author Contributions: Conceptualization, Weicheng Wu, Mingxing Hu and Yalan Liu; writing—original draft preparation, Mingxing Hu, Yalan Liu; writing—review and editing, Weicheng Wu. All authors have read and agreed to the published version of the manuscript.

Funding: This research received no external funding.

Institutional Review Board Statement: Not applicable.

Informed Consent Statement: Not applicable.

Data Availability Statement: Not applicable.

Acknowledgments: The guest editors would like to thank the authors who contributed to this Special Issue and to the reviewers who dedicated their time for providing the authors with valuable and constructive recommendations.

Conflicts of Interest: The authors declare no conflict of interest.

References

1. Zheng, Z.; Morimoto, T.; Murayama, Y. A GIS-Based Bivariate Logistic Regression Model for the Site-Suitability Analysis of Parcel-Pickup Lockers: A Case Study of Guangzhou, China. *ISPRS Int. J. Geo-Inf.* **2021**, *10*, 648. [CrossRef]
2. Han, B.; Hu, M.; Wang, J. Site Selection for Pre-Hospital Emergency Stations Based on the Actual Spatiotemporal Demand: A Case Study of Nanjing City, China. *ISPRS Int. J. Geo-Inf.* **2020**, *9*, 559. [CrossRef]
3. Han, B.; Hu, M.; Zheng, J.; Tang, T. Site Selection of Fire Stations in Large Cities Based on Actual Spatiotemporal Demands: A Case Study of Nanjing City. *ISPRS Int. J. Geo-Inf.* **2021**, *10*, 542. [CrossRef]
4. Liu, B.; Shi, Y.; Li, D.-J.; Wang, Y.-D.; Fernandez, G.; Tsou, M.-H. An Economic Development Evaluation Based on the OpenStreetMap Road Network Density: The Case Study of 85 Cities in China. *ISPRS Int. J. Geo-Inf.* **2020**, *9*, 517. [CrossRef]
5. Chen, L.; Yao, X.; Liu, Y.; Zhu, Y.; Chen, W.; Zhao, X.; Chi, T. Measuring Impacts of Urban Environmental Elements on Housing Prices Based on Multisource Data—A Case Study of Shanghai, China. *ISPRS Int. J. Geo-Inf.* **2020**, *9*, 106. [CrossRef]
6. Zhao, X.; Liu, J.; Hao, H.; Yang, Y. Quantifying the Spatial Heterogeneity and Driving Factors of Aboveground Forest Biomass in the Urban Area of Xi'an, China. *ISPRS Int. J. Geo-Inf.* **2020**, *9*, 744. [CrossRef]
7. Castanho, R.A.; Naranjo Gomez, J.M.; Vulevic, A.; Couto, G. The Land-Use Change Dynamics Based on the CORINE Data in the Period 1990–2018 in the European Archipelagos of the Macaronesia Region: Azores, Canary Islands, and Madeira. *ISPRS Int. J. Geo-Inf.* **2021**, *10*, 342. [CrossRef]
8. Wang, M.; Fu, J.; Wu, Z.; Pang, Z. Spatiotemporal Variation of NDVI in the Vegetation Growing Season in the Source Region of the Yellow River, China. *ISPRS Int. J. Geo-Inf.* **2020**, *9*, 282. [CrossRef]
9. Yin, J.; Qiu, Y.; Zhang, B. Identification of Poverty Areas by Remote Sensing and Machine Learning: A Case Study in Guizhou, Southwest China. *ISPRS Int. J. Geo-Inf.* **2021**, *10*, 11. [CrossRef]
10. Zhang, Y.; Wu, W.; Qin, Y.; Lin, Z.; Zhang, G.; Chen, R.; Song, Y.; Lang, T.; Zhou, X.; Huangfu, W.; et al. Mapping Landslide Hazard Risk Using Random Forest Algorithm in Guixi, Jiangxi, China. *ISPRS Int. J. Geo-Inf.* **2020**, *9*, 695. [CrossRef]
11. Chen, T.; He, H.; Li, D.; An, P.; Hui, Z. Damage Signature Generation of Revetment Surface along Urban Rivers Using UAV-Based Mapping. *ISPRS Int. J. Geo-Inf.* **2020**, *9*, 283. [CrossRef]
12. Liu, B.; Liu, X.; Li, D.; Shi, Y.; Fernandez, G.; Wang, Y. A Vector Line Simplification Algorithm Based on the Douglas–Peucker Algorithm, Monotonic Chains and Dichotomy. *ISPRS Int. J. Geo-Inf.* **2020b**, *9*, 251. [CrossRef]
13. Guo, H.; He, G.; Jiang, W.; Yin, R.; Yan, L.; Leng, W. A Multi-Scale Water Extraction Convolutional Neural Network (MWEN) Method for GaoFen-1 Remote Sensing Images. *ISPRS Int. J. Geo-Inf.* **2020**, *9*, 189. [CrossRef]
14. Li, K.; Liang, C.; Lu, W.; Li, C.; Zhao, S.; Wang, B. Forecasting of Short-Term Daily Tourist Flow Based on Seasonal Clustering Method and PSO-LSSVM. *ISPRS Int. J. Geo-Inf.* **2020**, *9*, 676. [CrossRef]

International Journal of
Geo-Information

MDPI

Article

Site Selection for Pre-Hospital Emergency Stations Based on the Actual Spatiotemporal Demand: A Case Study of Nanjing City, China

Bing Han, Mingxing Hu * and Jialing Wang

School of Architecture, Si Pailou Campus of Southeast University, Nanjing 210096, China; 230159008@seu.edu.cn (B.H.); 220160040@seu.edu.cn (J.W.)

* Correspondence: 101009930@seu.edu.cn

Received: 1 August 2020; Accepted: 23 September 2020; Published: 25 September 2020

Abstract: Rapid economic and social development has been accompanied by the occurrence of many major issues throughout the world. Specifically, there is an ever-increasing demand for emergent medical services among the public. In order to ensure timely responses to emergency demands, it is critical to reasonably configure the emergency stations. In general, emergency stations are mostly distributed according to the distribution of emergency demands and response time, which, however, fails to consider the negative impacts of randomly occurring emergency demands and traffic delays. In this study, first aid demands are combined with traffic states based on the spatiotemporal big data set covering model, which alleviates the negative impacts of randomly occurring first aid demands and traffic delay time on the planning of pre-hospital first aid stations. Moreover, the accuracy of the site selection model is improved, which meets the requirements that all randomly occurring simulated first aid demands can be approached within the target time under planning conditions and actual traffic constraints. Taking Nanjing City as an example, this study obtains multi-source big data, such as ambulance-carried GPS data from June 2018 to June 2019, Gaode Map-recorded traffic congestion data, and survey data of emergency rescue facilities. Basing on the processing and analysis of these data, it shows that first aid demands in Nanjing City are highly region-specific with high time delay. Various required factors are determined based on modeling and analysis, and the target time is agreed to be 8 min. The average vehicle speed on each road is calculated, accompanied by the establishment of an actual road network model. In this context, the transit time from the randomly distributed first aid stations to the hospital can be calculated, which are set to satisfy the model conditions so as to obtain the solution. Finally, such a solution is adjusted and verified according to the land-use situation. The results of this study, based on spatiotemporal big data, are expected to provide insights into improving the site selection model and enhancing efficiency while providing new technical methods.

Keywords: pre-hospital emergency; spatiotemporal demand; GPS data

---

1. Introduction

With the advancement in the global economy and people's living standards, conflicts between resource exploitation and environmental protection have intensified. Emergencies such as natural disasters, accidents, and infectious diseases continuously occur, causing huge economic losses and serious life threats. Specifically, emergency-related economic loss has reached 6% of the total GDP in China, which, together with the public's growing demand for emergency facilities, reveal the prominent importance and urgency of effective urban emergency management. The outbreak of COVID-19 not only threatens the safety and properties of individuals but also disrupts the global economic order. Moreover, anxiety has been spreading due to the economic slowdown and infrastructure imbalance.

Although all countries have taken active countermeasures, there are still many exposed problems in emergency management, especially when it comes to the emergency medical service system.

Pre-hospital care refers to medical activities prior to the arrival of patients at hospitals, including on-site care and monitoring during transit. This should be performed by professional emergency groups that are equipped with communication tools, transportation gears, and medical instruments. The pre-hospital care system, the development of which is highly emphasized in countries all over the world, is an integral part of both the urban public security emergency system and the public health system. It does not only directly impact the actual demands of health protection and safety of people, but also is an important representation of public service equalization. The response time to an emergency call is a critical measure of the efficacy of the pre-hospital care system, and moreover, a reasonable placement of emergency facilities is key to effectively reducing the response time to emergency calls [1]. Construction of emergency facilities is guided and controlled during urban planning, via allocation and configuration of city space and specifically the use of land resources [2]. It is necessary to study the reasonability of the emergency facility's site selection to build a fair and accessible pre-hospital emergency network, so that any patient can receive timely and effective help when they need emergency treatment.

2. Literature Review

On the site selection of first aid stations and other emergency facilities, previous research on planning methods was often based on the graph theory, in which each demand zone is represented by a point in a network and the site selection problem is equivalent to the problem regarding the minimum distance or overall coverage of multiple point sets to/on specific facilities. For the preliminary site selection models represented by the coverage model, p-center model, and p-median model and its modified model, the geographic/gravity center of each demand zone is used to replace the set of demand points [3–7]. Correspondingly, site selection considers only the decreasing coverage level with the increasing distance and ignores the effects of the specific spatial distribution of demands within a certain zone upon site selection. Accordingly, a considerable part of demands is in fact ineffectively covered, especially in the case of a large demand zone. Estimation of demand coverage suffers from great errors, which leads to inappropriate station configuration in actual applications of many cities.

Over the recent years, the detailed spatial information provided by the expanded application of the Geographic Information System (GIS) and big data offers people more accurate measurement approaches, and researchers begin to realize the constraints of the previous data and technical approaches. Under such circumstances, a more precise and accurate description of the temporal-spatial variables of demands has gradually become a research hot spot. Some scholars attempt to integrate the first aid demand temporal-spatial variable into the site selection model based on the minimum distance solution, and thus overcome the failure of homogeneous distribution of demand zone points to capture differentiation of emergency demands. For instance, Degal et al. [8] and Chen et al. [9], with the help of GIS, make attempts on meshing the research area and statistically summarize the temporal-spatial emergency demand by grids, which however fails to efficiently reduce the impacts of the random spatial distribution of demands. This is because some grids may present no demands for several months, and moreover grids in corners are also required to fulfill the constraint of coverage levels. Then, Fan [10] applies the K-means clustering data mining to site selection of emergency response centers, in which facilities are constructed at the centroids of the current data point clusters. Zhou [11] and Dou [12] simplify zones with high medical emergency activity densities into demand points, by mapping medical contacts of pre-hospital care, and these demand points are preferably placed with first aid stations. Although one is able to simulate the complex demand distribution within a certain area via methods such as kernel density estimation and artificial neural network, it is hard to ensure that the obtained distribution can remain stable for a long period of time. Furthermore, the occurrence of medical emergency activities is uncertain. Given this, some researchers further incorporate randomness of first aid demands into their studies. Recently, to accurately describe the random spatial distribution of first aid demands for the site selection has become a key aim. For example, Beraldi and Bruni [13] propose a probabilistic model to optimize the emergency service vehicle location. Nonetheless, they describe the probable coverage of demands simply by introducing a sole randomness parameter and do not probe deep into the specific effects of spatial randomness upon demand distribution. Su [14] applies the Gaussian mixture model (using a soft clustering approach) into the depiction of spatial randomness of demand. Based on the emergency service data of the Songjiang District of Shanghai, China, in 2013, he manages to decompose the massive spatially random first aid demands into the mixture of multiple Gaussian-distribution clusters, and constrain that each cluster has only one primary station and one alternative station. In essence, the demand zone is represented by a point as before, which roots in low guidance provided by this research upon practical site selection.

Neglecting the spatiotemporal status of traffic can be fatal, in terms of ambulances that chase time. The ever-intensifying traffic jam in cities may result in delayed delivery of medical emergency service and elongated emergency response distance, which greatly threatens people's lives. In previously adopted models, the division of grids or the linear distance cannot represent the actual road route. Fortunately, this has been noticed, and some scholars have presented the spatial distance calculation method based on obstacles.

In the urban traffic system, the straight-line distance and the actual route distance between the demand point and the facility location have huge gaps. With the deepening of research work, it is found that demanders prefer to choose the least time-consuming route rather than the shortest-distance route. Therefore, time consumption should be introduced into the location model of emergency facilities as the transportation cost. However, since it is difficult to acquire the actual transit time of specific roads, the design speed for various roads [15] or the running speed of vehicles [16–18] are used to estimate the minimum travelling time in most research. Due to urban traffic congestion, vehicles cannot run at the designed speed most of the time, and thus the actual transit time is mostly longer than the designed counterpart. The development of wireless positioning technology and telecommunication technology provides a more precise way to acquire traffic congestion conditions and corresponding transit time. Recently, some studies have begun to use Gaode Map or Baidu Map to calculate the travelling time [19–22], acquiring the least transit time to the emergency facility with various traffic modes through the route-planning interface in API. This method sufficiently exploits road data provided and updated by the map developers to estimate the travelling time using reliable real-time vehicle speed data.

To conclude, much progress has been made in addressing the emergency site selection issue based on models that consider temporal-spatial demands. However, there are two prominent shortcomings in these models. The first one is the insufficient investigation into the randomly occurring first aid demands. Specifically, existing site selection models mainly focus on the statistical significance of the spatial distribution of first aid demands, which, however, is only a description of the historical data. In this context, these site selection models are not guaranteed to always be valid within the planning period. Although certain efforts have been made to incorporate the spatial randomness factor into the site selection model, there is still a lack of a random simulation method for spatial demands. Specifically, such a method is expected to not only accurately describe the demand distribution, but also effectively optimize the model based on the description. In this context, the emergency site selection issue can be better solved. The second shortcoming of existing models refers to unconsidered impacts of travel temporal distance. Although there have been studies that obtain traffic data through API and confirm the impact of traffic conditions on the accessibility of emergency facilities, the traffic factor has not been quantitatively incorporated in the site selection methods.

This paper proposes a set covering model for site selection based on spatiotemporal big data. The massive simulated emergency service demands that are spatially random and the traffic status are integrated into the set covering model, which overcomes the difficulties in the description of emergency service demands, due to their complicated distribution, and also eliminates errors induced by the idealization of the traffic status. With Nanjing City taken as an example, multi-source data, such as the GPS data carried by the ambulances as well as the real-time traffic data and road network data obtained through the OPEN API of Gaode Map, are integrated into a spatiotemporal database. Based on this database, the annual first aid service data of the city in 2018-2019 are put into the clustering analysis. The clustering analysis simulation results and the characteristic speed of municipal roads of Nanjing City are input into the actual road network model, and substituted into the location-set model constrained by the planned land use. The models are solved, which can provide guidance on other similar emergency plans such as fire-fighting facilities and public security agencies while helping to optimize the urban emergency site selection in the context of the big data era. This paper is organized as below: The first part is the statement of the adopted model and method; the second is the basic analysis of the temporal-spatial data; the third part presents testing results of data and validation; and the last part states our research conclusion.

3. Introduction of the Model and Method

3.1. Set Covering Model

The set covering model is, in essence, a model for the solution of optimal location of discrete points. Discrete points generally refer to points with known demands, and the goal of the model is to find a solution fulfilling all known demand points. During the solving process, the quantity and location of placed facilities as well as the economic benefit shall be considered comprehensively. Depending on the solving method, the set covering model is divided into two types: the set covering model, and the maximum covering model. The set covering model requires minimum costs of facilities or construction, under the premise that all demand points are covered, while the maximum covering model pursues a maximum quantity of demands that can be fulfilled by appropriately placing service facilities, in the case of defined service station quantity and service range. The biggest difference between the maximum covering model and the set covering model is whether or not the quantity of facilities is considered. The former method emphasizes the fulfillment of demands, and yet the latter pays more attention to lowering cost (Figure 1). Given that the first aid station planning in China aims at achieving the full coverage of the medical emergency service network upon urban and suburban areas of the whole city and actual conditions of the Nanjing City, the set covering model is used in this paper.

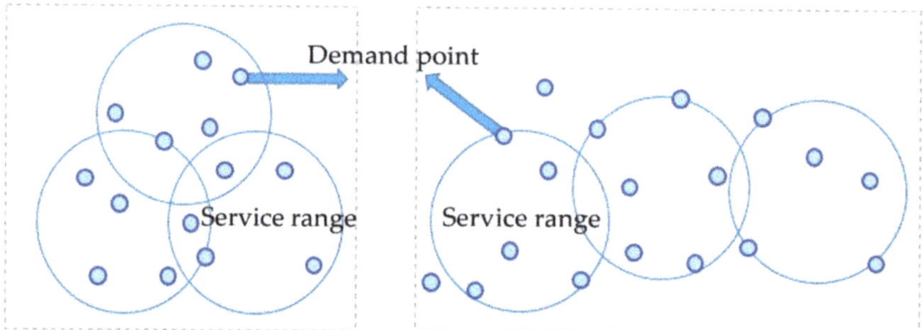

Figure 1. Schematic diagrams of the set covering model and the maximum covering model.

The location coverage model is one of the most important site selection models for emergency response stations [23]. Toregas et al. [4] apply, for the first time, the location set covering problem (LSCP) model to determine the layout of fire stations, with the goal to cover all objects (points, lines, and planes) with emergency service demands while minimizing the emergency service facilities [24]. The basic model is shown below:

$$\min \sum_{j \in W} x_j \tag{1}$$

$$\text{s.t}: \sum_{j \in N_i} x_j \geq 1, \forall i \in V \tag{2}$$

$$x_j \in \{0, 1\}, \forall j \in W \tag{3}$$

where V is the set of demand points, W is the set of facility service stations, i and j denote the different demand points and service stations, N_i is the number of stations covering demand points i, and x_j is a bool variable to evaluate the necessity of the j-th station.

The objective function Equation (1) is to achieve a minimum amount of newly established stations. According to constraint Equation (2), each demand point is covered by at least one station. Equation (3) specifies that the variable x_j is a bool type. Equations (1)–(3) are discrete models, requiring the input of a series of spatial demand elements (points, lines, and planes) and potential location sets of facilities. x_j is a node, and j reflects whether or not this node is chosen to build a facility (for being chosen, the value is 1; otherwise, 0). The facility planning with minimum quantities of planned facilities at certain locations enables the LSCP model to cover any continuous space, which also requires the value determination of the variables, namely the demand and candidate points, and solving x_j. The detailed site selection optimization procedure of the model is presented below.

3.2. Steps of the Algorithm

The framework of the method illustrating the steps of the algorithm is presented in Figure 2. Firstly, ensure the variable values of the set covering model, candidate station sites are selected according to land use planning, and written in the set M, clustering analysis is performed for first aid service data and simulation of first aid demands is carried out as set V. Then build an optimization algorithm based on the site covering model and solve the model. The final planning sites are collected and written in the Set W, requiring that the ideal travelling time from the first aid station set M to the service demand zone V, t_{ij} is lower than the specified time t_s^r for any simulated demand point in the context of the road network model T. The steps are described in detail below.

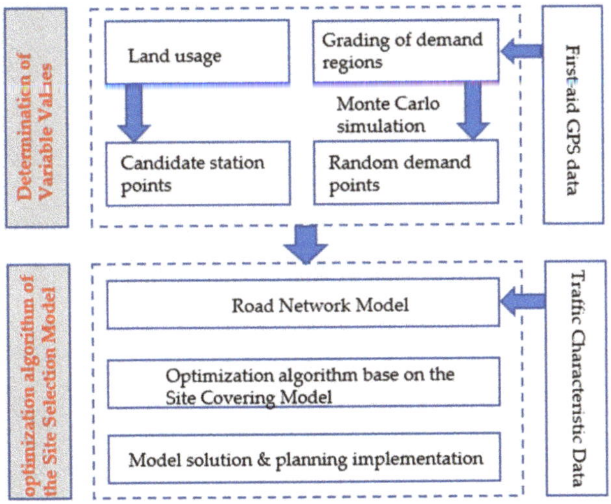

Figure 2. The framework of the method.

3.2.1. Determination of Variable Values (Demand Points and Candidate Points)

Identifying candidate station points according to the planning condition: Currently, there are 6 main modes for pre-hospital care in China due to differences among cities in their economic development and local medical emergency systems, including independent, hospital-built-in, government-operated, affiliated, fire-fighting department-integrated, and coordinative modes. For each mode, candidate site points are identified, according to the planned land use condition, and the set of candidate points is defined as M. For the demand temporal-spatial point presenting first aid service demands in 2019, the K-means clustering analysis is carried out. Subsequently, the Monte Carlo simulation of demand points is conducted, based on the results of the data clustering and Gaussian distribution fitting, which generates several random demand points for calculating the confidence level (under the premise that it follows the Gaussian distribution).

As a common method based on the probability statistics theory, a Monte Carlo simulation, also referred to as random sampling or the statistical test method, can be used to estimate the probability of an event based on the probability of the event in a large number of experiments. It works particularly well in estimating the dynamic nature of risk event systems such as first aid, and thus it is able to solve uncertain and complex problems. Its assumed function is as follows:

$$\mathrm{Y} = \mathrm{F}(\mathrm{X}_1, \mathrm{X}_2, \ldots \mathrm{X}_\mathrm{n}) \tag{4}$$

where $(\mathrm{X}_1, \mathrm{X}_2, \ldots \mathrm{X}_\mathrm{n})$ represents the probability of the spatial distribution of each group of demand points in the known demand point clusters in 2019. It is generally difficult to use analytical methods to solve the probability distribution of Y and its mathematical characteristics. The Monte Carlo method uses a random number generator to directly or indirectly sample the value of each group of random variables, and performs a large number of repeated independent random samplings of the random variables to generate a probability distribution of the function Y (i.e., the simulated first aid demand) that is close to the actual situation. The sampled values will then be substituted into the function in step (2) group by group until the final result is obtained.

3.2.2. Building the Site Selection Model

Now the model can be solved for x_j, and the set of the ultimately planned stations W can be obtained by substituting the demand and candidate station points into the model. In terms of the set of the existing stations H, the default practice is to directly merge it with Set W with no change. Given the predictability of medical emergency service demands, the optimization algorithm of the set covering site selection model is used to determine whether or not each station *j* of Set M shall be included in Set *W*, and consequently, the minimized station quantity and distribution with maximized first aid service efficiency are obtained. The algorithm is shown below:

$$\min \sum_{j \in W} x_j \tag{5}$$

$$\mathrm{s.t}: \Pr\left(\mathrm{t_{ij}} \leq \mathrm{t_s^r}\right) \geq \alpha \tag{6}$$

$$t_{ij} \leq t_s^r, \forall i \in V, \exists j \in W \tag{7}$$

$$x_j \in \{0,1\}, \cdot \forall j \in W \tag{8}$$

where *i* is the label of a random simulated demand point, j is the label for a candidate station, *V* is the set of demand regions, M is the set of the new candidate sites, *W* is the set of planned emergency stations, H is the set of existing stations, t_{ij} is the travelling time from station *j* to demand region *i*, according to the actual road network model T, and x_j is a 0–1 variable, which equals 1 if an alternative site *j* is enabled and 0 otherwise.

Objective Function (5) requires that the quantity of constructed service facilities is minimized. Moreover, Objective Function (6) is a global constraint condition, requiring that the ideal travelling time from the first aid station to the service demand zone t_{ij} is lower than the specified time t_s^r for any simulated demand point in the context of the road network model T incorporating traffic jam. Calculation of the actual traveling time t_{ij} from the first aid station j to a random simulated demand point i within the research area needs the values of distance and vehicle speed. Here, first, we convert the coordinates of the random demand points, and candidate and current first aid stations defined in Section 3.1 into a unified coordinate system and then the temporal-spatial database. Meanwhile, with the help of the route planning program based on the Dijkstra algorithm, the shortest route between two element points in the weighted graph (namely the road network T) is calculated, which is divided by the transportation characteristic speed of the corresponding road to yield the traveling time that is converted into the time matrix t_{ij}. It is required that more than 97% of the emergency services meet the maximum travelling time criterion (considering errors, α is artificially specified as 97%). This paper, taking the Nanjing City as an example, provides more accurate measurement methods for each variable in model equations, since it is based on actual demand points of emergency service events and road network data. Therefore, this study is able to more accurately capture the temporal-spatial characteristics of first aid service events, and ultimately generates more optimal site selection decisions. The next part describes the data sources obtained in Nanjing City.

4. Data Sources

4.1. Research Area

The present pre-hospital care mode in Nanjing city, the capital of Jiangsu Province, falls in the category of the affiliated mode, which means first aid stations are supported by general hospitals and community health services or township hospitals.

According to the analysis of a survey questionnaire, a total of 52 (i.e., the value of H) first aid stations are available in Nanjing City by June 2019. The allocation of these operating stations exhibits a general pattern with a higher density in the central urban area and a lower density in outer new districts or towns. Specifically, the density is obviously higher in districts such as the Xuanwu, Gulou, Qinhuai, Yuhuatai, Jianye, Qixia, and Pukou Districts (Figure 3). According to the Nanjing Regulation of Pre-Hospital Medical Emergency Service, urbanized regions should be equipped with stations within an ambulance action radius of 3~5 km, regions with dense population should be allocated with at least one station per 200,000 citizens, and each designated town or sub-district should have a station. It provides an estimation method for the coverage of emergency services in various areas. The coverage rates are 32% and 49% for service radii of 3 and 5 km, respectively.

4.2. Data Sources and Processing

4.2.1. GPS Data of Ambulances

The ambulance is a critical transportation tool for medical groups and services in case of emergency, and the installed GPS device on it records its moving trajectory and status within a certain period, which provides information on positions of accidents and first aid stations. GPS data of ambulances in Nanjing City during the period from 1 June 2018 to 1 June 2019 are obtained under the help from the Nanjing Emergence Center. In total, 99,598 records are valid, including critical information such as emergency call positions, responsive ambulance positions, and response time (Figure 3). The data consist of 85,332 records of patient information, corresponding to a service rate of 85.7%. The total number of station-based and empty-run responses is 67,802 accounting for 79.5%, and the rest is for inter-hospital-transfer response and is beyond the scope of this study. The attribute data are processed and spatialized using the PostGIS software, and the attribute fields after processing include the ambulance plate number, departure time, site arrival time, site longitudes and latitudes, and name of the target place and its longitudes and latitudes (Table 1). It is seen that medical emergency service demands in Nanjing City is highly heterogeneous, in terms of the spatial distribution, and characterized by series delay, from a temporal point of view, which further validates the necessity of clustering simulation of the demand zone and obtaining the transportation characteristic speed in our model.

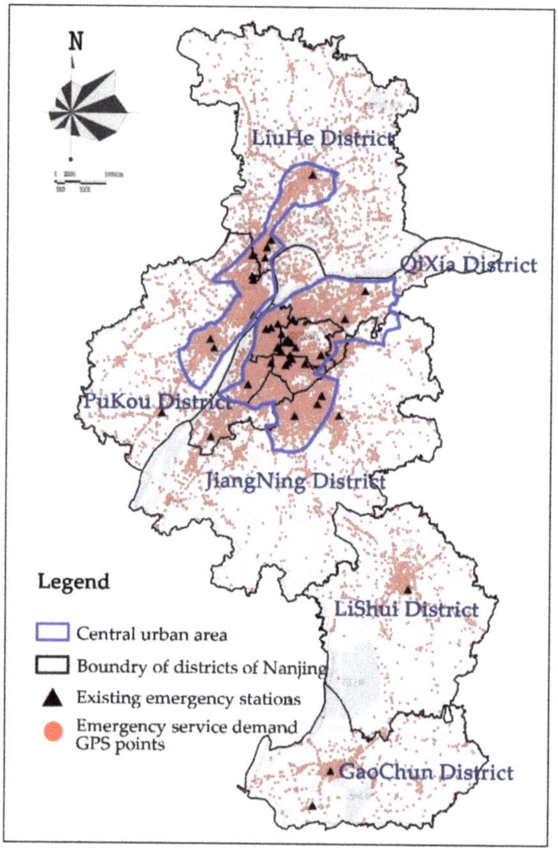

Figure 3. Map of pre-hospital emergency stations in Nanjing.

Table 1. Example of GPS data for emergency service events.

Ambulance Plate Number	Departure Time	Arrival Time at the Demand Site	Site Longitude and Latitude	Arrival Time at the Hospital	Destination Hospital
SUXXXXXX	2 June 2016 08:25	08:37:48	118.7907; 32.0241	08:45:28	Chengnan Sub-Station
SUXXXXXX	2 June 2016 14:18	14:35:20	118.7787; 32.01897	14:52:30	Brain Hospital Sub-station
SUXXXXXX	2 June 2016 16:22	16:34:40	118.7687; 32.02224	16:38:10	Chengnan Sub-Station
SUXXXXXX	2 June 2016 17:35	17:53:18	118.7959; 32.0029	18:02:38	Chengnan Sub-Station
SUXXXXXX	2 June 2016 19:15	19:20:46	118.7722; 32.0146	19:27:24	Chengnan Sub-Station

(1) According to the autocorrelation analysis of the whole area after spatial visualization, the global Moran's I coefficient has a value of 0.28, and z, a value of 3.60, suggesting the presence of high-demand cluster regions, and prominent spatial positive correlation and concentration characteristics of first aid

service demand distribution. Generally speaking, the heterogeneous demand distribution features the central urban area, with intensively high demands and the circumferential newly emerging municipal area with relatively low demands. Moreover, the visualization analysis of the OD (original-destination) service distance from the first aid demand point to the first aid station in the ArcGIS software confirms the regional differentiation of first aid service demands. The first aid stations in the central urban areas are close to each other, resulting in highly overlapped service regions among neighboring stations and a consequently short service radius for each station. It indicates that several stations are commonly available for an individual position, without a clear boundary of the service area (Figure 4). The outer suburban areas have a relatively low density of first aid stations, which thus have a large service region per station. The emergency routes converge to single points, as shown in Figure 4. In the suburban areas, emergency demands in Liuhe District mainly rely on Liuhe People's Hospital, those in Gaochun District on Gaochun People's Hospital station, and those in Lishui District on Lishui People's Hospital. Similar rules are also true for Gaochun District Dongba Station, Pukou Central Hospital Linqiao Station, and Meishan Hospital Sub-Station. Quantitative statistics show that the average response distance for first aid services is 3.8 km, while the peak probability is located from 3 to 4 km. The cases with response distances over 4 km account for 37.3%, which is a considerably large fraction (Figure 5).

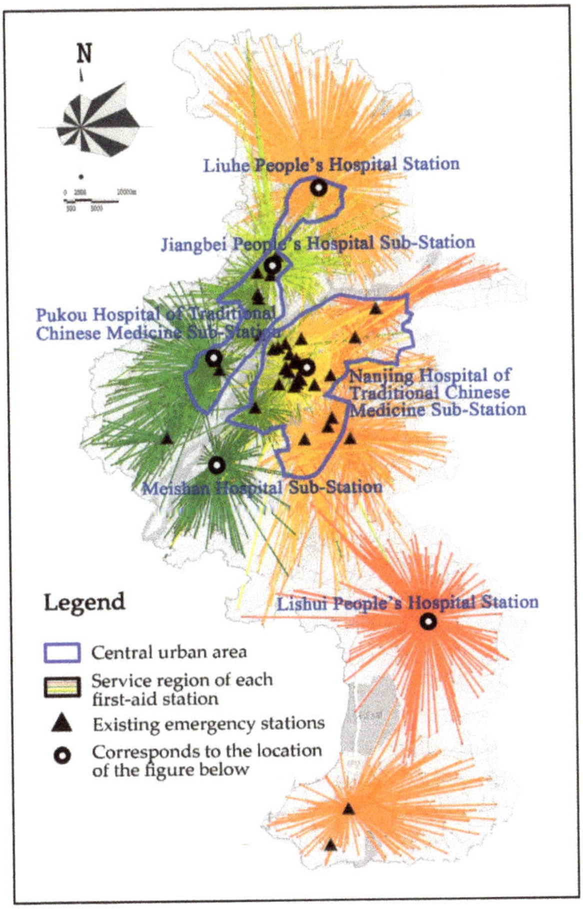

Figure 4. *Cont.*

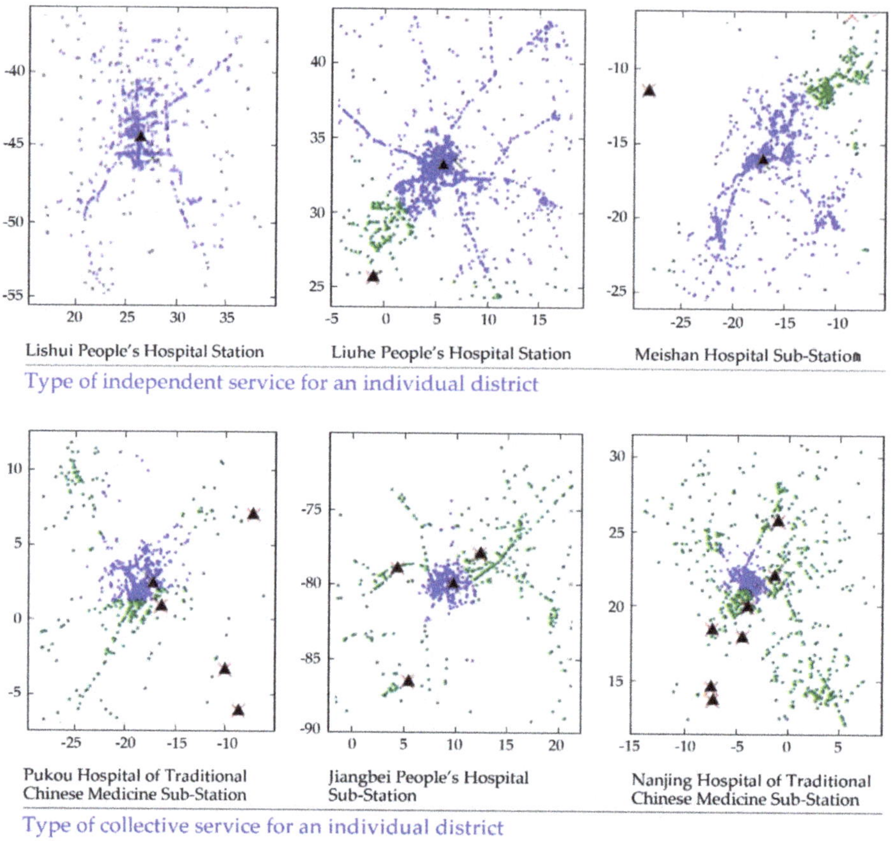

Figure 4. Map of service range of each first aid station (the right figure represents first aid stations, blue circles are service demands of this station, and green circles are those of other stations).

(2) According to the time difference between the ambulance departure time and the site arrival time, the average ambulance travelling time is 14.6 min. Spatial statistics of seriously-delayed cases with travelling time over 20 min show that the distribution of highly-delayed demand points generally coincides with that of the overall service demand points, which therefore demonstrates that the response delay is a city-wide systematic issue. Although the central urban area has a high first aid station density, it suffers from response delays due to the low traffic efficiency. On the contrary, the circumferential suburban area is prone to medical emergency service delay, owing to the low density of first aid stations and thus long travelling distance. According to the average travelling time per unit distance (Figure 6), the travelling time consumed by the same distance is highly differentiated, and the response time of ambulances during rush hours is longer than that during non-rush hours. These indicate that the travelling time of ambulances is tremendously restrained by traffic efficiency, and thus obtaining the characteristic road speed of Nanjing City through the API of Gaode Map is necessary for the estimation of ambulance response time.

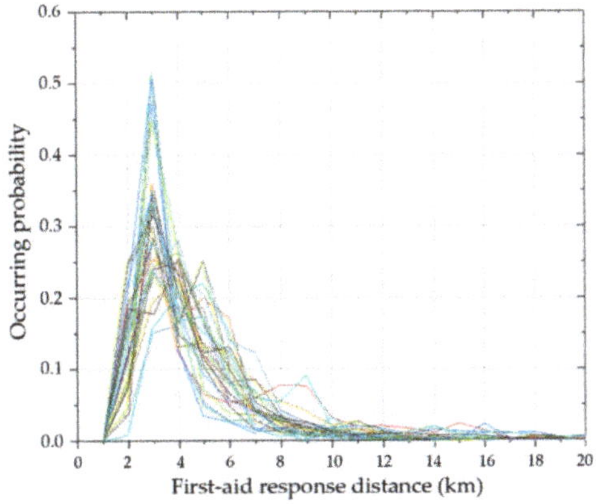

Figure 5. First aid response distance distribution.

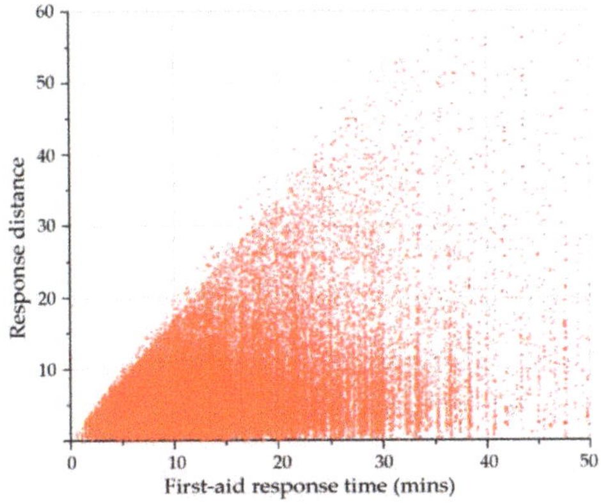

Figure 6. Consumed time vs. response distance.

4.2.2. Obtaining Real-Time Traffic Data through the Gaode Map Open API

It is noted that a more accurate calculation for transportation costs would contribute to a more accurate evaluation for medical service efficiency and capability, during the investigation on emergency service accessibility. Thus, through the analysis of big data concerning transportation, road congestion is introduced into the optimization of site selection in this study, for the sake of more rational planning. Gaode Map offers travel data of both the floating car of the transportation industry and over 700 million users of this app, and it is available in 40 cities of China including Nanjing. The traffic situation data for roads of all levels within the research area are requested and obtained via the Gaode web-service OpenAPI, specifically including the road name, road geographic coordinate geometry, road speed, congestion status, and so on. The overall idea can be summarized as follows: To begin with, data for model parameters are requested via the traffic situation API of Gaode. Since it is required that the request zone shall be rectangular, the urban road network is first divided into several units when obtaining the actual road network model T. Then, the traffic situation acquisition software runs, and the original traffic information within the longitude and latitude range of each unit is extracted and imported into the database. Finally, the units are merged into the actual regional road network, and the original data are pre-processed and linked to the effective traffic information. Furthermore, it is set that the traffic information database acquisition software runs automatically for 28 days in a row at the interval of one hour (3600 s), which achieves the intelligent batch processing of streaming data and thus automatic acquisition, pre-processing, spatialization, and saving (in the format of shapefile) of the city-wide traffic situation data and building the traffic situation temporal-spatial database. After dimensionality reduction, the characteristic speed of each road is extracted and mapped. Statistics of all traffic data of Nanjing City from May 28th to June 10th 2019, provided by Gaode Web API show that the city-wide daily traffic congestion delay index of Nanjing is about 1.55, and the daily average speed is around 28.29 km/h. During rush hours, the congestion delay index grows to 1.81, with the average speed slowed down to 24.21 km/h. The region within ~5 km from the central urban area is characterized by overall slow (jammed) traffic and local presence of unblocked traffic. Meanwhile, in the region far away from the urban area, traffic in most roads is smooth, while some congested road sections exist in the central areas of this region and cross-region main roads. Urban traffic congestion of Nanjing has evolved into a malady and severely impacts emergency response and rescue.

5. Factor Analysis and Results

5.1. Determination of t_s^r

t_s^r refers to the target response time for first aid services in Nanjing City, and is typically defined as the time consumption between the emergency call and the arrival of the ambulance at the demand location, composed of the time for emergency call answering, staff response, and ambulance travelling. Therefore, before determining the target time, one first needs to analyze the current temporal characteristics of medical emergency service in Nanjing Cities, using the GPS data of ambulances. According to the statistics of Nanjing Emergency Center, the current average emergency call answering time is 1 min, and the average staff response time is 3 min. The analysis of ambulance GPS data reveals an average travelling time of 14.6 min. Hence, the total average response time is 18.6 min, with a median of 17.03 min, which is slightly longer than the Chinese national standard of 15 min [25–27]. This paper highlights the planning and site selection of first aid stations constrained by the emergency response time, so as to reduce the time consumed by the process to transfer patients to hospitals for medical emergency services after answering the emergency call and sending out the ambulance. At present, the average emergency call response time for developed countries is within 8–12 min [28–30], based on which the target emergency response time is set as 12 min, also with consideration of illness and megacity. Given the current pre-hospital care reality of Nanjing City (namely the average emergency call answering time of 1 min and the average staff response time of 3 min), the ambulance travelling time t_s^r in this paper targets 8 min, and this time target is used for planning the first aid station network of Nanjing City.

5.2. Generated Results after Substituting Data into the Model

5.2.1. Selecting Candidate Station Sites According to Land Use Planning and Writing the Selected Sites in the Set M

The pre-hospital care mode in Nanjing City falls into the category of affiliated mode, i.e., hospitals are involved in the co-building of their respective emergency stations. According to Nanjing Medical Treatment and Public Health Facility Planning, a total of 2083 sites are selected as candidates from three kinds of land usages, including general hospitals (A5), community public health service centers (Aa), and community public health service stations (Rc) (Figure 7).

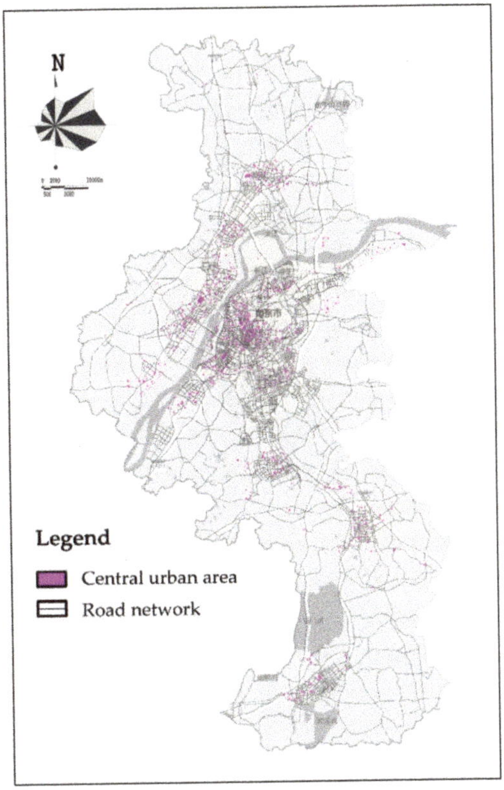

Figure 7. Candidate site set M.

5.2.2. Clustering Analysis for First aid Service Data and Simulation of First aid Demands

The K-means clustering analysis is conducted for the first aid service demand spatiotemporal points n in 2018-2019, with the cluster number of K. The averages of the lateral and vertical coordinates of points of each cluster relative to the cluster center μ and the covariance COV are calculated and satisfy the test function of the normal distribution. The Monte Carlo simulation of demand points, based on the results of clustering and Gaussian distribution fitting of data for 2019, is performed to generate several random demand points, in order to calculate the confidence level. It should be noted that the K-means clustering requires the input of the cluster number K (in this paper, 200). The clustering process mainly involves three steps: First, K random initial points are taken as the centroids of clusters. Then, according to the distance to the centroid, data in the data set are allocated to each cluster, and the average value of the data within each cluster is computed and taken as the new centroid. At last, the previous step is repeated until all clusters remain un-changed. The analysis of the clustering results in cases of different K values shows that K = 200 presents the optimal performance in data grouping based on temporal-spatial characteristics, specifically with neither shadowing of characteristics nor chaotic classification (Figure 8).

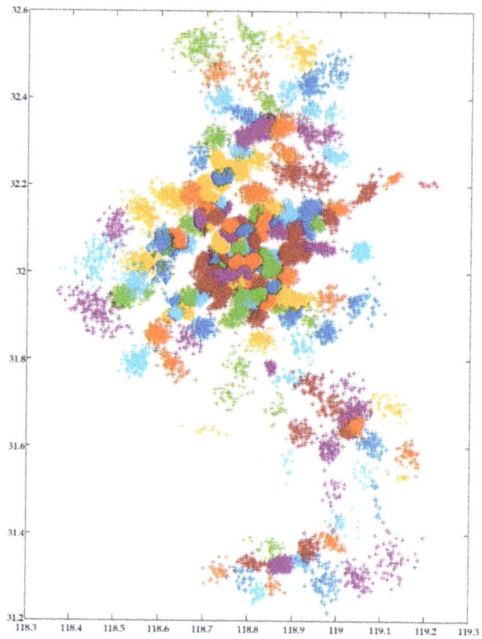

Figure 8. K-means clustering results.

5.2.3. Building the Road Network Model and Calculating the Minimum Time Matrix t_{ij}

The road network data set is constructed and used as the traffic network model, based on the processed *Shapefile*-format road network file, after topological processing such as intersection breaking and interface merging. The data set includes road sections and intersections. The road section is the line element for the road network, and it is represented by the arc in ArcGIS. Each road section possesses attributes such as the congestion vehicle speed (km/h), average vehicle speed (km/h), traveling time in the case of the average vehicle speed (s) and length (km). Road intersections are represented by nodes in ArcGIS and combined with the table of turning instruction at road intersections. Thus, they are able to truly mimic the actual traffic scenarios such as waiting for the red light, no-straight-through, no-left-turning, and driving through an overpass. In ArcGIS, the time required for each road section is calculated, with respect to the actual road network and corresponding characteristic speed. In this research, the characteristic speed refers to the actual average travelling time. Vehicle speed data between two consecutive weeks within the research period are compared, and the covariance calculation results show that the distribution curves for the average speeds at a certain o'clock sharp can be well fitted, with a correlation coefficient of 0.973. Thus, it is safe to say that the data of the two weeks used in this research are periodically representative.

5.2.4. Collecting Final Planning Sites and Writing Them in the Set W

The existing 52 sites of Set *H* are directly added to Set *W*. Via the algorithm presented in the first part of this paper, the calculation is performed to determine whether or not each station *j* in Set *M* should be included in Set *W*. The actual factors of Nanjing City are substituted into the global condition constrain equation one after another. As stated above, t_s^r is set as 8 min, and α is 0.97. For any random simulated demand point *j* generated by clustering, it is required that the travelling time to the station *i* in Set M (plus the OD time matrix *t*) should be less than 8 min. Accordingly, the least station quantity x_j and distribution of *j*, with the maximized emergency response efficiency, can be obtained.

5.2.5. Iterative Computation in Matlab

Results show that at the 120th iteration the optimum solution is obtained (Figures 9 and 10). The optimal quantity of constructed stations is 134. After mapping the results into space, the 52 existing facilities can be simply identified, and 82 potential station sites can be selected among the other candidate points.

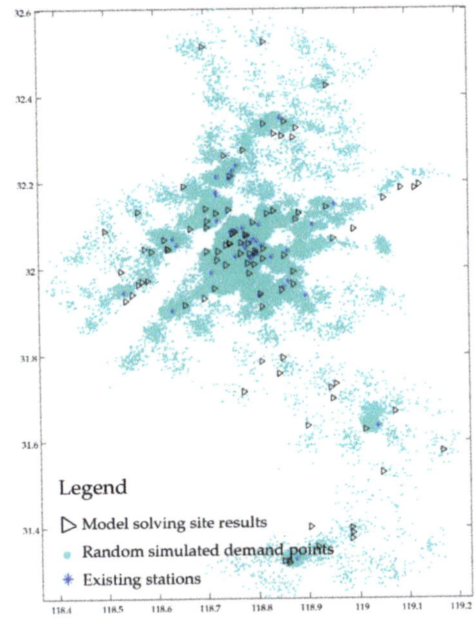

Figure 9. Site selection results.

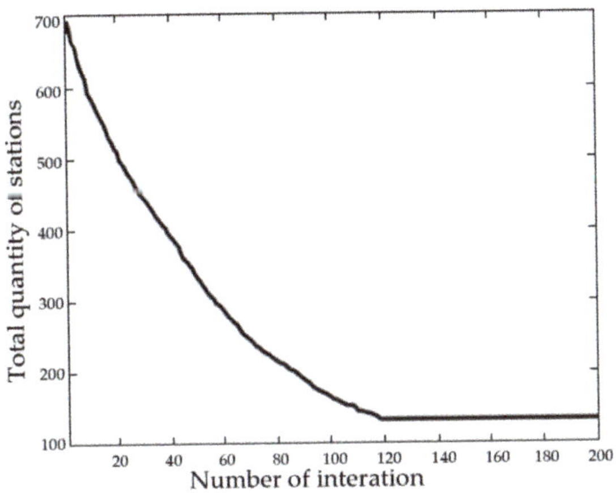

Figure 10. Iterative computation in Matlab.

5.2.6. Examining and Adjusting Simulated Results

First, the land usage corresponding to the location of the planned station site should be clarified. Fine adjustment is conducted according to the following preference sequence, general hospitals (A5) > community public health service centers (Aa) > community public health stations (Rc), to facilitate the implementation of the planning.

The eventual planning of the allocation is shown in Figure 11. The planning involves a total of 136 emergency stations, with 48 stations reserved, 88 newly established, and 8 adjusted. A total of 99 stations mainly serve the central urban regions while other 37 stations serve the suburban regions. With the planning, the coverage rate of the 3-km service radius is increased from 32% to 63%, and that of the 5-km service radius is increased from 49% to 95%. Each town in the suburban region has at least one emergency station. The average response time is decreased from 18.6 min to 12 min.

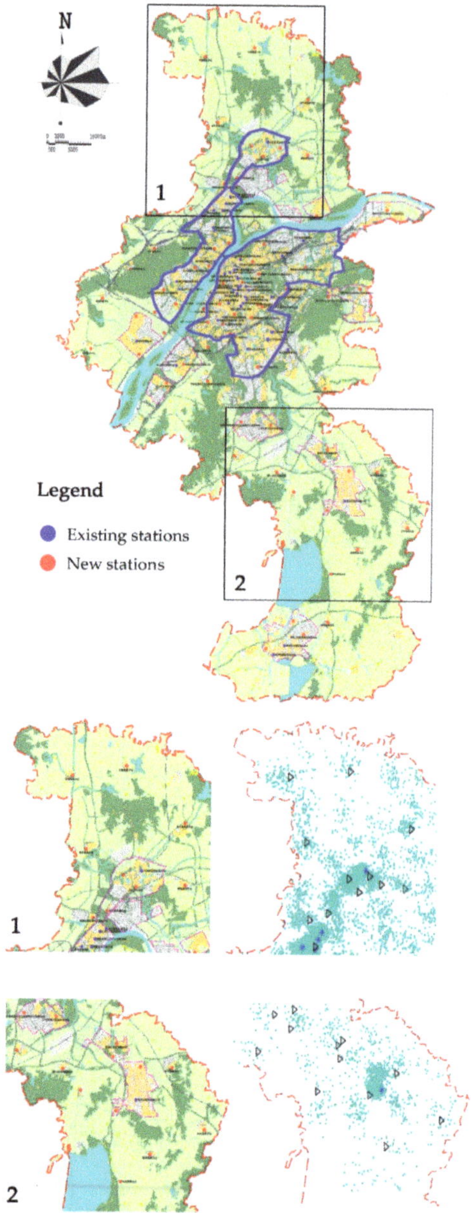

Figure 11. The final layout of first aid stations.

6. Discussion

(1) The optimization of emergency facility sites has been a classic research topic in the field of geography. Compared with the traditional model, the optimized location set coverage model is demonstrated to greatly improve the coverage and effectively shorten the emergency response time. This confirms the presence of obvious shortcomings when it comes to the distribution of pre-hospital emergency stations in Nanjing City. Specifically, current emergency facilities are far from meeting emergency demands due to insufficient and unbalanced allocation of medical resources in various districts. The overwhelmingly concentrated medical resources in Gulou and Xuanwu Districts contradict the goal of Nanjing's multi-center development plan of "one center and three sub-cities". Moreover, the unbalanced allocation of emergency resources is also present within each district. For instance, emergency resources in the Xuanwu and Gulou Districts are mainly concentrated on the south and east of Xuanwu Lake, respectively.

(2) The time target parameter in the proposed model is selected according to the existing data analysis and the actual situation of Nanjing City with references to the requirements of other cities in China and other countries. In the future, with the further improvement of the first aid legislation and various supporting facilities, the government may be able to further shorten the target emergency call response time. Moreover, the emergency response time and personnel response time may be incrementally reduced with promoted standardized management. In addition, the traffic cost based on the road network model may be changed as a result of road reconstruction and expansion. Therefore, it is crucial to adjust and update the dynamic factors in the model in a timely manner. Accordingly, traffic data over a longer period of time will be collected to observe the changes in vehicle speeds during different seasons and months, which will help obtain more accurate time cost factors that can be substituted into the model. Furthermore, the emergency data in 2020 can be acquired and substituted into the proposed model for further comparison and verification.

7. Conclusions

Digitalization in the new era has penetrated all aspects of urban life, transportation, and medical care. The development of spatiotemporal big data provides new opportunities for optimizing the emergency site selection. This study first introduces the basic model of emergency site selection based on actual traffic conditions and the simulation method of random spatial demands. Subsequently, it is proposed that these random spatial demands should be processed based on the actual data, accompanied by the application of the K-means clustering method to quantitatively describe and simulate the distribution characteristics of emergency demands. On this basis, an optimization algorithm that integrates the actual speed of the road network into the set coverage model is established, which contributes to a shorter time than the target counterpart from the simulated demand point to the emergency station at the actual speed. In the practical analysis, the first aid data of Nanjing City from 1 June 2018 to 1 June 2019 are first analyzed in terms of their spatiotemporal characteristics, followed by the determination of various necessary factors based on the modeling analysis. Eventually, the model is solved under the land-use constraints, with an agreed target time of 8 min. This paper attempts to apply the location set covering model integrated with the temporal-spatial big data into the urban planning, so as to provide a novel model for site selection of emergency service facilities that incorporates the randomness of emergency service events and traffic situations that have previously been neglected in research. The availability of emergency service and traffic spatiotemporal big data provides a more accurate data basis for the model, which is vital for the site selection of pre-hospital care facilities that requires high accuracy and timeliness. This research was commissioned to the Nanjing emergency center, and the research result provided theoretical support to the "layout planning of Nanjing first aid station" which is a legal planning implemented by the local government since 2019. The methodology used in this research provides a digital perspective of urban inherent law and a new branch of emergency facility renewal which could be learned and popularized. Moreover, this study also offers a digitalized perspective for probing into the intrinsic regularity of cities and modifying the conventional layout planning method.

Author Contributions: Conceptualization, Bing Han, Mingxing Hu and Jialing Wang; Methodology, Bing Han and Jialing Wang; Software, Bing Han and Jialing Wang; Validation, Mingxing Hu; Formal Analysis, Bing Han and Mingxing Hu; Investigation, Bing Han; Data Curation, Mingxing Hu and Jialing Wang; Writing—Original Draft Preparation, Bing Han and Mingxing Hu; Writing—Review and Editing, Bing Han and Mingxing Hu; Visualization, Bing Han and Jialing Wang; All authors have read and agreed to the published version of the manuscript.

Funding: This work was funded by the National Natural Science Foundation of China, grant number 51778127.

Acknowledgments: Thanks for the original data provided by Nanjing Municipal Planning Bureau and full cooperation.

Conflicts of Interest: The authors declare no conflict of interest.

References

1. Yang, X.H.; Ren, Y.R. Research on the layout of urban emergency medical service network. *Chin. J. Hosp. Manag.* **1994**, *10*, 673–677.
2. Lin, W.P.; Yan, Z. Medical and health system reform and urban medical and health facilities planning. *Urban Plan.* **2006**, *30*, 47–50.
3. Adenso-Díaz, B.; Rodríguez, F. A simple search heuristic for the MCLP: Application to the location of ambulance bases in a rural region. *Omega* **1997**, *25*, 181–187. [CrossRef]
4. Toregas, C.; Swain, R.; ReVelle, C.; Bergman, L. The Location of Emergency Service Facilities. *Oper. Res.* **1971**, *19*, 1363–1373. [CrossRef]
5. Church, R.; ReVelle, C. The maximal covering location problem. *Pap. Reg. Sci. Assoc.* **1974**, *32*, 101–118. [CrossRef]
6. Murray, A.T.; O'Kelly, M.E. Assessing representation error in point-based coverage modeling. *J. Geogr. Syst.* **2002**, *4*, 171–191. [CrossRef]
7. Pitt, E.; Pusponegoro, A. Prehospital care in Indonesia. *Emerg. Med. J.* **2005**, *22*, 144–147. [CrossRef]
8. Degel, D.; Wiesche, L.; Rachuba, S.; Werners, B. Time-dependent ambulance allocation considering data-driven empirically required coverage. *Health Care Manag. Sci.* **2015**, *18*, 444–458. [CrossRef]
9. Chen, A.Y.; Lu, T.Y.; Ma, M.H.M.; Sun, W.Z. Demand Forecast Using Data Analytics for the Preallocation of Ambulances. *IEEE J. Biomed. Health Inform.* **2016**, *20*, 1178–1187. [CrossRef]
10. Fan, B. Spatial clustering mining method for site selection problem of emergency response center. *J. Manag. Sci.* **2008**, *11*, 20–28, 30–32.
11. Zhou, Y.N.; Lu, X.; Dai, Z.; Shen, H.; Zhu, Q.Z.; Luo, L. Planning method and empirical study of emergency site. *China Health Policy Res.* **2016**, *9*, 69–73.
12. Dou, Q.Z.; Zhang, W.W.; Zhu, H.D.; Xu, J.; Chen, M.F.; Li, C.; Duan, W.Q.; Miao, J.G. Nonuniform and relatively stable pre-hospital first-aid model based on spatial big data. *Comput. Technol. Appl.* **2018**, *044*, 130–133.
13. Beraldi, P.; Bruni, M.E. A probabilistic model applied to emergency service vehicle location. *Eur. J. Oper. Res.* **2009**, *196*, 323–331. [CrossRef]
14. Su, Q.; Yang, Q.; Wang, Q.G. Ambulance location planning considering the spatial randomness of demand. *Chin. J. Manag. Sci.* **2019**, *27*, 110–119.
15. Ni, J.; Wang, J.; Rui, Y.; Qian, T.; Wang, J. An enhanced variable two-step floating catchment area method for measuring spatial accessibility to residential care facilities in Nanjing. *Int. J. Environ. Res. Public Health* **2015**, *12*, 14490–14504. [CrossRef]
16. Tao, Z.; Cheng, Y.; Dai, T.; Rosenberg, M.W. Spatial optimization of residential care facility locations in Beijing, China: Maximum equity in accessibility. *Int. J. Health Geogr.* **2014**, *13*, 33. [CrossRef]
17. Chen, J.; Zhou, S.; Liu, L.; Xiao, L.; Song, G. Impact of traffic congestion on time-space accessibility of emergency medical services—Take Guangzhou as an example. *Prog. Geogr. Sci.* **2016**, *35*, 431–439. [CrossRef]
18. Xia, T.; Song, X.; Zhang, H.; Song, X.; Kanasugi, H.; Shibasaki, R. Measuring spatio-temporal accessibility to emergency medical services through big GPS data. *Health Place* **2019**, *56*, 53–62. [CrossRef]
19. Cheng, G.; Zeng, X.; Duan, L.; Lu, X.; Sun, H.; Jiang, T.; Li, Y. Spatial difference analysis for accessibility to high level hospitals based on travel time in Shenzhen, China. *Habitat Int.* **2016**, *53*, 485–494. [CrossRef]

20. Wang, F.; Xu, Y. Estimating O-D travel time matrix by Google Maps API: Implementation, advantages, and implications. *Ann. GIS* **2011**, *17*, 199–209. [CrossRef]
21. Gu, W.; Wang, X.; McGregor, S.E. Optimization of preventive health care facility locations. *Int. J. Health Geogr.* **2010**, *9*, 17. [CrossRef] [PubMed]
22. Tao, Z.; Yao, Z.; Kong, H.; Duan, F.; Li, G. Spatial accessibility to healthcare services in Shenzhen, China: Improving the multi-modal two-step floating catchment area method by estimating travel time via online map APIs. *BMC Health Serv. Res.* **2018**, *18*, 345. [CrossRef]
23. Murray, A.T.; Tong, D.; Kim, K. Enhancing classic coverage location models. *Int. Reg. Sci. Rev.* **2010**, *33*, 115–133. [CrossRef]
24. ReVelle, C. Review, extension and prediction in emergency service siting models. *Eur. J. Oper. Res.* **1989**, *40*, 58–69. [CrossRef]
25. Liu, J.; Hao, Y.H.; Wu, Q.H.; Dong, X.; Xu, J.; Wu, Z.Y.; Chen, H.P.; Sun, Y.H. An international comparison of pre-hospital emergency mode and job training of emergency personnel. *Chin. Health Resour.* **2013**, *16*, 30–32.
26. Zhang, A.H.; Tao, H.; Gui, L. Review and prospect of the development of national and international Emergency Medical Service System. *J. Nurs. Adm.* **2004**, *4*, 23–25.
27. General Office of the State Council. *Notice of The General Office of the State Council on Forwarding the Construction Plan of Medical Treatment System for Public Health Emergencies of the Ministry of Health (Development and Reform Commission No.82 [2003])*; General Office of the State Council: Beijing, China, 2003.
28. Chen, K.H. Investigation report of European emergency service system. *Chin. Hosp.* **2006**, *10*, 79–81.
29. Sun, M. Progress of American emergency medical service. *Chin. Hosp. Manag.* **1987**, *7*, 53–55.
30. Legislative Council of the Hong Kong. *Legislative Council of the Hong Kong Special Administration Region of the People's Republic of China*; LC Paper No. CB(2)114/00-01; Legislative Council of the Hong Kong: Hong Kong, China, 2000.

International Journal of *Geo-Information*

MDPI

Article

Site Selection of Fire Stations in Large Cities Based on Actual Spatiotemporal Demands: A Case Study of Nanjing City

Bing Han [1], Mingxing Hu [1,*], Jiemin Zheng [1] and Tan Tang [2]

[1] School of Architecture, Si Pailou Campus, Southeast University, Nanjing 210096, China; 230159008@seu.edu.cn (B.H.); 220190065@seu.edu.cn (J.Z.)
[2] Nanjing Yunzhu Urban & Rural Planning Limited Company, Nanjing 210096, China; tangtan@njyzcxggsjwwgc.onexmail.com
* Correspondence: humingxing@seu.edu.cn

Abstract: The rapid expansion of cities brings in new challenges for the urban firefighting security, while the increasing fire frequency poses serious threats to the life, property, and safety of individuals living in cities. Firefighting in cities is a challenging task, and the optimal spatial arrangement of fire stations is critical to firefighting security. However, existing researches lack any consideration of the negative effects of the spatial randomness of fire outbreaks and delayed response time due to traffic jams upon the site selection. Based on the set cover location model integrated with the spatiotemporal big data, this paper combines the fire outbreak point with the traffic situation. The presented site selection strategy manages to ensure the arrival of the firefighting task force at random simulated fire outbreak points within the required time, under the constraints of the actual city planning and traffic situation. Taking Nanjing city as an example, this paper collects multi-source big data for the comprehensive analysis, including the full data of the fire outbreak history from June 2014 to June 2018, the traffic jam data based on the Amap, and the investigation data of the firefighting facilities in Nanjing. The regularity behind fire outbreaks is analyzed, the factors related to fire risks are identified, and the risk score is calculated. The previous fire outbreak points are put through the clustering analysis, the spatial distribution probability at points in each cluster is calculated according to the clustering score, and the random fire outbreak points are generated via the Monte Carlo simulation. Meanwhile, the objective emergency response time is set as five minutes. The average vehicle speed for each road in the urban area is calculated, and the actual traffic network model is built to compute the travel time from massive randomly-distributed simulated fire points. The problem is solved by making the travel time for all simulated demand points below five minutes. At last, the site selection result based on our model is adjusted and validated, according to the planned land use. The presented method incorporates the view of the spatiotemporal big data and provides a new idea and technical method for the modification and efficiency improvement of the fire station site selection model, contributing to a service cover ratio increase from 58% to 90%.

Keywords: fire station; spatiotemporal demand; fire risk evaluation

Citation: Han, B.; Hu, M.; Zheng, J.; Tang, T. Site Selection of Fire Stations in Large Cities Based on Actual Spatiotemporal Demands: A Case Study of Nanjing City. *ISPRS Int. J. Geo-Inf.* **2021**, *10*, 542. https://doi.org/10.3390/ijgi10080542

Academic Editors: Wolfgang Kainz and Marco Helbich

Received: 14 June 2021
Accepted: 10 August 2021
Published: 12 August 2021

1. Introduction

Urban fires have been ever-increasingly frequent in recent years due to the deepened conflicts between urban population, resources, and the environment. The number of fire cases in China reached 252,000 in 2020, which is 65% higher than that in 2012. The consequent direct property loss in 2020 was 4.0 billion yuan, which is 83% higher than that in 2012. These fires seriously threaten the safety and property of individual lives, and they severely influence normal economic activities. In this context, it is of great importance to enhance urban firefighting efficiency. The fire risk of large cities is tremendously complex. The rapid development of large cities has resulted in massive accumulated risks, while the continued development brings in various new fire rescue risks. New-type disaster drivers interact with conventional ones; high-rise and large-volume buildings greatly increase; new materials and new products are extensively applied; population movement is ever diversified; and uncertainties and uncontrollable factors of outbreaks of fires grow significantly. Thus, prevention and control difficulties and extreme requirements are enhanced upon response time. Given this, the layout of fire stations is of utmost importance. Rational layout of fire station is of great significance for improving firefighting security and disaster prevention.

As is commissioned by the Nanjing Public Security Fire Bureau, this study aims to apply the location set cover model that is integrated with spatio-temporal big data to urban planning, based on the five-year fire data from June 2014 to June 2018. In this way, a novel site selection model for emergency service facilities can be built. It can then be integrated with random spatial demand simulation methods and real-time traffic networks to consider the influences of event randomness and traffic conditions that are often ignored in previous models. The availability of firefighting and traffic spatiotemporal big data lays an accurate data foundation for the modeling, which is of crucial importance for the site selection of fire station. The research findings provide sufficient theoretical support for "Nanjing fire station layout planning" that has been implemented since 2020 by the local government.

In terms of the structure, this article consists of five parts: the first part is a literature review; the second part is the research method and modeling; the third part introduces the research scope and data sources; the forth part presents the detailed analysis process, test results, and verification; and the fifth part is discussion and conclusions.

2. Literature Review

It has been a continuously important topic to investigate fire station arrangements in cities. The U.S. and Europe have started studying the layout of fire stations and developing relevant firefighting plans and regulations since the 1970s. Code for Planning Urban Fire Control of China stipulates that, within the urban development land, the principle of the ordinary fire station arrangement is to ensure the arrival of firefighting trucks within five minutes. Moreover, taking the road and traffic conditions in the downtown and marginal areas of cities into consideration, the layout of fire stations shall enable a firefighting administration zone of 4–7 km^2 in the downtown urban area with a marginal urban area no bigger than 15 km^2. The conventional fire station arrangement theory is mostly based on the graph theory. The planning area is divided into the basic firefighting units, and the center of each unit represents a node in the corresponding fire outbreak position network. By doing so, the site selection problem is converted into the problem of coverage or the minimum distance of facilities to multiple point sets. The early location models, represented by the covering model, the P-center model, as well as the P-median model and its modified models, use the geographic center or key points of each zone to replace the demand point set. Consequently, the site selection process considers only the expansion of coverage with the shortened distance [1–3]. It has been pinpointed that planning based on the travel time or travel distance of firefighting trucks cannot keep pace with the city development and that the fire station layout shall consider more factors [4]. Moreover, multi-objective programming models are proposed to involve more factors such as the water supply for firefighting and political interference [5]. Chen and Zhao [6] introduce the constraints such as the weather and terrain into their case simulation and build the comprehensive objective function of facility site selection via the analytic hierarchy process. Similarly, some scholars apply the multi-objective model to the fire station selection in Samia, Canada. These models attempt to introduce the factors that may trigger fires, and yet they still optimize the fire station layout with respect to the average demand. This may lead to massive firefighting activities in high-risk zones and, in the meantime, much fewer activities in the low-risk zones. The firefighting response time also varies with fire risk levels as well as road and traffic conditions in different zones. Therefore, the aforementioned fire station site selection process would tend to result in high randomness in serve zone demarcation and sparse distribution of fire stations.

With progress in developing multi-objective models, random factors are gradually incorporated in site selection for emergency responses [7,8]. Specifically, random factors generally derive from the demand and traffic uncertainties induced by the spatiotemporal environmental dynamic variation. Hence, how to simulate the actual demand to evaluate the fire risk and how to overcome the traffic network restraint have become the key spots of research. Efforts have been made to investigate the index system and methodology of the fire risk evaluation [7,9], which is normally achieved from four aspects, namely the fire hazard source, the firefighting facility construction status, the firefighting management status, and the regional disaster relief capability [8,10–16]. The evaluation results are then applied to optimize the site selection model of fire stations [17].

At present, the evaluation methodology mainly includes the expert scoring method, fire-grade hierarchy method, risk matrix method, and the fire risk index method [18–21]. Among applications of such methods, the studies carried out by Wang et al. [22] and Xu et al. [23] to perform risk grading by land uses via an analytical hierarchy process are relatively representative. Recently, it has been common to identify the factors related to fire risks from the historic fire data owing to the rapid development of GIS, data mining, and visualization techniques. Xu et al. [24] applies the kernel density analysis to the city fire risk ranking based on fire outbreaks in 2013–2016, which confirms the superiority of the site selection results based on the evaluation index of fire risk level to those of the conventional model. Zhang [25] and Lin et al. [26] demarcate the urban risk zones and perform comprehensive evaluations of the city protection grade, using the fire data and kernel density of the key objects for safety; the zone with a higher grade is given priority in fire station site selection. Wang [27] established the classification of fire risk level based on the nuclear density of POI facilities, and substituted the SAVEE model for site selection planning. Fire station selection based on the evaluation results of fire risks greatly improves the effectiveness and service rates of fire stations. Nonetheless, fire outbreaks as random events are theoretically subjected to uncertainty. In other words, every space or building has the possibility of being on fire and shall be regarded as the potential fire outbreak point. Furthermore, fire outbreaks are attributed to numerous factors, and thus the hot-spot zone of fire outbreaks cannot fully manifest the outbreak risks, though it is, to some extent, representative. Moreover, the existing fire station selection models generally fail to consider the effects of traffic jams. The traffic network is in most cases abstracted into the road topological map, and the response time or service zone is calculated using the minimum weighted distance model [28–30]. It should be noted that the urban firefighting access way is the important premise for the rapid arrival of firefighting taskforces to the scene of the fire and thus the reduction and relief of fire disasters. The urban traffic network construction and traffic condition are important factors affecting the fire station arrangement and its optimization. Although there are studies considering the traffic capacity of the city road upon the travel speed of emergency rescue trucks, which is defined as the road design speed or assuming speed in these studies [31,32], they ignore the delay that may be caused by the ever-growing traffic jam in the city and the great threats to life and properties derived from the extended travel distance of firefighting taskforces. Wang [33] collects the data of the traffic performance index of Shenzhen over three years, and maps the daily traffic jam. Mao [34] and Ming [35] collected traffic data for one hour in the morning of June 2020 and one week in May 2020, respectively, to evaluate fire vehicle speeds based on Amap API. Specifically, the data collected by Ming [35] only have three constant speeds for three different traffic congestion states, and thus they are not representative.

To conclude, much progress has been made in addressing the fire station site selection issue based on models that consider temporal–spatial demands. However, there are two prominent shortcomings in these models. (1) The randomness of the spatial demand distribution is not sufficiently discussed. For random events, such as a fire, the existing site selection model mainly uses fire-related factors to divide the risk level. However, this is only an analysis of historical data, and the obtained site selection plan cannot be ensured to be valid during the whole planning period. Although certain efforts have been made to incorporate the spatial randomness factor into the site selection model, there is still a lack of a random simulation method for spatial demands. Specifically, such a method is expected to not only accurately describe the demand distribution but also effectively optimize the model based on the description. In this context, the site selection issue can be better solved. (2) Existing site selection models generally lack consideration of travel time and distance. Some studies have demonstrated the impact of traffic conditions on the accessibility of emergency stations by obtaining traffic data through APIs, and they have also have tried to include short-term data into the models. However, there are still no sufficiently efficient site selection methods that can stream long-term data and include traffic factors quantitatively into the model.

3. Introduction of the Model and Method

3.1. Methodology

This study targets the urban area of Nanjing City and builds the spatiotemporal dataset of multi-source data, including the fire history from June 2014 to June 2018 as well as the real-time traffic data and traffic network data via the Amap OPEN API. Moreover, this study investigates the regularity behind fire outbreaks and identifies and selectively incorporates the risk factors into the fire risk evaluation system. Based on the entropy weight method, the risk factors of the different fire types are normalized and the weight of each factor is obtained. After meshing the urban area of Nanjing, the fire risk value distribution across the grid cells is identified via superposition. Meanwhile, the fire data of Nanjing in 2014–2018 are put through the clustering analysis, and the probability distribution at each point of the cluster is calculated using the average score. The Monte Carlo simulation is performed to map the random fire outbreak points, and then these massive randomly distributed simulated fire outbreak points are inputted into the actual traffic network model, which is finally transferred into the location set cover model constrained by the land use plan to solve the site selection problem (Figure 1). The presented method overcomes the difficulty in mapping the fires caused by the complex fire distribution and meanwhile eliminates the error induced by the idealization of the traffic situation. It is worth noting that this research project is granted by the Nanjing Fire Department and the Nanjing Bureau of Planning and Natural Resources. The research findings provide important guidance on the actual site selection of fire stations of Nanjing, and the presented method, a novel method for urban fire station site selection in the big data era, is practical to provide references for analogous specialized planning of other cities. Fire station site selection based on fire risk evaluation can greatly improve the effectiveness and service rate of fire stations.

3.2. The Improved LSCP Model

At present, the frequently used site selection models include the set cover model, the P-median model, and the P-center model. The set cover model is in essence a model dealing with the optimal site selection for discrete points, which are often the identified distributed demand points. The site selection based on the set cover model needs to comprehensively consider multiple factors such as the quantity and position of the facility placement and the economic effectiveness. The set cover model can be divided into two types by the corresponding objectives, namely the set cover model and the maximal covering location model. The former is first proposed by Toregas et al. [36,37] and aims at the minimum facility or construction costs under the premise that all demand points are covered. Subsequently, Church and ReVelle [38] develop the maximal covering location model, based on the location set cover model. The maximal covering location model targets the facility layout that facilitates the maximum served demands under the premise of the known service station quantity and service range. The most important difference between the maximal covering location model and the set cover model is whether or not the facility quantity is considered. In addition, the former highlights serving demands, while the latter emphasizes the minimum cost. The P-median model, proposed by Hakimi [39], aims at minimizing the total distance between each demand point and the corresponding facility, so as to realize the best overall service performance of fire stations. Then, Hakimi modifies the P-median model and develops the P-center model [40]; the optimization objective of which is to minimize the maximal distance for all demand points to the corresponding facilities. The P-center model can result in a more distance-balanced layout of facilities. Given the goal of the fire station planning in China to realize full coverage of the rescue and disaster relief network across the urban and suburban areas and the reality, this paper adopts the location set cover (problem) model (LSCP) as the basic model.

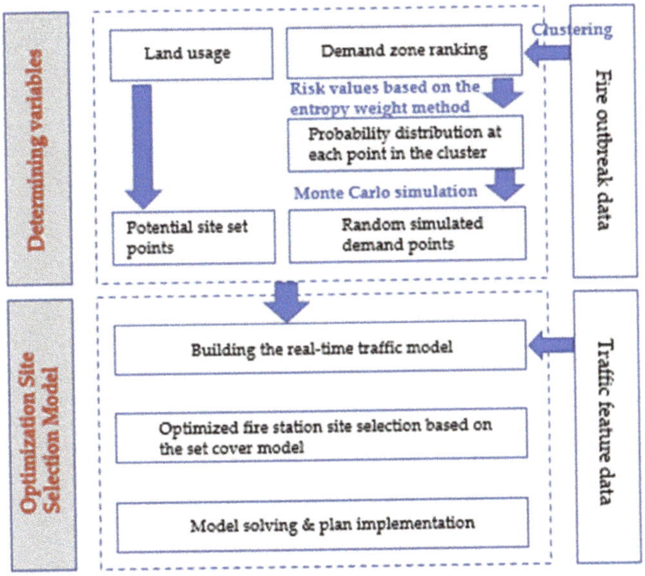

Figure 1. Workflow of our research methodology.

LSCP is one of the most important site selection models for locating emergency facilities [41]. Toregas et al. [36] for the first time apply LSCP in locating fire stations, which is shown below:

$$\min \sum_{j \in W} x_j \tag{1}$$

$$\text{s.t.}: \sum_{j \in N_i} x_j \geq 1, \forall i \in V \tag{2}$$

$$x_j \in \{0,1\}, \forall j \in W \tag{3}$$

where V is the set of the demand zone; W is the set of facility points; i is the demand point sequence number; j is the facility point sequence number; N_i is the set of the facilities that can serve the demand point i; x_j is a variable equal to one or zero, representing whether or not to build the j-th facility.

The objective function Equation (1) requires a minimal quantity of the service facilities. The constraint Equation (2) states that each demand point shall be served by at least one facility. Equation (3) is constrained by the values of the variable. Equations (1)–(3) constituent a discrete model, requiring the input of a series of spatial demand elements (including points, lines, and planes) and the location set of potential facility sites. x_j represents the node, and j refers to being chosen to build a facility ($x_j = 1$) or not ($x_j = 0$). The LSCP model realizes coverage of any continuous space by placing a minimal number of facilities in some locations, which thus requires determining the variable value (namely the demand point) and solving x_j. However, the basic location set coverage model is not fully applicable to the fire requirement characterization. The model is improved according to the characteristics of fire needs. (1) A random simulation point i is used to replace the original event point. (2) The ideal travel time from the fire station to the demand area (t_{ij}) should be shorter than the target time (t_s^r) in the road network model (T) based on traffic congestion corresponding to any simulated demand point. The specific steps of the optimized location set cover model algorithm are as follows (Figure 2).

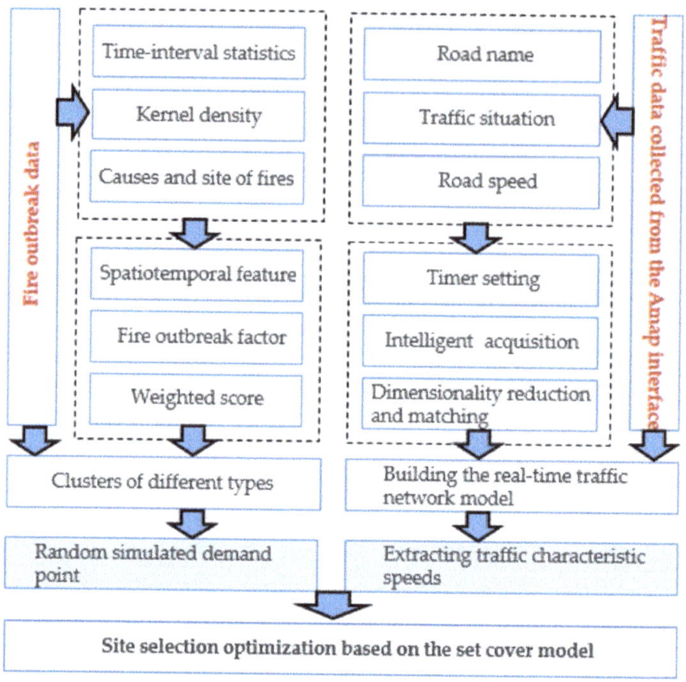

Figure 2. Technical workflow of data processing.

3.2.1. Generation of Random Demand Points

The overall methodology of this research is summarized below. First, the spatiotemporal features are analyzed using the historic fire outbreak data, the fire outbreak factors are identified, and each index is weighted using the entropy weight method. The study area is ten meshed into spatial grid cells, and the weights are assigned to each grid cell to obtain the fire outbreak risk value of each grid cell. Second, the fire outbreak points are put through the k-means clustering analysis, which generates several clusters, each with a cluster center; the point coordinates in the cluster are checked for compliance with the Gaussian distribution; the mean value and variance are calculated; the average score across a cluster is computed; and the probability distribution at the point in the cluster (interval) and the number of points generated by each cluster are calculated. Finally, random fire outbreak points are generated via the Monte Carlo simulation based on the mean value and variance to calculate the confidence level.

Further explanation is needed. (1) The meshing method. The study area needs to be meshed to study the potential distribution of different fire risk factors in space. Traditional gridding methods are normally based on regular quadrilateral units. In contrast, this study meshes the study area into 2784 closely connected honeycomb units that are regular hexagons with side lengths of 1000 m in ArcGIS10.6. Although the hexagonal shape is more complicated, its advantages are also very prominent: (i) it can reduce the sample deviation caused by the boundary effect of the grid shape; (ii) it is the most circular polygon and can be inlaid to form a uniform grid; and (iii) its pattern can be accurately recognized.

(2) The entropy method. In order to minimize and avoid subjective factors and some objective limitations in the process of weight determination, this paper uses the entropy method to assign weight to each index. Entropy is originally a thermodynamic concept in physics, and it can reflect the degree of chaos in the system. In the information theory, entropy is a measure of the degree of chaos in the system, while information is a measure of the degree of order. Entropy and information have the same absolute values but different signs. In the index data matrix $X = \{x_{ij}\}_{n*m}$ that consists of n plans to be evaluated and m evaluation indexes, a large degree of data dispersion corresponds to smaller information entropy, which means a larger amount of information, thus higher importance to the comprehensive evaluation, and consequently a larger weight. In this way, the index weight can be scientifically assigned to solve the problem of information overlap among multiple indexes. Practically, this study first evaluates the degree of dispersion of each sample data, then uses information entropy to determine the index weights, and finally assesses the fire risk factors in the urban space. The calculation steps are as follows:

(a) Standardizing the original positive index data: $x_{ij}' = (x_{ij} - \overline{x})/s_j$ where x_{ij} is the original value of the i-th sample and the j-th index, x_{ij}' is the standardized index value, $\overline{x}$ and s_j are the average and standard deviation of the j-th index, respectively.
(b) Quantifying all indexes in the same way and calculating the weight of the i-th factor in the j-th index (p_{ij}):

$$p_{ij} = Z_{ij}/\sum_{i=1}^{n} Z_{ij}(i = 1,2,\ldots,n;\ j = 1,2,\ldots,m)$$

where n is the number of samples (indexes) and m is the number of indexes.
(c) Calculating the entropy value of the j-th index (e_j): $e_j = -k\sum_{i=1}^{n} p_{ij}\ln(p_{ij})$ where $k = 1/\ln(n)$ and $e_j \geq 0$.
(d) Calculating the difference coefficient (g_j) of the j-th index: $g_j = 1 - e_j$
(e) Normalizing the difference coefficient and calculating the weight of the j-th index

$$(g_j) : w_j = g_j/\sum_{i=1}^{m} g_j(j = 1,2,\ldots,m)$$

(3) The Monte Carlo simulation. This is a common computation method based on probability statistics, also called the random sampling/statistical testing method. Its principle is that the probability of an event can be estimated using the occurring probability of this event in a large number of tests. There are several reasons for choosing this method: (i) it is convenient to perform a large number of repeated sampling of all spatial data, and it can simulate the dynamic relationship between variables randomly, thereby solving uncertain and complex problems; (ii) it is of high applicability and is less constrained by the problem conditions than other numerical methods; and (iii) when performing numerical calculations, its convergence speed is not related to the problem dimension. The assumption function is presented below:

$$Y = F(X_1, X_2, \ldots X_n) \tag{4}$$

where $(X_1, X_2, \ldots X_n)$ represents the known spatial distribution probability of fire outbreak points in each cluster after performing the clustering analysis of fire outbreak points over five years.

In most cases, it is very difficult to calculate the probability distribution and its mathematical features of Y via an analytical process. The Monte Carlo method using a random number generator can perform a great amount of repeated independent random sampling for the random variable, by generating a set of values of the random variables via direct or indirect sampling, and can produce the probability distribution of the function Y (namely, the simulated fire outbreaks) that is close to the reality. At last, these sampled values are substituted into Equation (6) in Step 3.2.3 one set after another until the ultimate results are obtained.

3.2.2. Traffic Model T Incorporating Traffic Jam

In site selection of firefighting facilities, precisely calculating the traffic cost can result in increased accuracy of evaluation upon the firefighting service efficiency and capacity. This paper, with the help of the big data analysis, incorporates the traffic jam factor into the site selection process, and thus produces an optimized planned fire station layout. Both data acquisition and processing are implemented using Python3.6. The obtained data are from the real-time traffic speed and congestion status data of Nanjing's road network provided by Amap Maps from 28 May to 10 June 2018, with a data collection interval of 1 h and thus 24 pieces of data per day. Amap Open Platform Web Service API function is used, and the request parameters include user authority identification, key, query road level, return data format type, callback function, and the longitude and latitude coordinate pairs of the lower left and upper right vertices of the rectangular area to be queried. Among these parameters, key refers to the authorization key that the user applies for on the official website of Amap Maps; query road level, return data format type, and callback function are all set as the default values. The innermost distance in the rectangular area to be queried is required to be less than 10 km, and due to this limitation, the Nanjing city area is segmented into 230 rectangular units (each 0.06° by 0.06°). Then, the units are merged into the actual regional traffic network, and the raw data are pre-processed to link the effective traffic situation information. In the meantime, the traffic situation acquisition program is set to auto-run at the interval of one hour (3600 s) for 28 days in a row, and the intelligent batch processing of streaming data is achieved, thus allowing for automatic acquisition, pre-processing, spatialization, and storage (readable in the shapefile format) of the traffic data of the whole city. By doing so, the spatiotemporal dataset of the traffic situation is built. At last, the characteristic speed of each road is extracted via dimensionality reduction and mapped for visualization.

As the speed obtained by the Amap Map API is based on real-time road conditions, it is vulnerable to influences of holidays, major traffic accidents, and traffic jams. However, it is critical during the traffic feature extraction by highlighting the stable traffic connections and travel time between facilities and demand points. Therefore, it is necessary to measure the difference and stability of the transit time obtained by the Amap Map API at different times. Accordingly, this study chooses to collect vehicle speed data for the two weeks from May 28 to 10 June 2018, and average them to each hour. Then, the vehicle speed data of these two weeks are compared with those of weekdays and weekends, respectively. The covariance calculation shows that the average speed distribution curve during the studied two weeks can be well fitted. Therefore, it can be inferred that the difference in the average speed between weekdays and weekends does not affect the analysis of this study. The all-day congestion delay index of Nanjing is calculated to be 1.55, and the average speed of the whole day is about 28.29 km/h. In contrast, the congestion delay index during the rush hour is up to 1.81, and the corresponding average speed is as low as 24.21 km/h. The urban traffic congestion problem in Nanjing has become a serious issue to be solved. However, it is not suitable to use the congestion speed for the site selection model, because that will bring inefficient resource allocation. As a countermeasure, the characteristic speed in this model is set as the two-week average speed of each road to represent the traffic difference in different regions. Multiple small fire stations will be considered in congested areas in our future research.

3.2.3. Site Selection Model Based on Random Simulated Demand Points and Traffic Characteristic Speeds

The determined demand points and the traffic network model T are then substituted into the location set cover model to compute x_j and the set of the ultimate planned station site set W. For the existing fire stations (corresponding to the set H), the default operation is to directly merge them into the set W (in other words, no demolition or relocation). Owing to the predictability of fire outbreaks, the optimized algorithm of the set cover location model is used to determine whether or not to include the station candidate j in the set M to the set W one after another and ultimately produce the fire station distribution with the minimal station quantity and maximal firefighting efficiency improvement. The mentioned algorithm is presented below:

$$\min \sum_{j \in W} x_j \tag{5}$$

$$\text{s.t.} : \Pr\left(t_{ij} \le t_s^r\right) \ge \alpha \tag{6}$$

$$t_{ij} \le t_s^r, \forall i \in V, \exists j \in W \tag{7}$$

$$x_j \in \{0,1\}, \cdot \forall j \in W \tag{8}$$

where i is the sequence number of the simulated demand point; j is the sequence number of the new site candidate; V is the set of the demand zone; M is the set of the new site candidates; W is set of the facility sites planned to put into service; H is the set of the existing stations; t_{ij} is the actual travel time from the fire station j to the random simulated point i in the demand zone, derived from the traffic network model T based on the actual traffic jam situation; x_j is a variable equal to one or zero, representing whether or not the site candidate j will be put into service (putting into service corresponds to $x_j = 1$, and otherwise $x_j = 0$).

The objective function Equation (5) requires a minimal quantity of the constructed service station. The objective function Equation (6) is a global constraint requiring that, for any simulated demand point, the ideal travel time (t_{ij}) from the corresponding fire station to its service zone shall be lower than a specified time (t_s^r), according to the traffic network model T considering the traffic jam. Calculating the actual travel time t_{ij} from the fire station j to the random simulated point i in the demand zone needs to determine the corresponding distance and travel speed. Here, we first change the random demand point generated in Step 3.1 with the potential and existing fire station, convert their coordinates into a unified system, and export them into the spatiotemporal dataset. Subsequently, the single-source shortest path between two points in the weighted graph (namely the traffic network model T) is calculated using the path solver based on the Dijkstra algorithm, which is divided by the traffic characteristic speed of the corresponding road to yield the travel time. The travel time results are converted into the time matrix t_{ij}. This constraint requires the response probability above 90% over the service range (considering errors, α is artificially set as 90%). This paper takes Nanjing city as an example, and accurately dissects the spatiotemporal characteristic of firefighting events, based on the actual demand points of firefighting events and the traffic network data. Thus, we manage to perform more precise measurements of each variable in the equations and develop a more reasonable site selection plan. The next chapter introduces the fire data of Nanjing and the data sources.

4. Data Sources

4.1. Study Area

Nanjing, the capital city of Jiangsu Province, has an urban area of 1364.85 km^2. By 2020, there are a total of 57 fire stations that have been built and put into service in Nanjing (including 2 special-duty fire stations, 54 normal fire stations, and 1 firefighting support fire station). A total of 39 stations are located in the urban area, while 18 lie in the suburban area. There are no underground fire stations. The current distribution of the effectively-operated fire stations is generally characterized by the high density in the central urban area and low density in the developing urban fringe area. Xuanwu, Gulou, Qinhuai, Yuhuatai, Jianye, Qixia, and Pukou districts are found with more existing fire stations (points), correspondingly associated with relatively concentrated distributions (Figure 3). The Standard for Urban Fire Station Construction (MOHURD Standard 152-2017, a national sector standard of China) stipulates that the service area of the normal fire station in cities should be no larger than 7 km^2, and that of the normal fire station in suburbs should be no larger than 15 km^2, which means there should be at least 114 fire stations in Nanjing. The quantity of the existing fire stations is far less than the stipulated one. The service area of the fire stations is, in most cases, overwhelmingly large, and the five-minute arrival required by the standard cannot be realized (at present, the average travel time to the fire site is about 11.6 min). The service area of an individual fire station far exceeds the upper limit of the service area. Moreover, the traffic jam of the city is intensified over recent years. Thus, the optimal response time for rescue and disaster relief cannot be guaranteed. It should also be noted that the land that can be used to build fire stations is in short in the city, which leads to the uneven distribution of fire stations and considerable firefighting dead zone. For example, the South New Town centered at the Nanjing South Railway Station and the Maqun area in the eastern Purple Mountain are found with no placement of fire stations.

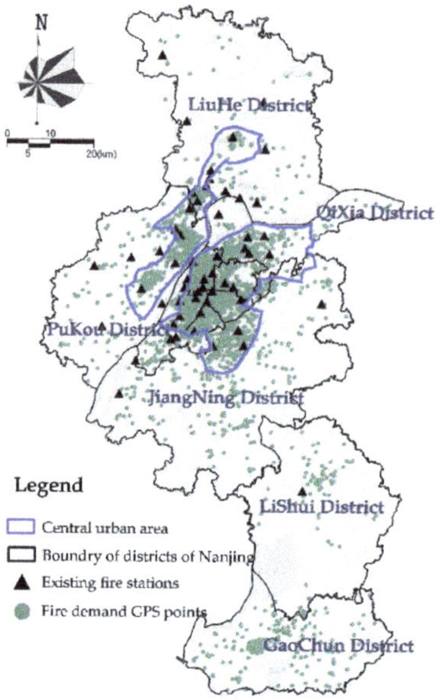

Figure 3. The fire outbreak locations and current fire stations in Nanjing.

4.2. Fire Outbreak Data

4.2.1. Basic Characteristics of Fire Outbreaks

Optimization of the fire station layout needs to consider the fire characteristics in Nanjing. With the help of the Nanjing Fire Department, we collect the fire outbreak time, fire location, fire site, and cause of fire recorded by the Emergency Call for Fires (119) from 1 June 2014 to 1 June 2018, and identify a total of 9561 pieces of fire data (Table 1). The location information of all fire events is translated into the latitudes and longitudes, based on which the fire events are spatially positioned in the hexagonal grid cells with radii of 500 m meshing the Nanjing urban area. A summary of kernel density analysis on the fire outbreak quantity in each grid cell reveal that the hot-spot zones of fires are mainly located in the core areas of the central urban area (including Gulou, Xuanwu, Qinhuai, and Jianye districts), which is the intensive core of fire outbreaks. The global spatial autocorrelation coefficient reaches 0.875, suggesting the extremely high spatial aggregation of fire outbreaks. The major peak occurs at the Xinjiekou area, around which the fire outbreaks gradually decline radially towards the opposite direction, and meanwhile, another peak of fire outbreaks occurs at the outward Gaochun area (Figures 4 and 5).

Table 1. Example of GPS data for emergency service events.

Administrative District	Fire Time	Fire Location	Fire Site	Cause of Fire
Yuhuatai District	31 December 2015 16:12:00	The fourth block of Langshilv	Others	Unknown
Jianye District	31 December 2015 14:50:00	The Eighth Bureau of Construction, Youth Olympic Village, Jianye District	Waste	Other-residual fire
Yuhuatai District	31 December 2015 13:53:00	Jindi Free City reed marshes	Others	Unknown
Xuanwu District	31 December 2015 12:30:00	East of Jiming Temple, Xuanwu District	Others	Electrical fire-electrical circuit failure-other
Yuhuatai District	31 December 2015 08:55:00	Old glass factory next to Oasis Machinery Factory	Others	Unknown
Gulou District	31 December 2015 03:03:00	North Gate of Workers' New Village, Gulou District	Others	Electrical fire-electrical circuit failure-other
Gulou District	30 December 2015 21:48:00	1st Floor, No.49 Yucai Apartment, Gulou District	Residence	Electrical fire-electrical circuit failure-other
Gulou District	30 December 2015 16:48:00	Room 101, Unit 7, Building 16, Xinyi Village, Jinling Community, Gulou District	Residence	Electrical fire-electrical circuit failure-other

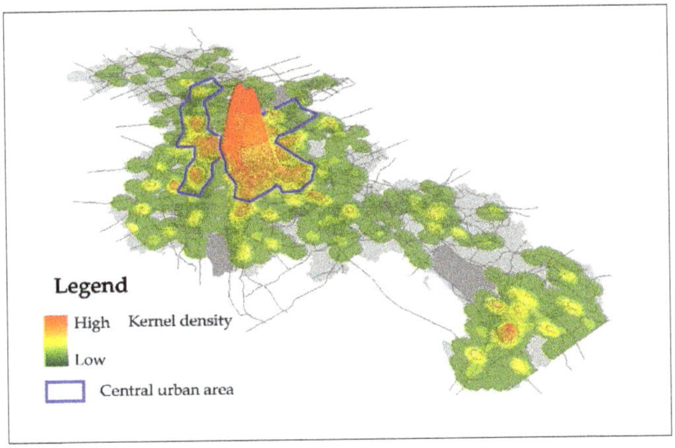

Figure 4. Fire outbreak distribution (3D view).

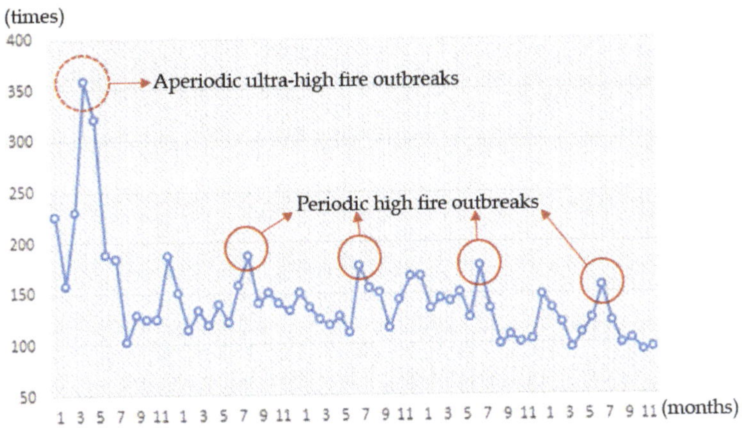

Figure 5. Fire outbreaks by month.

From a temporal point of view, the fire outbreaks are obviously periodic. Regarding months, July and August are associated with periodic high fire outbreaks, resulting in nearly 200 fire events over the last three years. However, July–August in 2014 presents itself as a rare valley of fire outbreaks, which is attributed to the fire security circles (for fire prevention and facility protection) set up by the Nanjing Municipal Government for the Youth Olympic Games held at that time in Nanjing. Besides, January and February in winter are the secondary periodic peak for fire outbreaks. April in 2014 is seen with aperiodic ultra-high fire outbreaks (360 outbreaks).

4.2.2. Fire-Triggering Factors

The cause and site of fires are analyzed using the dominance index. The electrical fire is the main cause of fires for all districts, accounting for 76.3%, followed by the careless living fire activity (11.1%), the production operation fire (4.5%), and self-ignition fires (2.5%). The electrical fire is mainly triggered by electrical circuit failure, electrical equipment failure, and electrical heating devices, and often occurs in summer and winter. The high ambient temperature (plus the self-heating of the running electrical equipment) and thunderstorms in summer promote occurrences of electrical fires. In winter, the overhead electrical lines are prone to contacting and connecting driven by intensive winds and discharging electricity to cause fires. Moreover, inappropriate applications of heating devices and burning inflammable materials are also the characteristic causes of fire in winter. These analysis results are consistent with the temporal characteristics of fire outbreaks mentioned above. Weather, hazardous goods, gas pipe network distribution, and population distribution are all main factors related to fire outbreaks.

Furthermore, the fire outbreak sites are divided into five types, namely the residence land, industrial land, facility land, commercial land, and land for plazas and plants. The uppermost outbreak site is the residence land, accounting for 71.2%, while the shares of the other four types of lands are 4.5%, 12.7%, and 2.4%, respectively, and other types of fire account for 9.2%. Residence fires are mainly caused by electrical failures, with a correlation coefficient of 0.725. Their frequency distribution of daily outbreaks is found to conform to the exponential law. In other words, for most days, the daily fire outbreaks are very low, and yet for some specific dates, the fire outbreaks dramatically grow. The presented analysis further reveals that fire outbreaks are highly dependent on the residential land and population distribution, and present a certain regularity.

5. Optimization and Application of the Site Selection Model

5.1. Underlying Fire Risk Evaluation

The fire risk evaluation of the urban area involves various aspects, such as the occurrences, development, control, and firefighting and rescue in the urban area, and is characterized by numerous and complex influential factors. The fire risk evaluation calculates the fire outbreak probability, predicts the disaster consequence, and quantifies the fire risk, by analyzing the factors affecting fires. It can provide scientific references for developing the urban firefighting plan and direct the urban fire safety management to improve the resistance of cities to fire disasters. By far, the previous urban fire risk evaluation cases in China and other countries mostly focus on the evaluation index system establishment, the evaluation model and its application, etc. The risk ranking of the U.S. considers risk factors such as the application scenario of architectures, building density, and fire separation, while the "urban ranking system" of Japan mainly considers the type and structure condition of buildings, climate condition, and firefighting system [8,10]. Ding and Wang [42] choose the population, firefighting infrastructure, firefighting capacity, public security condition, and major hazardous source as the five primary factors. Zhang [11] further refines the risk factors into the firefighting key area, population density, high-rise building distribution, large crowded underground space, and firefighting taskforce to build the fire risk evaluation framework, from the perspective of the urban space, and identifies the high, medium, and low-risk areas. In general, when it comes to building the evaluation system, previous studies often focus on investigating and refining the fire hazardous source, firefighting capacity development, firefighting management status, and regional disaster-resistant capacity. The relevant studies are increasingly mature, and yet with the ever-complicated development of large cities, the urban fire risk evaluation shall highlight the spatial analysis of each underlying factor, quantify the risk fire evaluation into the material space, and precisely rank the underlying fire risks of each urban land blocks, in order to realize refined management of the urban firefighting security. Hence, based on the previous studies, the features of this research, and the characteristics of the fire outbreaks in Nanjing, the population density, high-rise building distribution, underground space distribution, site distributions of gas facilities, and hazardous chemical substances, and historic fire outbreak frequency are determined as the six major underlying factors for evaluation. The weight and score of each factor and the comprehensive score distribution across the urban area are calculated using the entropy weight method (Table 2, Figures 6 and 7). This research simulates the underlying fire outbreaks and assumes that the firefighting capacity is limited to putting out fires and rescue. Thus, the firefighting capacity is not included as an evaluation factor.

Table 2. Fire risk ranking.

Underlying Fire Risk Ranking		
Risk Factor	**Evaluation Factors**	**Weight**
Fire outbreak probability	Historic fire outbreak frequency	0.14
	Population density	0.15
Fire hazardous source	Gas pipe networks	0.24
	Hazardous chemical substances	0.14
Regional disaster resistance	Underground space distribution	0.12
	High-rise building distribution	0.17

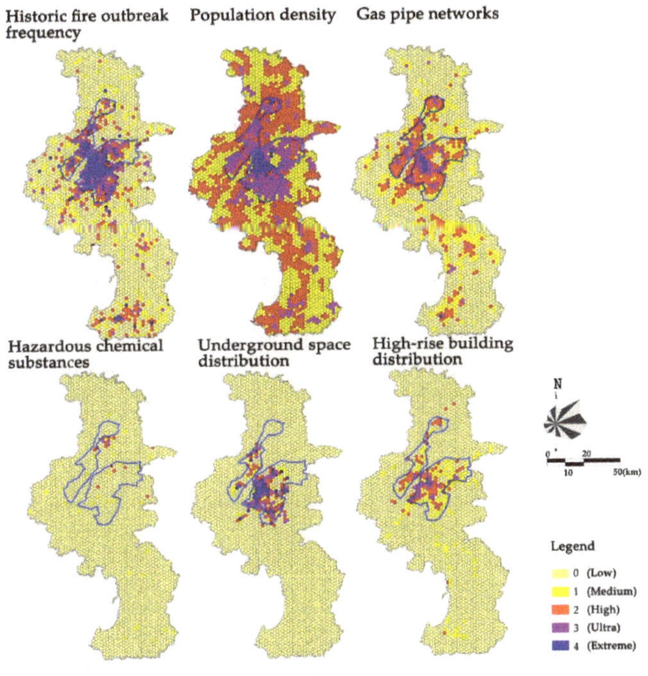

Figure 6. Evaluation factors for fire risks.

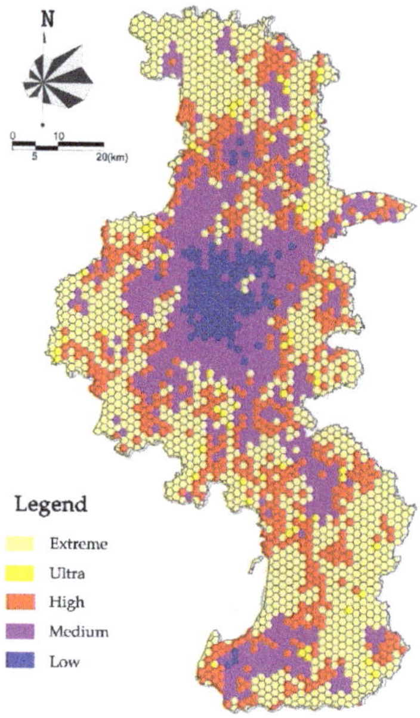

Figure 7. Fire risk scoring.

5.2. Ranking of Demand Zones

5.2.1. Determining the Fire Station Site Candidates with the Urban Planning Taken into Consideration

By comprehensively considering the urban land use plan and the current land use and construction status and eliminating the positions that are unfavorable for building fire stations (e.g., hills and lakes), the land blocks over 2000 m^2 (according to Standard for Urban Fire Station Construction, a national sector standard of China) are screened out. There are a total of 4084 region blocks, which are defined as the site candidate set M.

5.2.2. Clustering and Simulation of Historic Firefighting Data

The three-dimensional K-means clustering analysis is carried out for the demand spatiotemporal points of fire outbreaks from 2014–2018, and the cluster quantity is K. The average horizontal and vertical coordinates μ of the relative cluster center for points of each cluster are calculated, which is combined with the fire risk value of the corresponding grid cell to form the 3D coordinates for clustering analysis. The clustering adopts the mean-variance normalization, and the clustering and Gaussian distribution fitting results of the data of 2014–2018 are used. It shall be noted that the K-means clustering requires determining the cluster quantity K, which is set as seven in this research. The main clustering process has three steps: first, K random starting points are chosen as the mass centers; second, the data in the dataset are assigned to each cluster according to the distance to the mass centers, and the averages of each cluster are calculated and set as the new mass centers; and third, the second step is repeated until there is no change for all clusters (Figures 8–10). According to the elbow method, K = 7 brings about the best classification performance for spatiotemporal features, which means no concealing features and in the meanwhile clear grouping. The data clustering and Gaussian fitting results are used for the Monte Carlo simulation to generate several random demand (fire outbreak) points.

5.2.3. Building the Traffic Network Model and Calculating the Minimal Time Matrix t_{ij}

Based on the processed traffic network shapefile-format files, the traffic network dataset is constructed as the traffic network model, via topological operations, such as road intersection breaking and interface merging. The model consists of the road intervals and intersections. The road interval is the line element of the traffic network, which is represented using the arc in ArcGIS. Each interval is assigned the attributes of the jam vehicle speed (km/h), average vehicle speed (km/h), travel time at the average speed (s), and length (km). The road intersection, represented by the node in ArcGIS, is combined with the turning table for road intersections to vividly mimic the actual on-the-road scenarios, such as waiting for the traffic light, no straight-through, no left turn, and passing through the elevated road. The travel time for each road interval is calculated in ArcGIS, according to the actual traffic network and corresponding characteristic speed, which is defined as the average actual travel time. The vehicle speeds between two adjacent weeks in the study period are compared and the covariance calculation results show that the distribution of the average vehicle speed at exact hours during the two weeks can be well fitted, with the correlation coefficient of 0.973. Thus, the data of every two weeks is of periodic representativeness in our research.

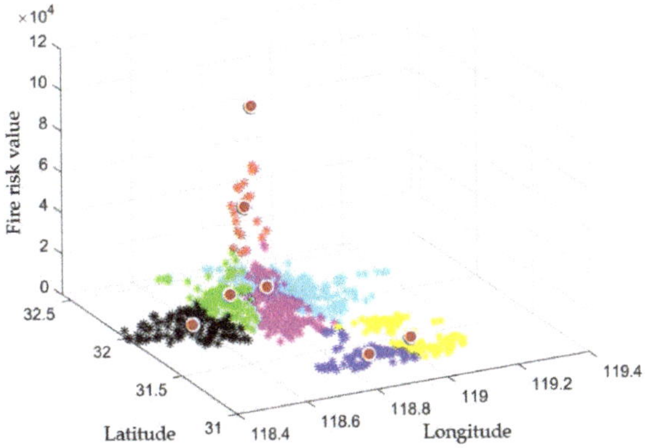

Figure 8. Clustering results of the historic fire data (K = 7).

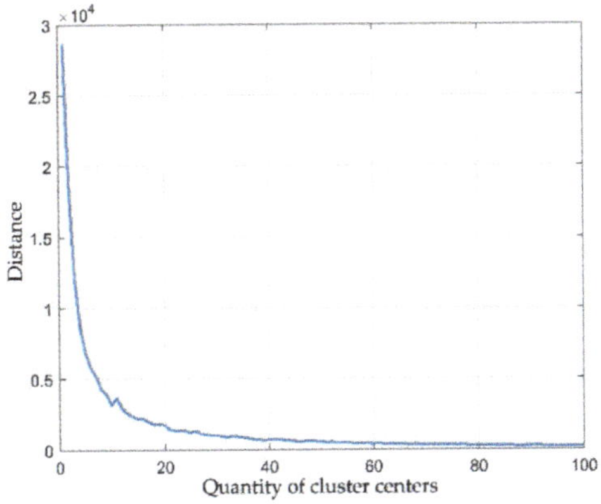

Figure 9. Elbow method.

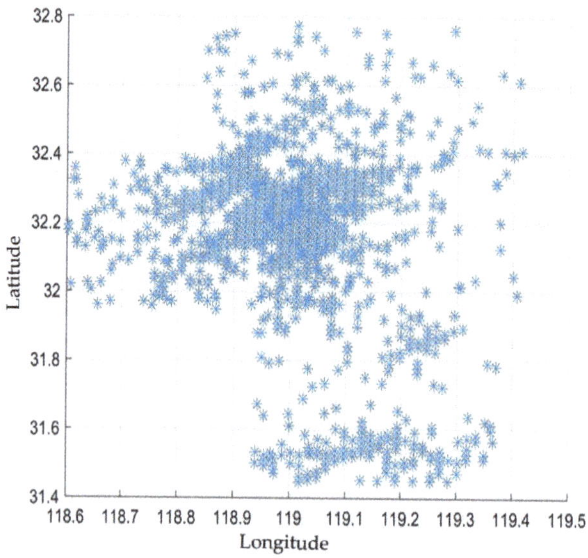

Figure 10. Random demand (fire outbreak) points.

5.2.4. Generating the Set of the Planned Sites W

It is determined that the set of the current 57 stations (the H set) is directly merged with the set W, with no station demolition and relocation. Based on the algorithm presented in Chapter 1, the calculation is made for each station site j to determine whether or not to be included in the set W, and the actual factors of Nanjing are all input into the global constraint equation. t_s^r is set as five minutes, and α is set as 0.90, which means that, for the clustering-based random simulated demand point j, the travel time to the station i included in the set M (the travel time is added with the OD time matrix) is less than five minutes. The problem is solved in Matlab R2021a using the genetic algorithm, and the minimal station quantity and station distribution (x_j and j) that can best improve the firefighting efficiency are obtained.

5.2.5. Adjusting the Model-Produced Results and Reviewing the Planned Land Use for the Fire Station Sites According to the Regulatory Plans of Each District to Ensure the Feasibility of the Ultimate Planned Fire Station Layout

There are 274 ultimate planned fire stations (Figure 11), including 57 existing stations and 217 to-be-built stations. It should be noted that 28 of the model-planned fire stations are adjusted. Among these fire stations, 125 stations mainly serve the central urban area and its surrounding, while 149 stations mainly serve the urban fringe area. The average service range for each fire station is 4.3 km^2, less than the specified upper limit of 7 km^2. Our calculation shows that in this plan, the service area that realizes the five-minute arrival of the firefighting taskforce in the central urban area and concentrated construction area accounts for 91% of the total construction land use, and 99% of the construction land use of the central urban area. The locations of 217 fire stations have been fed back to the regulatory plan development group for approval of their land uses. The fire stations are all placed in the street-facing sections of the main and secondary main roads, more than 200 m far away from the sites storing hazardous goods, such as gas and LNG stations, and 50 m from crowded sites. The firefighting response time is decreased from the current 11.6 min to 5 min.

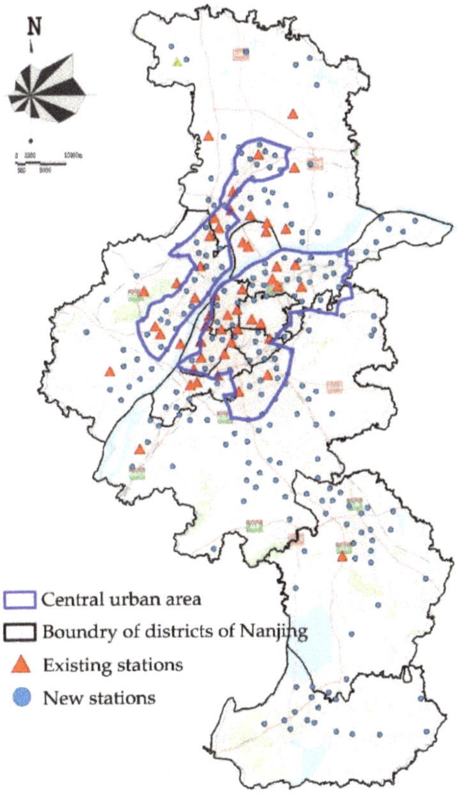

Figure 11. The final layout of fire stations.

6. Discussion

(1) Optimization of the emergency facility site selection has always been a classic research topic in geography. Given that the fire outbreak probabilities are different for demand points of the varied fire risk zones, this research proposes a site selection method that manages to realize the arrival of firefighting taskforces within the target time to all random simulated fire outbreak points under the constraints of the administrative regulatory plan and actual traffic situation. The solution workflow based on the set cover model is designed, and a case study of Nanjing city is performed. Compared with the conventional model, the optimized location set cover model greatly improves the covering range and effectively reduces the firefighting response time. Our research also reveals the considerable flaws of the layout of the fire stations in Nanjing. Specifically, due to the insufficient and unbalanced distribution of firefighting resources in the districts, the current firefighting facility is far from satisfying the emergency response demand. The overall cover ratio of the fire station service is only 58%. The firefighting resources are excessively concentrated in the Gulou and Xuanwu districts. Consequently, the firefighting service has formed a large regional cover across the central urban area, while multiple firefighting dead zones exist in other current urban built-up areas. A lack of fire stations is the main reason for the large firefighting dead zone.

(2) Various fire risk factors have increasingly emerged due to the continuous concentration of population and resources in large cities. Moreover, the accessibility and rescue efficiency of the urban road network have been seriously affected by the overwhelming large traffic flow and, thus, the unsmooth fire control passages in the central and new urban

areas. Furthermore, fire control facilities in many large cities of China are not sufficient and outdated. All these factors bring about severe challenges to fire safety management in large cities. This study manages to accurately assess the distribution of urban fire risks, road traffic conditions, and other situations by analyzing the spatial data. In this way, much progress has been made in effectively suppressing the negative impacts of random fire occurrences and traffic delays on site planning, thereby saving urban public resources. As more and more cities are entering a digital era, our future research will consider and integrate different tools into the ArcGIS toolbox, which will be more conducive for the relevant department to optimizing fire station planning. These results will provide references for similar cities in China and thus help enhance the fire control capacities of cities.

(3) The site selection of fire stations in this paper assumes that facilities all belong to the same level. However, firefighting facilities are in fact subjected to differentiation in types, levels, scales, and service ranges. Such an assumption to some extent results in errors, which shall be corrected in further studies. In addition, the traffic cost based on the traffic network model may change due to road modification and widening. Therefore, it is critical to adjust and update the dynamic factors in this model in a timely manner. The next step is to integrate different tools into the ArcGIS toolbox and visualize it, so as to facilitate the timely correction of the road network and update the data. Given this, we plan to collect the traffic data over a prolonged period of time and investigate the vehicle speed variation with different seasons and months to more accurately calculate the time-cost factors and substitute them into our model. We also plan to collect the fire outbreak data in 2019–2020 and import them into our model for further comparison and validation. The firefighting response time is decreased from the current 11.6 min to 5 min.

7. Conclusions

Digitalization in the new era has penetrated into all aspects of urban life, transportation, and medical care. In this context, the emergence and development of spatio-temporal big data provide new opportunities for the optimization of emergency site selection. This study first introduces the basic model of emergency site selection based on actual traffic conditions and the simulation method based on random demand in space. Then, this study employs the K-means clustering method to quantitatively describe and simulate fire demand based on actual data. Subsequently, an optimization algorithm is established that integrates the actual speed of the road network into the location set cover model, which contributes to the shortest travel time from the fire station to the simulated demand point at the actual speed. In order to evaluate the validity of the proposed model, this study analyzes the spatio-temporal characteristics of fire data from 1 June 2014 to 1 June 2018 in Nanjing City, identifies various necessary factors based on the modeling analysis, and solves the model with a set target time of 5 min under the conditions of land-use constraints. Compared with traditional models, the optimized location set cover model greatly improves the coverage area and effectively shortens the fire response time. This study is commissioned by the Nanjing Public Security Fire Bureau, and the results of this study provide important theoretical support for the statutory plan, i.e. the Nanjing Fire Station Layout Plan, implemented by the local government in 2020. Nonetheless, congestion status might change as the city develops, which will have an impact on the existing research results. In future work, we will regularly monitor changes in the road network status and adjust the fire station layout accordingly, including adding some small fire stations where necessary.

Author Contributions: Conceptualization, Bing Han and Mingxing Hu; Methodology, Bing Han; Software, Bing Han; Validation, Mingxing Hu; Formal Analysis, Bing Han and Mingxing Hu; Investigation, Bing Han and Tan Tang; Data Curation, Mingxing Hu, Writing—Original Draft Preparation, Bing Han; Writing—Review and Editing, Bing Han, Jiemin Zheng and Mingxing Hu; Visualization, Bing Han. All authors have read and agreed to the published version of the manuscript.

Funding: This research was funded by the National Natural Science Foundation of China, grant number 51778127.

Data Availability Statement: The original data were supported by Nanjing Fire Brigade.

Acknowledgments: The authors are grateful for the original data provided by Nanjing Fire Brigade and full cooperation.

Conflicts of Interest: The authors declare no conflict of interest. The funders had no role in the design of the study; in the collection, analyses, or interpretation of data; in the writing of the manuscript, or in the decision to publish the results.

References

1. Shin, S.C. Study on the emergency mechanism of major public events. *China Saf. Prod. Sci. Technol.* **2005**, *1*, 24–26.
2. Chen, C.; Ren, A.Z. Optimization of fire station locations using computer. *J. Tsinghua Univ. Nat. Sci. Ed.* **2003**, *43*, 94–97. [CrossRef]
3. Chen, H. *Study on Optimization of the Spatial Distribution of Urban Fire Station*; Tongji University: Shanghai, China, 2007.
4. Feng, F.B. *Study on Set Coverage Problem*; Shandong University: Jinan, China, 2014.
5. Badri, M.A.; Mortagy, A.K.; Alsayed, C.A. A multi-objective model for locating fire stations. *Eur. J. Oper. Res.* **1998**, *110*, 243–260. [CrossRef]
6. Chen, T.; Zhao, X.F. Creating & application of facility location model. *J. Nanjing Univ. Inf. Sci. Technol. Nat. Sci. Ed.* **2010**, *2*, 211–215.
7. Barr, R.C.; Caputo, A.P. Planning fire station locations. In *Fire Protection Handbook*; Cote, A.E., Ed.; National Fire Protection Association: Quincy, MA, USA, 1996; pp. 311–318.
8. Yoshida, Y. Evaluating building fire safety through egress prediction: A standard application in Japan. *Fire Technol.* **1995**, *31*, 158–174. [CrossRef]
9. Bukowski, R.W.; Clarke, F.B.; Hall, J.R.J.; Stiefel, W.S. *Fire Risk Assessment Method: Description of Methodology*; National Fire Protection Association: Quincy, MA, USA, 1999.
10. Insurance Services Office. *The Fire Suppression Schedule*; ISO Press: New York, NY, USA, 1999.
11. Zhang, Y.X.; Bian, Z.H. Initial probing the fire risks grades of Soozhou's Old City Region. *Fire Technol. Prod. Inf.* **2003**, *02*, 10–12.
12. Du, X.; Zhang, X.; Liu, T.Q.; Ma, Y.H. The current status and applying of city fire risk assessment technology in foreign country. *Fire Sci. Technol.* **2004**, *23*, 137–139. [CrossRef]
13. Lian, D.J.; Dong, X.L.; Wu, L.Z. A summary of fire risk assessment of city area. *Fire Sci. Technol.* **2004**, *23*, 240–242. [CrossRef]
14. Li, B.H. *Research of Urban Fire Risk Assessment Based on Fuzz Information Optimization Method*; University of Science and Technology of China: Hefei, China, 2010.
15. Chen, G.L.; Hu, R.; Wei, G.Z. Study on comprehensive assessment index system of urban fire risk in Beijing City. *China Saf. Sci. J.* **2007**, *17*, 119–124. [CrossRef]
16. Wu, X.T.; Wu, L.P. Evaluation of the Fire Emergency Rescue Capability in Urban Community. *Procedia Eng.* **2011**, *11*, 536–540. [CrossRef]
17. Zhang, C.W.; Xia, C.H.; Zhong, S.B. GIS application framework of urban fire risk regionalization. *Fire Sci. Technol.* **2012**, *31*, 1233–1237. [CrossRef]
18. Chevalier, P.; Thomas, I.; Geraets, D.; Goetghebeur, E.; Janssens, O.; Peeters, D.; Plastria, F. Locating fire stations: An integrated approach for Belgium. *Socio Econ. Plan. Sci.* **2012**, *46*, 173–182. [CrossRef]
19. Chen, Z.F.; Chen, J.; Huang, C.F.; Tan, M.Y. Fire risk assessment index system for large-scale public places(II): Indexes and their weights. *J. Nat. Disasters* **2006**, *15*, 164–168. [CrossRef]
20. Hu, C.P. *Research on Regional Fire Risk Assessment and Flighting Rescue Strength Layout Optimization*; Tongji University: Shanghai, China, 2006.
21. Wang, D.B. *Practice and Thought of the Applies of Fire Risk Evaluation to Fire Station Layout*; Northeast Normal University: Changchun, China, 2009.
22. Wang, Q.H.; Zhang, J.; Deng, J.; Zhu, H.Y. Optimization location of Fire Fighting stations based on fire risk assessment theory—A case study of Yaofeng town of Xia city in Shanxi province. *J. Xi'an Univ. Sci. Technol.* **2014**, *34*, 681–685.
23. Xu, Z.B.; Zhou, L. Hierarchical coverage location model for fire station based on urban fire risk: A case study on urban area in Jinan. *Geogr. Sci. Prog.* **2018**, *4*, 87–98. [CrossRef]
24. Zhibang, X.; Liang, Z.; Ting, L.; Zhonghui, W.; Li, S.; Rongwei, W. Spatial optimization of mega-city fire station distribution based on Point of Interest data: A case study within the 5th Ring Road in Beijing. *Geogr. Sci. Prog.* **2018**, *37*, 535–546. [CrossRef]
25. Zhang, G. Urban fire risk evaluation and its application based on spatial analysis: A case study of Xi'an. *City Plan.* **2016**, *40*, 59–64. [CrossRef]
26. Lin, J.X.; Jiang, X.; Zhu, J.G.; Cao, C.W.; Kong, Y. Application of GIS model in urban fire station layout planning. *City Plan.* **2018**, *42*, 65–70. [CrossRef]
27. Wang, W.; Xu, Z.; Sun, D.; Lan, T. Spatial Optimization of Mega-City Fire Stations Based on Multi-Source Geospatial Data: A Case Study in Beijing. *ISPRS Int. J. Geo Inf.* **2021**, *10*, 282. [CrossRef]

28. Li, X.; Claramunt, C.; Kung, H.-T.; Guo, Z.; Wu, J. A Decentralized and Continuity-Based Algorithm for Delineating Capacitated Shelters' Service Areas. *Environ. Plan. B Plan. Des.* **2008**, *35*, 593–608. [CrossRef]
29. Ni, J.; Wang, J.; Rui, Y.; Qian, T.; Wang, J. An Enhanced Variable Two-Step Floating Catchment Area Method for Measuring Spatial Accessibility to Residential Care Facilities in Nanjing. *Int. J. Environ. Res. Public Health* **2015**, *12*, 14490–14504. [CrossRef] [PubMed]
30. Chen, Z.F.; Li, J.W.; Lu, F.X.; Li, Q. Optimizing location of fire stations and it' s enlightenments for Xiongan New Area. *China Saf. Prod. Sci. Technol.* **2018**, *14*, 12–17. [CrossRef]
31. Tao, Z.; Cheng, Y.; Dai, T.; Rosenberg, M.W. Spatial optimization of residential care facility locations in Beijing, China: Maximum equity in accessibility. *Int. J. Health Geogr.* **2014**, *13*, 33. [CrossRef]
32. Xu, Q.; He, X.S. New fire station location problem based on AHP and set coverage model. *Inf. Comput.* **2019**, 26–28.
33. Wang, X.J. *Research on Urban Fire Station Location Optimization Based on Key Factor Analysis–ShenZhen Scenarios*; South China University of Technology: Guangzhou, China, 2013.
34. Mao, K.; Chen, Y.; Wu, G.; Huang, J.; Yang, W.; Xia, Z. Measuring Spatial Accessibility of Urban Fire Services Using Historical Fire Incidents in Nanjing, China. *ISPRS Int. J. Geo Inf.* **2020**, *9*, 585. [CrossRef]
35. Ming, J.; Richard, J.-P.P.; Zhu, J. A Facility Location and Allocation Model for Cooperative Fire Services. *IEEE Access* **2021**, *9*, 90908–90918. [CrossRef]
36. Toregas, C.; Swain, R.; Revelle, C.; Bergman, L. The Location of Emergency Service Facilities. *Oper. Res.* **1971**, *19*, 1363–1373. [CrossRef]
37. Friedrich, C.J.; Weber, A. *Alfred Weber's Theory of the Location of Industries*; University of Chicago Press: Chicago, IL, USA, 1929.
38. Church, R.; Revelle, C. The maximal covering location problem. *Pap. Reg. Sci. Assoc.* **1974**, *32*, 101–118. [CrossRef]
39. Hakimi, S.L. Optimum Locations of Switching Centers and the Absolute Centers and Medians of a Graph. *Oper. Res.* **1964**, *12*, 450–459. [CrossRef]
40. Hakimi, S.L. Optimum Distribution of Switching Centers in a Communication Network and Some Related Graph Theoretic Problems. *Oper. Res.* **1965**, *13*, 462–475. [CrossRef]
41. Murray, A.T.; Tong, D.; Kim, K. Enhancing Classic Coverage Location Models. *Int. Reg. Sci. Rev.* **2009**, *33*, 115–133. [CrossRef]
42. Ding, W.B.; Wang, J.P. Urban regional fire risk assessment with the method of BP neural network. *J. Yunnan Univ. Nat. Sci. Ed.* **2009**, *S2*, 232–235.

International Journal of
Geo-Information

MDPI

Article

A GIS-Based Bivariate Logistic Regression Model for the Site-Suitability Analysis of Parcel-Pickup Lockers: A Case Study of Guangzhou, China

Zilai Zheng *, Takehiro Morimoto and Yuji Murayama

Faculty of Life and Environmental Sciences, Graduate School of Life and Environmental Sciences, University of Tsukuba, 1-1-1 Tennodai, Tsukuba 305-8572, Ibaraki, Japan; tmrmt@geoenv.tsukuba.ac.jp (T.M.); mura@geoenv.tsukuba.ac.jp (Y.M.)

* Correspondence: zhengzilai23@gmail.com; Tel.: +81-80-4068-7266

Abstract: The site-suitability analysis (SSA) of parcel-pickup lockers (PPLs) is becoming a critical problem in last-mile logistics. Most studies have focused on the site-selection problem to identify the best site from given potential sites in specific areas, while few have solved the site-search problem to determine the boundary of the suitable area. A GIS-based bivariate logistic regression (LR) model using the supervised machine-learning (ML) algorithm was developed for suitability classification in this study. Eight crucial factors were selected from 27 candidate variables using stepwise methods with a training dataset in the best LR model. The variable of the proximity to residential buildings was more important than that to various commercial buildings, transport services, and roads. Among the four types of residential buildings, the most crucial factor was the proximity to residential quarters. A test dataset was employed for the validation process, showing that the best LR model had excellent performance. The results identified the suitable areas for PPLs, accounting for 8% of the total area of Guangzhou (GZ). A decision-maker can focus on these suitable areas as the site-selection ranges for PPLs, which significantly reduces the difficulty of analysis and time costs. This method can quickly decompose a large-scale area into several small-scale suitable areas, with relevance to the problem of selecting sites from various candidate sites.

Keywords: parcel-pickup lockers; site-suitability analysis; GIS-based; bivariate logistic regression model; suitability classification

Citation: Zheng, Z.; Morimoto, T.; Murayama, Y. A GIS-Based Bivariate Logistic Regression Model for the Site-Suitability Analysis of Parcel-Pickup Lockers: A Case Study of Guangzhou, China. *ISPRS Int. J. Geo-Inf.* **2021**, *10*, 648. https://doi.org/10.3390/ijgi10100648

Academic Editor: Wolfgang Kainz

Received: 30 June 2021
Accepted: 21 September 2021
Published: 26 September 2021

1. Introduction

The rapid development of e-commerce has severely impacted parcel distribution, and the last-mile delivery problem restricts logistics development. Many e-commerce companies, logistics service providers, and other stakeholders considered effective systems for last-mile delivery to be essential competitive advantages and attempted to tackle the bottleneck by innovative methods, such as parcel-pickup points (PPPs, also called collection and delivery points), drone delivery, and autonomous ground vehicle delivery [1–4]. PPP is the most widely used novel solution that helps firms reduce costs through consolidated shipments and provide customers with a flexible, convenient, and comfortable means of receiving parcels.

PPPs have garnered significant interest in logistics research. Studies address the advantages of PPPs such as economic efficiency, environmental friendliness, and high service quality [5–8]. There are two types of PPPs: parcel-pickup shops (PPSs) and parcel-pickup lockers (PPLs). PPLs rely on intelligent technology without human interaction, whereas PPSs cooperate with commercial facilities. PPLs exhibit the advantages of long opening hours, flexible collection times, and anonymity. Consumers are allowed to collect their parcels without being bound to shop opening hours. In addition, parcels can be retrieved anonymously because no human interaction is required [9,10]. Given that PPSs cooperate with existing facilities and PPLs are built-in uncertain locations, the location planning problem for PPLs is more complex and challenging to solve. In the 13th Five-Year Plan for the logistics development of China, the installation of PPLs was accelerated, and the percentage of PPL delivery was projected to reach 10% by the end of 2020 [11]. However, there are no clear guidelines on the location planning for PPLs from the government. Furthermore, the coronavirus (COVID-19) pandemic impacts people's lifestyles, and social distancing limits face-to-face contact with others, resulting in more online shopping and larger parcel volumes. PPLs play a specific role in the prevention and control of the COVID-19 pandemic. Therefore, the need for PPLs is most urgent.

Site-suitability analysis (SSA) is conducted to identify the most appropriate spatial locations or patterns for planning according to specific requirements, preferences, or predictors of a certain activity [12–14]. There are two types of SSA: site-selection analysis and site-search analysis. Site-selection analysis determines the best site from a given set of potential sites, while site-search analysis identifies the area or location of the best site [15]. SSA is becoming a particularly critical topic for PPL planning. Most studies of PPL have focused on the site-selection problem in SSA to identify the best sites by ranking or rating candidates based on different indicators [16,17]. However, they present the limitation of requiring specific areas with sets of predetermined candidates. For the planning of large-scale areas, decision-makers rarely have specific lists of predetermined candidates. First, they need to search for suitable areas and then further identify specific candidates within these suitable areas to select the most appropriate points. Determining how to search for suitable areas quickly is critical. It can help decision-makers to be significantly more efficient at the beginning of planning. It can also serve to quickly break down a large-scale area of study into several small-scale areas of study, with relevance to the site-selection topics of most current studies on PPL planning.

Thus, the main aim of this study was to develop a GIS-based bivariate logistic regression (LR) model with supervised classification algorithms to search for areas suitable for PPLs in a large-scale area. The selection criteria were chosen from many potential variables and their weights were determined using a data-driven Machine Learning (ML) algorithm. A decision-maker can focus on these suitable areas as site-selection ranges for PPLs, which significantly reduces the difficulty of analysis and time costs.

2. Literature Review

The three core issues related to the location analysis for PPPs in previous studies are (1) influencing factors, (2) spatial distribution patterns, and (3) site selection. For the influencing factors, some studies state that the distribution of PPPs is strongly related to the population density, land-use types, urban development, and spatial accessibility according to their agglomeration pattern [18–21]. Some studies have found that residents' behavior also has a relationship with PPP layout, and thus developed methods for measuring customers' spatial access to PPPs, considering differentiated supply and demand [22]. For spatial distribution patterns, the patterns of PPSs in several cities of China (Changsha, Wuhan, and Xi'an) were investigated using point of interest (POI) data [21,23,24]. The results showed that there are more PPSs in the central regions and fewer in the periphery regions, and there are multi-core agglomerations in general. For the site selection, research determined the best sites by ranking or rating candidates based on different indicators [16,17]. In general, previous studies related to location analysis for PPPs only analyzed the location characteristics and impact factors. Few studies addressed the site-search problem in SSA to identify the boundaries of the suitable sites in a large-scale area, such as a metropolis.

GIS-based SSA techniques are widely applied in urban, regional, and environmental planning activities, such as labeling potential hazards, ecological resources, habitats, and geological favorability, or locating advantageous sites for facilities, agricultural activities, and urban development [15,25–29]. The challenging aspect of GIS-based SSA is determining the important factors and their weights. Three major groups of approaches to

GIS-based SSA are computer-assisted overlay mapping, multi-criteria evaluation (MCE), and ML algorithms [15]. However, a criticism of the computer-assisted overlap map approach is that it is often used without verifying independent assumptions regarding the suitability criteria, nor is it standardized using appropriate methods [30]. In the MCE approach, the weights of the suitability criteria are determined subjectively, which is imprecise and ambiguous. Different multi-criteria evaluation rules generate remarkably different suitability patterns [15,31]. As a new data-driven technique, ML could overcome the limitations of the aforementioned approaches and better address problems involving enormous datasets. There are two types of models for the ML algorithm: white-box models are the explainable-type modes that allow an interpretation of the model parameters; black-box models, such as support vector machines or artificial neural networks, do not allow such an interpretation and can only be verified externally [32]. The LR model is the most common and useful white-box model for supervised classification algorithms due to its easy and efficient operation. The data types of the variables can be continuous or categorical. The result of the LR model is measured as a probability from 0 to 1, which can be considered as the suitability index. Thus, the large-scale area in this study was subdivided into a micro-scale raster to form the basic units of observation, and the classification of each raster was conducted according to its suitability index.

3. Materials and Methods

3.1. Study Area and Data

China has the largest e-commerce market globally, with over 40% of global e-commerce transactions originating from the country as of 2017. Guangzhou (GZ) is one of the four most developed metropolises in China, where PPLs occupy the market in the early stage. Furthermore, GZ has been ranked first for parcel receipts in China for seven consecutive years, from 2014 to 2020 [33]. As shown in Figure 1, GZ (112°57′E–114°30′ E; 22°26′ N–23°56′ N), located in the center of Guangdong Province in south China, had a population of 15.3 million in 2019 and covered an area of 7434 km^2. GZ is the third-largest metropolis in China, containing 11 administrative districts.

In this study, the suitability modeling for PPLs was conducted using five types of data: POI data, road-network data, population data with a resolution of 100 m, land-price data, and a digital elevation model (DEM) with a resolution of 30 m, as shown in Table 1. Given the large quantity and wide distribution of PPLs and the related facilities, manual data acquisition was time-consuming and inaccurate, hindering the progress of PPL research. POI data—a novel form of data incorporating information such as latitudinal and longitudinal coordinates, specific locations, place names, and other attribute information—played an essential role in the analysis of macro-scale spatial distribution characteristics. POI data had the advantages of comprehensive coverage, high recognition accuracy, and high accessibility. Thus, POI big data improved the quality of micro-scale studies on PPL locations. In this study, POI data were obtained from Gaode Map, which was an everyday navigation application popular in China. It used three-level classification codes to classify objects of POI data. From the open application programming interface (API) of Gaode Map, developers could extract data for a specific area, a specific category, or a keyword for the name. According to the literature, PPL distributions were strongly related to traffic convenience and residential and commercial areas. The influential factors from the POI data were chosen from two major categories with several subcategories: transportation service and commercial/house, as shown in Table 2. The locations of PPLs were searched for using the keywords 'parcel locker' or 'self-pickup locker'. A total of 679 PPLs were extracted from Gaode Map in 2019. The road network data were collected from OpenStreetMap (OSM).

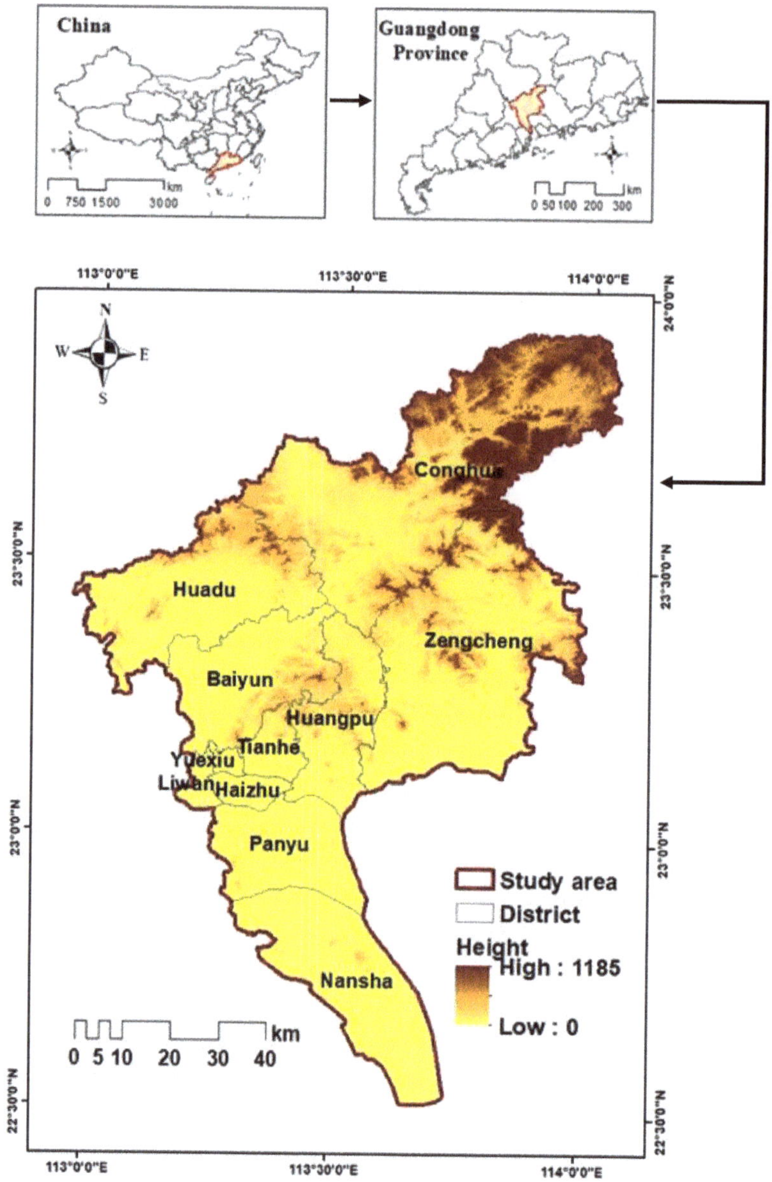

Figure 1. The study area.

3.2. Methodology

Figure 2 shows the methodology used in this research. It mainly consisted of five parts: (1) the conversion of multi-source data to the same scale, (2) the preparation of the observation data, (3) the diagnosis of the assumptions of the LR model, (4) the determination of the best combination of explanatory variables, (5) the evaluation of the model's performance, and (6) the generation of the suitability map using the best model.

Variables X1 to X27 are explained in Table 3, and their distribution maps are shown in Figure 3.

Table 1. List of data used.

Layer	Description	Source	Data Type
Road Network		OSM (2019) https://www.openstreetmap.org/ (accessed on 20 December 2019)	Vector (line)
POI		Gaode Maps (2019) https://ditu.amap.com/ (accessed on 20 December 2019)	Vector (point)
DEM	DEM-GDEMV2 30 m	ASTER GDEM Project (2019) https://www.gscloud.cn/ (accessed on 20 December 2019)	Raster
Population	Resolution of 100 m	WorldPop Project (2019) https://www.worldpop.org/geodata/summary?id=6275 (accessed on 11 September 2021)	Raster
Standard Land Price (Housing)	12 Levels of Price	Guangzhou Municipal Planning and Natural Resources Bureau 2019	Vector (polygon)

Table 2. List of POI data used.

Big Category	Mid Category	Subcategory	Number
Commercial House	Building	Business Office Building	5658
		Commercial-residential Building	825
		Villa	280
	Residential Area	Residential Quarter	7619
		Dormitory	2031
		Community Center	353
Transportation Service	Subway Station	Exit	808
	Bus Station	Bus Station Related (The bus stops for the airport bus or stopping operation were not considered.)	6778
	Parking Lot	Parking Lot Related	9882

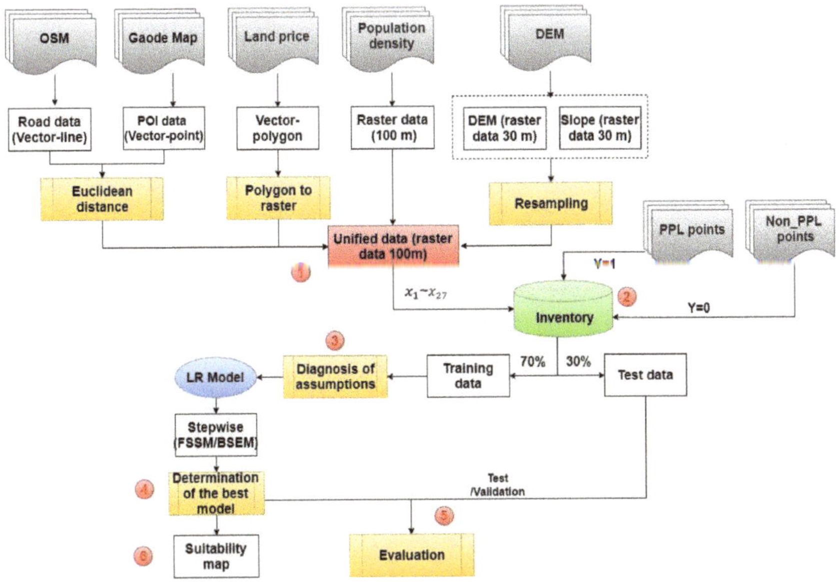

Figure 2. Methodological framework. Note: ① conversion of multi-source data to the same scale; ② preparation of the observation data; ③ diagnosis of the assumptions of LR model; ④ determination of the best model; ⑤ evaluation of the model performance; and ⑥ generation of the suitability map using the best model.

Table 3. The abbreviations of the 27 candidate variables.

No.	Potential Explanatory Variable	Variable Code	Type
X1	DEM	DEM	Topographic factors
X2	Slope	Slope	
X3	Population density	POP	Social factors
X4	Standard land price	SLPrice	
X5	Euclidean distance to the nearest residential quarter	Dist_Res_Qua	Accessibility factors: Proximity to various types of building
X6	Euclidean distance to the nearest residential community center	Dist_Res_CC	
X7	Euclidean distance to the nearest residential villa	Dist_Res_Vil	
X8	Euclidean distance to the nearest residential dormitory	Dist_Res_Dor	
X9	Euclidean distance to the nearest commercial and residential building	Dist_Com_ResB	
X10	Euclidean distance to the nearest commercial office building	Dist_Com_OffB	
X11	Euclidean distance to the nearest primary road	Dist_Road_Pri	Accessibility factors: Proximity to various types of road
X12	Euclidean distance to the nearest secondary road	Dist_Road_Sec	
X13	Euclidean distance to the nearest tertiary road	Dist_Road_Ter	
X14	Euclidean distance to the nearest unclassified road	Dist_Road_Unc	
X15	Euclidean distance to the nearest residential road	Dist_Road_Res	
X16	Euclidean distance to the nearest special type of road	Dist_Road_Spe	
X17	Euclidean distance to the nearest path road	Dist_Road_Path	
X18	Euclidean distance to the nearest metro exit	Dist_MetroExit	Accessibility factors: Proximity to various types of transport
X19	Euclidean distance to the nearest bus stop	Dist_BusStop	
X20	Euclidean distance to the nearest parking lot	Dist_ParkingLot	
X21	Euclidean distance to the nearest water area	Dist_WaterArea	
X22	Kernel density of parking lot	Dens_ParkingLot	Urban development factors: Density of various types of POI
X23	Kernel density of metro exit	Dens_MetroExit	
X24	Kernel density of bus stop	Dens_BusStop	
X25	Kernel density of commercial building	Dens_ComB	
X26	Kernel density of residential building	Dens_ResB	
X27	Kernel density of road	Dens_Road	

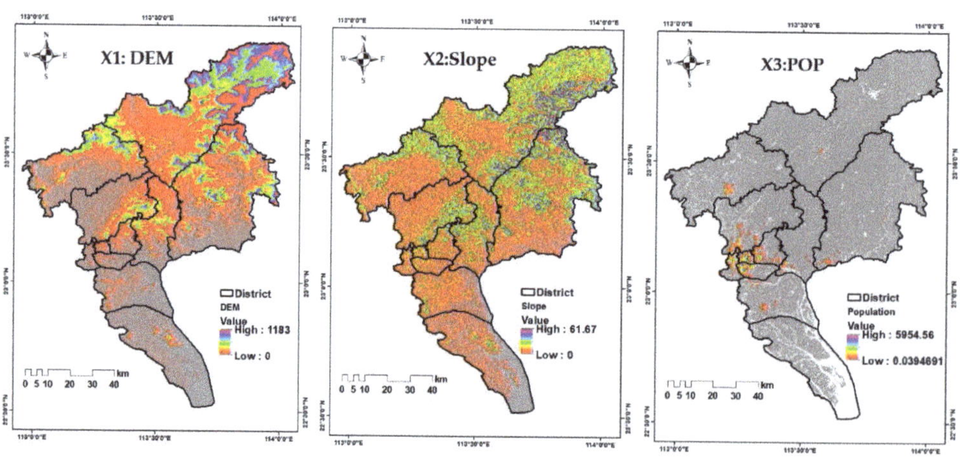

Figure 3. *Cont.*

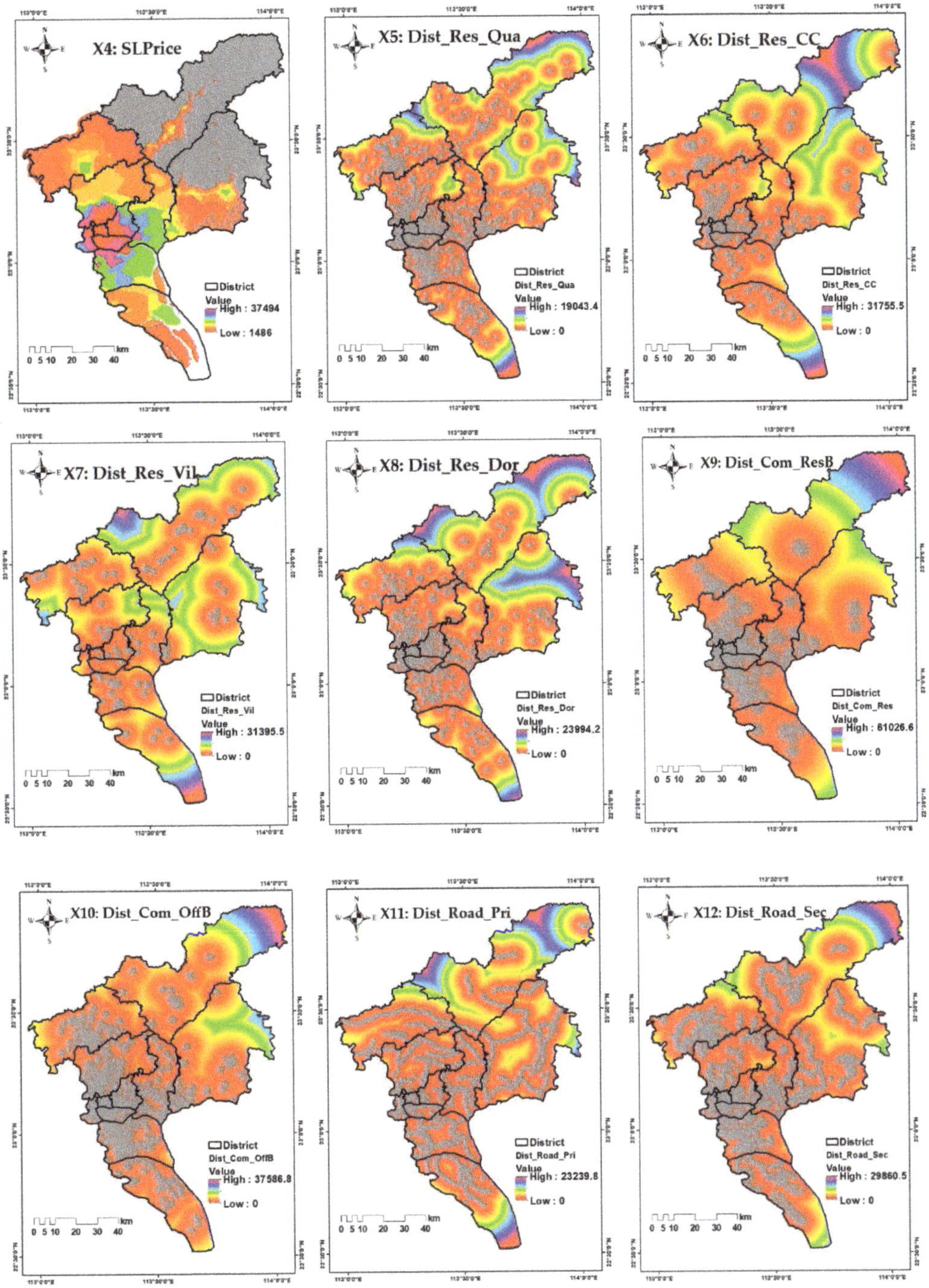

Figure 3. *Cont.*

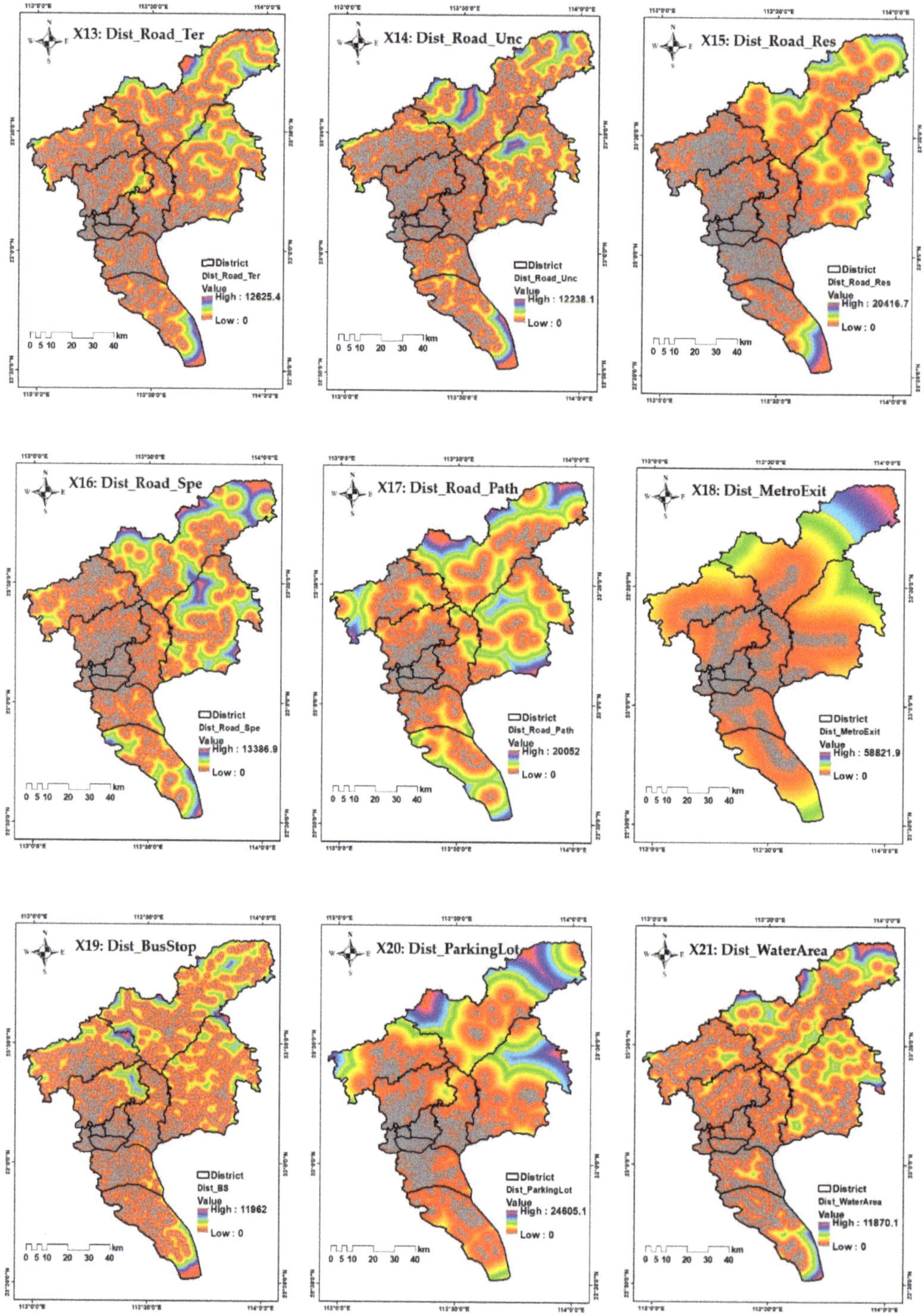

Figure 3. *Cont.*

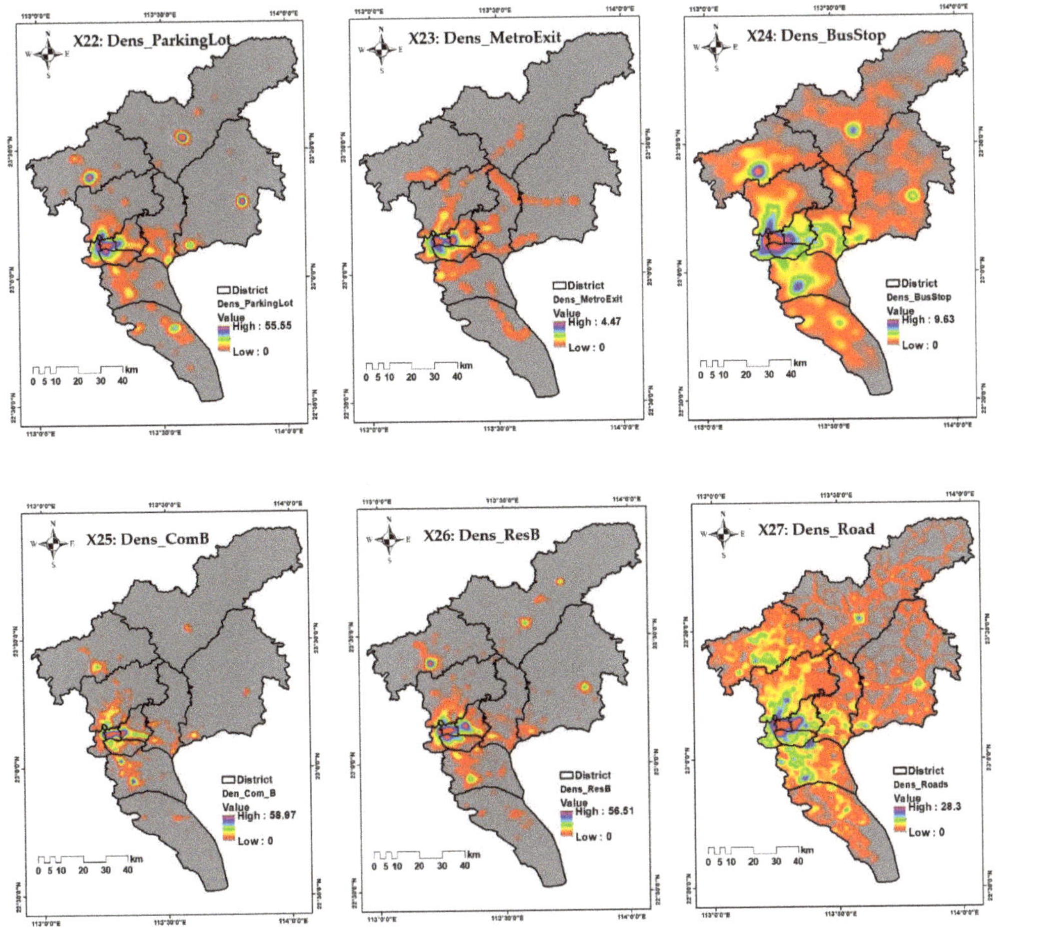

Figure 3. The distribution maps of the 27 candidate variables used in the model.

3.2.1. Conversion of the Multi-Source Data to the Same Scale

The challenges associated with multi-source data were attributable to the different types and scales of the data. The multi-source data should be unified to the same type and unit in the preprocessing stage. This study used four different data types—vector-line, vector-point, vector-polygon, and raster data—with different resolutions. As this study aimed to identify suitable areas at the pixel level, all the data needed to be converted to the same data type (raster) with the same resolution. The vector-line and point data were converted using the Euclidean distance and kernel density method. The vector-polygon data were directly converted to raster data. Higher-resolution raster data were converted to a lower resolution using the resampling tool of the ArcGIS 10.6 software. A total of 27 conversion results with a resolution of 100 m were candidate variables in the modeling, as shown in Table 3 and Figure 3.

3.2.2. Preparation of the Observation Data

An observation database was prepared for the LR model to learn the data features, including suitable and unsuitable location points, with the values of their explanatory variables. The location points of PPLs were collected from the POI data from Gaode

Map. This study assumed that ranges of 500 m around the existing locations of PPLs were suitable (approximate walking distance of 5 min) [17]. After erasing the water and assumed suitable areas, the non-PPL points were randomly sampled in the remained area. The classification by the LR model using ML algorithms should have avoided the class-imbalance problem [34]. In order to make the sample sizes of the positive and negative datasets similar, 690 non-PPL points were randomly selected. Figure 4 shows the locations of all the observation points.

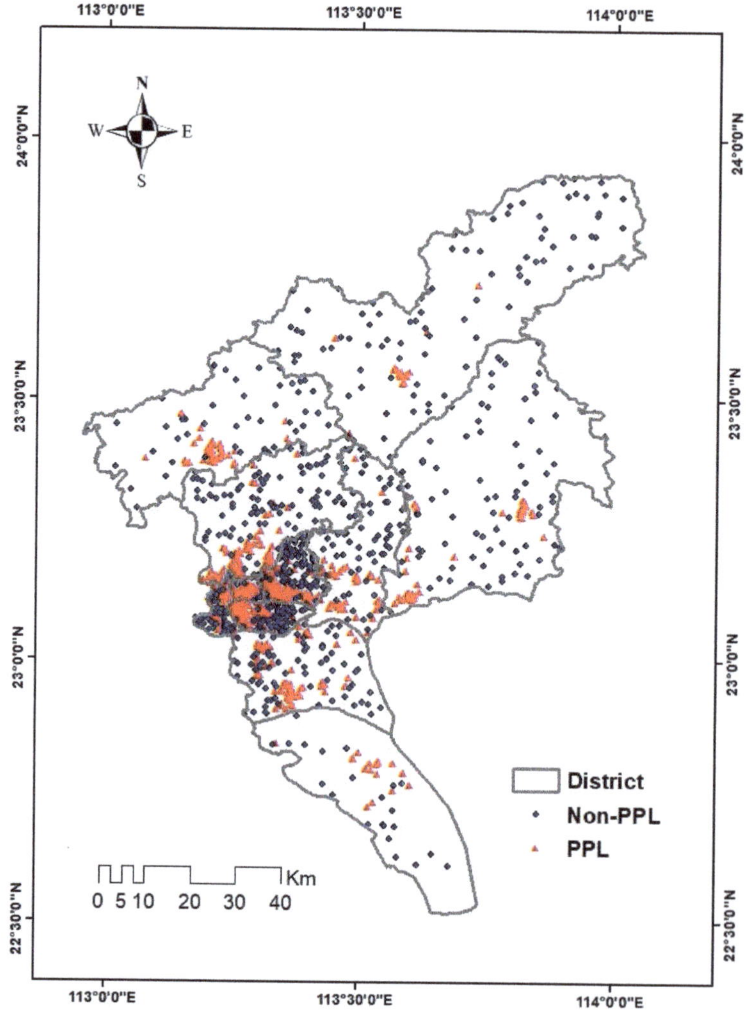

Figure 4. The locations of PPL and non-PPL points.

Next, the values of all the observation points were extracted from the raster layers of 27 candidate variables to create the reference database. There were several points that extracted the null values from the raster layers. These abnormal points were neglected to reduce the model bias. Empirical studies showed that the best results were obtained by training and testing data with a ratio of 70:30 or 80:20 [35]. In order to employ more data to test the performance of the model, this study chose the ratio of 70:30. The data were randomly split into a training dataset and a test dataset.

3.2.3. Diagnosis of the Assumptions of LR Model

Before applying the LR model, it was necessary to examine the assumptions shown in Table 4. The data for modeling satisfied the requirements for the first four assumptions during the dataset design, but the last three had to be examined using other methods. Here, the diagnosis was conducted using Version 25 of the IBM SPSS statistics software.

Table 4. Assumptions of the LR model.

No	Assumptions	Explanation	Examination
1	Dependent variable is required to be a binary variable.	1: PPL presence; 0: PPL absence.	Y
2	Observations were required to be independent of each other.	The observations come from different measurements or matched data.	Y
3	There is at least one dependent variable. The independent variable can be a continuous variable or a categorical variable.	There is one dependent variable and 27 independent variables.	Y
4	A large size of the sample is required. In general, the minimum sample quantity should be more than ten times the number of the independent variables.	There are 1205 points of PPL data. The sample quantity is more than 270.	Y
5	The linearity of independent variables and log odds is assumed.	Box–Tidwell method	?
6	There is little or no multicollinearity among the independent variables.	Multicollinearity diagnosis	?
7	There are no obvious outliers.		?

Note: In the examination column, 'Y' indicates that the assumption met the requirement and '?' indicates that the assumption needed to be verified.

- Diagnosis of the linearity of independent variables and log-odds

The Box–Tidwell method was employed here. It incorporated the interaction term between the continuous independent variable and its natural logarithmic value into the regression equation [36]. First, the natural logarithms of all the continuous independent variables were calculated using the compute variable function in SPSS. Then, the interaction terms between the continuous independent variables and their logs were included in the binary LR analysis using SPSS. The statistical significance of this predictor suggested a non-linear logit. When the interaction term was statistically significant (p-value < 0.05), there was no linear relationship between the corresponding continuous independent variable and the logit conversion value of the dependent variable. It was recommended that all the items in the analysis (including the intercept term) be corrected using the Bonferroni method when testing the multiple significance of the linearity hypothesis [37]. In this study, 55 items were included in the model analysis: 27 continuous independent variables, 27 interaction terms with their independent variables and their natural logs, and the intercept term (constant). A p-value less than the corrected value (i.e., $0.05 \div 55 = 0.000091$) was taken to indicate nonlinearity. There was no observed p-value less than the corrected value. Hence, linear relationships existed between all the continuous independent variables and the log-conversion value of the dependent variable.

- Diagnosis of multicollinearity

A good LR model exhibits low noise and is statistically robust. It means that the explanatory variables are highly correlated with the dependent variable but minimally correlated with each other [38]. Multicollinearity occurred when explanatory variables exhibited strong correlations or associations with each other. When the degree of correlation was extremely high, the standard errors of the coefficients increased, which caused some variables to appear statistically insignificant in the results, even though they were significant. Multicollinearity made the coefficients unstable [39] and reduced the precision or interfered with the result when fitting the model [40]. This was mainly detected with the help of the tolerance (Tol) and reciprocal, called the variance inflation factor (VIF) [41]. The formulae are defined as follows:

$$\mathrm{Tol} = 1 - \mathrm{R}^2 \quad (1)$$

$$\mathrm{VIF} = \frac{1}{\mathrm{Tol}} \quad (2)$$

where R^2 is the coefficient of determination for the regression of the explanatory variable on all the remaining independent variables.

VIF > 10 and Tol < 0.1 were common thresholds for assessing multicollinearity between explanatory variables [38,42]. There were several ways to address the multicollinearity problem. First, multiple variables with collinearity could be combined into a single variable. Second, the sample size could be increased to decrease standard errors. Third, some variables causing multicollinearity could be omitted from the model. Omitting some variables was the most direct, simple, and effective way. In order to retain as many variables as possible, the most correlated variable was neglected each time until the collinearity problem was not severe. Table 5 shows the VIF values of all the variables after omitting the variable with multicollinearity in the model.

Table 5. VIF values of all variables after omitting the variable with the multicollinearity problem.

		Step0		Step1		Step2	
No.	Variable	Tol	VIF	Tol	VIF	Tol	VIF
X1	DEM	0.316	3.164	0.316	3.163	0.316	3.161
X2	Slope	0.594	1.685	0.594	1.683	0.595	1.681
X3	POP	0.682	1.466	0.692	1.445	0.692	1.444
X4	SLPrice	0.165	6.044	0.166	6.022	0.203	4.918
X5	Dist_Res_Qua	0.142	7.021	0.143	6.984	0.143	6.969
X6	Dist_Res_CC	0.292	3.419	0.293	3.419	0.296	3.376
X7	Dist_Res_Vil	0.533	1.878	0.533	1.877	0.557	1.797
X8	Dist_Res_Dor	0.183	5.475	0.183	5.458	0.183	5.454
X9	Dist_Com_ResB	0.151	6.637	0.151	6.626	0.154	6.49
X10	Dist_Com_OffB	0.14	7.142	0.14	7.13	0.14	7.128
X11	Dist_Road_Pri	0.452	2.212	0.458	2.182	0.458	2.181
X12	Dist_Road_Sec	0.305	3.282	0.307	3.256	0.309	3.241
X13	Dist_Road_Ter	0.375	2.664	0.377	2.65	0.378	2.643
X14	Dist_Road_Unc	0.559	1.788	0.56	1.786	0.56	1.785
X15	Dist_Road_Res	0.424	2.359	0.425	2.356	0.425	2.353
X16	Dist_Road_Spe	0.326	3.066	0.327	3.062	0.328	3.052
X17	Dist_Road_Path	0.32	3.123	0.324	3.091	0.325	3.079
X18	Dist_MetroExit	0.155	6.452	0.155	6.452	0.159	6.271
X19	Dist_BusStop	0.38	2.63	0.381	2.623	0.382	2.618
X20	Dist_ParkingLot	0.11	9.098	0.11	9.096	0.11	9.095
X21	Dist_WaterArea	0.693	1.444	0.695	1.438	0.698	1.433
X22	Dens_ParkingLot	0.139	7.196	0.173	5.769	0.175	5.704
X23	Dens_MetroExit	0.097	10.286	0.097	10.283	Omitted	
X24	Dens_BusStop	0.098	10.235	0.108	9.241	0.114	8.756
X25	Dens_ComB	0.16	6.251	0.178	5.632	0.211	4.741
X26	Dens_ResB	0.086	11.685	Omitted			
X27	Dens_Road	0.106	9.401	0.11	9.087	0.119	8.395

- Diagnosis of obvious outliers

An outlier is an exceptional value that is very different from the others in a dataset. The LR model is sensitive to outliers. The usual approach to detecting outliers is based on the values of standardized residuals. If its absolute value is larger than three, it is usually considered an outlier [36]. After deleting the outliers, model fitting was conducted for the training dataset of 961 samples.

3.2.4. Determination of the Best Model Using the Stepwise Methods

There were many candidate variables in the model. It was important to detect the best variable combination for model fitting. A good model should adequately fit the data, and the predictor variables should not be too complicated. It was challenging to select the smallest number of candidate variables that could predict the dependent variable sufficiently while considering sample size constraints [36]. The forward and backward stepwise methods were frequently applied in previous studies of the LR model [43].

The forward stepwise selection method (FSSM) selected several significant predictors for the final model. Model optimization was performed using the least-squares criteria. It started with a blank model with no predictors. Variables were sequentially added one at a time to an empty model to predict the best output variable. Subsequently, a second variable that could best improve the model fitting was sought. The process was continued until a stopping rule was satisfied. In FSSM, variables added early in the process could be removed at a later stage because they became unimportant when other variables were added to the model. FSSM used a systematic method for adding variables based on their statistical significance in a regression. The process started with no explanatory variables in the model and then compared the incremental explanatory power of larger models [44]. Using the FSSM technique, the variables could be ranked by importance according to the priority of the added variables.

Unlike FSSM, the backward stepwise elimination method (BSEM) started with all the predictors of the least-squares model and then eliminated the least effective predictors one at a time. This method was continued until a stopping rule was satisfied. In the literature, the recommended stopping rule was a p-value of ~0.15 [45,46]. In the SPSS software, the default values for FSSM and BSEM were 0.05 and 0.1, respectively.

3.2.5. Evaluation of the Model's Performance

The performance of the LR models was evaluated based on their discrimination and calibration. Discrimination referred to the ability of the model to correctly distinguish between the two suitability classes based on prediction values. The capacity of discrimination was often measured using a confusion matrix and by calculating indices of classification performance [47]. The LR model used the logistic function to map the predictions to probabilities between 0 and 1. The default threshold of 0.5 was commonly used. It assumed that a PPL was present if the probability was above 0.5; otherwise, it was absent. The classification accuracy was determined by comparing the predictions with the real values. The classification table was divided into four types. True positives (TPs) and true negatives (TNs) indicated the number of correctly predicted PPLs and non-PPLs; false positives (FPs) and false negatives (FNs) denoted the numbers of incorrect predictions. Several further indications were used to measure the performance of a model or predictors. The accuracy was the total number of correct predictions divided by the total number of predictions made for a dataset. However, even unskillful models could show high accuracy scores when the class imbalance was severe. An alternative to using the classification accuracy was to use precision and recall. Unfortunately, precision and recall may sometimes contradict each other. The F-Measure (also known as the F-Score) was the most common method for balancing both indications in a single score. The mathematical basis was the same as in Equations (3)–(6). Here, the classification accuracy and F-Measure represented the index of the discrimination.

$$\text{Precision} = \frac{\text{TP}}{(\text{TP} + \text{FP})} \tag{3}$$

$$\text{Recall} = \frac{\text{TP}}{(\text{TP} + \text{FN})} \tag{4}$$

$$\text{Accuracy} = \frac{(\text{TP} + \text{TN})}{(\text{TP} + \text{FP} + \text{FN} + \text{TN})} \tag{5}$$

$$\text{F} - \text{Measure} = \frac{2 \times \text{Presision} \times \text{Recall}}{(\text{Precision} + \text{Recall})} \tag{6}$$

The discrimination only compared the predicted probability value with a certain threshold of 0.5. However, it ignored how far the predicted value was from the true value. Calibration resolved this shortcoming, and it described how close the predicted value was to the actual value. The Brier score was an important calibration index that measured the accuracy of probabilistic predictions. It was applicable to tasks in which predictions assigned probabilities to a set of mutually exclusive discrete outcomes. The set of possible outcomes could be either binary or categorical in nature, and the probabilities assigned to this set of outcomes must have summed to 1, where each individual probability ranged from 0 to 1 [48]. The lower the Brier score for a set of predictions, the better the predictions were calibrated. In this study, the reduction ratio for the variables involved in modeling (the model optimization rate) was added to evaluate the model's performance:

$$\text{B} = \frac{\sum_{i=1}^{n} (x_i - q_i)^2}{n} \tag{7}$$

where x is the real dependent variable, and q is the predicted probability.

The receiver operating characteristic (ROC) curve was also a popular method for testing a model's accuracy and describing the quality of a probabilistic prediction system [49]. The area under the ROC curve (AUC) was a common metric for the level of discriminative ability; the larger the area, the better the performance of the model. The following classification using the AUC was considered for accuracy: 0.90–1 (excellent), 0.80–0.90 (good), 0.70–0.80 (fair), 0.60–0.70 (poor), and 0.50–0.60 (fail) [50,51].

3.2.6. Generation of the Suitability Map

The coefficient of the selected optimum variables and the constant of the best LR model was substituted into Equation (9). The suitability index of Equation (10) was applied in each raster of the whole study area for prediction. According to the classification threshold of the LR model, the suitability map for PPLs consisted of two categories. The raster with a predicted value between 0.5 and 1 was reclassified as a suitable area, and the raster with a value between 0 and 0.5 was reclassified as an unsuitable area.

$$\text{Z} = \sum_{i=1}^{n} w_i x_i + \text{Constant} \tag{8}$$

$$\text{y} = \frac{1}{1 + e^{-(z)}} \tag{9}$$

4. Results

4.1. The Optimum Variable Combination for the Best Model

Table 6 shows the model's performance with the combination of variables selected by the FSSM and BSEM. The discrimination and calibration of the BSEM are also slightly better than those of the FSSM. However, the optimization rate for the FSSM is 20% higher. It indicates that the two methods for selecting the optimal variable combination show a similar model accuracy and bias. In terms of the index of model optimization, the FSSM performed better than the BSEM. Table 7 shows the coefficient of the best explanatory variable combination as determined by the BSEM. The Wald value indicates the significance of the variables. Eight significant variables were selected from the 25 variables without multicollinearity. Among these eight variables, five were selected from the accessibility factors, and one each was selected from the social factors, topographic factors, and urban development factors. Among the five selected accessibility factors, three were from the variables of proximity to various types of buildings. According to the Wald value, the most crucial factor was Dist_Res_Qua, with a value of 45.5, followed by SLPrice (29), Dist_BusStop (28.4), and Dens_ComBs (20.7). According to the signs of the coefficients, the variables of Dist_Res_Quar, Dist_BusStop, Dist_Com_OffB, Dist_Road_Sec, Dist_Res_Vil, and SLPrice were negatively correlated with the suitability for PPLs in the raster unit. The DEM and Dens_ComB were positively correlated. Thus, a PPL site may be situated close to residential quarters, commercial offices, and residential villas. The areas were near bus stops or secondary roads with relatively low land prices, and in high-density zones of commercial buildings.

Table 6. Performance of FSSM and BSEM with the training dataset.

Method	Discrimination		Calibration	Optimization
	Accuracy	F-Measure	Brier Score	Reduction Ratio
FSSM	88.20%	88.50%	0.088	68.00%
BSEM	88.40%	88.70%	0.085	48.00%

Table 7. Coefficient of the best explanatory variable combination in the standard LR model.

Variable Type	Variable Code	Selected	Coefficient	Wald
Topographic Factors	DEM	Y	0.019	10.5
	Slope	N	-	-
Social Factors	POP	N	-	-
	SLPrice	Y	−0.0001	29
Accessibility Factors: Proximity to various types of building	Dist_Res_Qua	Y	−0.0032	45.5
	Dist_Res_CC	N	-	-
	Dist_Res_Vil	Y	−0.0002	6.7
	Dist_Res_Dor	N	-	-
	Dist_Com_ResB	N	-	-
	Dist_Com_OffB	Y	−0.0013	18.6
Accessibility Factors: Proximity to various types of road	Dist_Road_Pri	N	-	-
	Dist_Road_Sec	Y	−0.0006	9.3
	Dist_Road_Ter	N	-	-
	Dist_Road_Unc	N	-	-
	Dist_Road_Res	N	-	-
	Dist_Road_Spe	N	-	-
	Dist_Road_Path	N	-	-
Accessibility Factors: Proximity to various types of transport	Dist_MetroExit	N	-	-
	Dist_BusStop	Y	−0.0039	28.4
	Dist_ParkingLot	N	-	-
	Dist_WaterArea	N	-	-
Urban development Factors: Density of various types of POI	Dens_ParkingLot	N	-	-
	Dens_BusStop	N	-	-
	Dens_ComB	Y	0.090	20.7
	Dens_Road	N	-	-

4.2. *Evaluation of the Classification Performance*

The test dataset was used to conduct an unbiased evaluation of the final model's fit on the training dataset. The final LR model with the best variable combination and coefficients was applied to the test dataset. The F-measure, Brier score, and AUC were the indicators used to evaluate the model's classification performance, as shown in Table 8. The larger the F-measure index, the higher the discrimination accuracy of the model's classification. The F-Measure values for both the training and test data, were all greater than 89%. The lower the Brier score, the smaller the deviation predicted and the higher the calibration degree of the model. The Brier scores were less than 0.09. The value of the AUC for both datasets was between 0.9 and 1, indicating excellent accuracy.

Table 8. Predicted performance of the best model.

	F-Measure	Brier Score	AUC
Training dataset	89.11%	0.088	0.954
Test dataset	91.69%	0.069	0.963

Overall, the predicted performance of the final LR model was effective. Additionally, the performance with the test dataset was better than that with the training dataset.

4.3. The Boundaries of the Suitable Areas

Figure 5 demonstrates the suitability for PPLs simulated using the best LR model. The suitability for PPLs is divided into two classes: the suitable area in orange and the unsuitable area in blue. Most of the suitable areas are concentrated in the central districts and dispersed in small areas in the outer districts. Figure 6 summarizes the sizes and percentages of the suitable area by the district. Panyu district has the greatest suitable area, while Liwan district has the smallest. Yuexiu district has the greatest proportion of suitable area, more than 80%, while Conghua has the smallest, only 1%. Overall, the suitable area is appropriately 614 sq. km, accounting for 8% of the total area of GZ. The site-selection range for PPLs can focus on these suitable areas, which significantly reduces the difficulty of analysis and time costs.

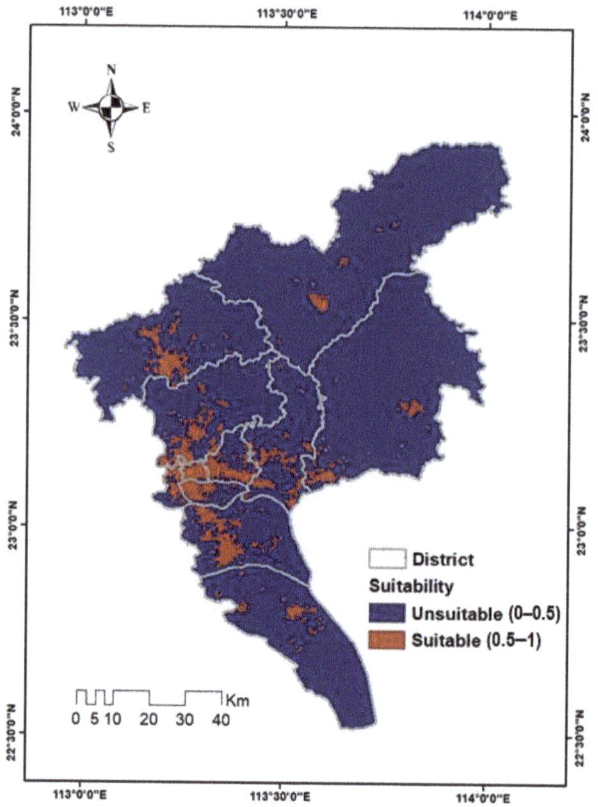

Figure 5. Suitability map for PPLs using the best LR model.

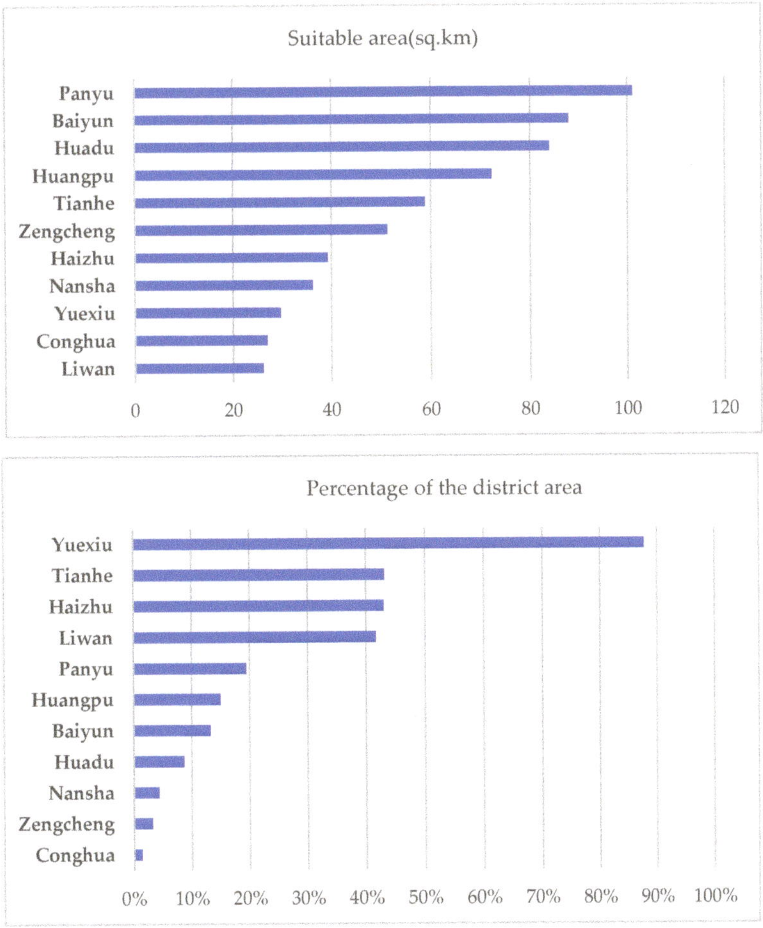

Figure 6. Summary of the suitable areas for PPLs.

5. Discussion

Big data make location analysis in a macro-scale area possible. POI data, an innovative data source with a low cost, can identify the existing locations of PPLs and other related facilities. Some studies used POI data to analyze the PPL distribution patterns in specific cities of China and found them to be strongly consistent with economic development levels, population density, and traffic convenience [21,23,24]. This study further developed a GIS-based LR classification model using an ML algorithm to identify suitable areas from bottom to top with massive, detailed data, which was different from previous studies conducted by the MCE approach. The optimum explanatory variables from the 27 candidates and their coefficients for LR models were determined using a training dataset with stepwise methods. The FSSM performed better than the BSEM in the optimization of variables. The most crucial variable was Dist_Res_Qua. It was much more important than the variables of the distance to various transport services/roads and the density of related points. This result was consistent with the preferences of customers for PPLs being located near their home addresses [52]. Furthermore, this study subdivided residential buildings into four types as candidate variables to analyze the relationships with PPLs. The results showed that the type of residential quarters was the most crucial variable; the types of dormitory and community center (CC) were not determining variables for the locations of PPLs. A

CC is a place providing recreational, cultural, and social activities for surrounding groups of residential neighborhoods. Although a CC is usually near the residential building of the community, it is difficult to combine the behavior of picking up parcels with entertainment or social activities. The residential buildings of dormitories are mainly located in colleges, factories, or institutions with closed management. The dormitory areas are usually far from the entrance. It takes a long time to distribute parcels to PPLs near a dormitory building, and the delivery vehicles have limited accessibility. Due to the safety of internal personnel and the long delivery times, parcels for dormitories are generally signed for and stored by guards or shops which offer parcel-pickup services. The population in the dormitory area is dense, and the capacity of PPLs is limited. Due to the high machine cost, it is not possible to set up several facilities of PPLs to meet the great demand there. Moreover, the nature of PPLs is more inclined toward that of a public service facility, and their economic benefit is limited. The dormitory management prefers to lease the land to commercial shops rather than PPLs, to obtain more rent.

Another interesting finding was that the variable of population density was not selected as the critical factor for determining the locations of PPLs in the study. It was somewhat different from the previous studies that proposed that the density of PPPs had a strong positive correlation with population density [20,21]. The reason for this may be that the research scales used were different. The previous studies were based on the unit of the administrative boundary. According to the characteristics of the existing locations of PPPs, the relationship between the density of PPPs and the density of various factors in each administrative unit of the study area was investigated using correlation analysis [21,23,24]. The analysis focused mainly on the quantitative relationship and ignored the relationship with the location distances of various factors. For distance analysis, the statistical method was widely used to determine the distance range between the location of most PPPs and the surrounding features. Unlike previous studies, this work attempted to model the locations using raster units. The locations of existing PPLs and random non-PPL points were used as the training and testing datasets. The features of existing PPL locations were extracted by the ML algorithm and generated the best model. Other unknown raster units were classified into suitable and unsuitable sites by the model. This method considered both the number and distance-related factors, and the model could further identify locations suitable for PPLs rather than only analyzing the characteristics of existing points. The model could distinguish variables that yielded locations suitable for PPLs from a large number of candidate factors using Wald values (importance). Another reason was that the raster population data were not highly accurate and only considered the nighttime population. The population density data source used here was the population prediction data in the WorldPop dataset developed by the WorldPop Project. Up-to-date raster data for population density with a high resolution were hard to obtain. The predicted population in the WorldPop dataset was simulated from the official census population data and nighttime satellite images [53]. In the best model of this study, the most critical variables that yielded suitable PPL locations included Dis_Res_Qua, Dist_Busstop, and Dist_Com_OffB. These factors also had a strong relationship with the population.

There are several assumptions and limitations in this study due to the insufficient data. First, the existing PPL locations are considered as locations suitable for PPL. These locations of points serve as the sample for the ML algorithm, which learns their features. However, they may not be consistent with the actual suitability. Only current POI data are available, not historical POI data. It is impossible to analyze the relationship between the historical PPL locations and the surrounding environment. In addition, because PPL usage data are not available, it is not possible to determine whether the existing PPL locations are realistically appropriate. Second, the competition of PPLs was not considered in this study. In reality, PPLs are operated by different companies, and they may compete with each other. Third, the 27 candidate factors in the model are social and location-related factors; market and user-behavior preference factors are not included in this study.

Moreover, a metropolis is a large city consisting of a densely populated urban core and less-populated surrounding territories under the same administrative jurisdiction [54]. The PPL density is also unbalanced in different areas of a metropolis. Future research could divide metropolitan areas into multiple zones according to population density for modeling and further analyze the differences in the variables chosen by the model in the various zones.

6. Conclusions

Previous studies of SSA for PPLs commonly addressed the site-selection problem with given sites in a specific area [16,17]. Few studies have focused on the site-search problem with quantitative models. GIS-based SSA techniques were widely applied in urban planning activities with multiple factors. ML method was superior to the other two approaches of GIS-based SSA and worked best for problems involving enormous datasets. The LR model was the most common and explainable model of the data-driven ML algorithms. This paper proposed a GIS-based bivariate LR model with supervised classification algorithms for the SSA of PPLs and explicitly identified the boundaries of suitable areas. The micro-scale raster provided the basic unit of observation, and the suitability classification was conducted in each raster. The crucial factors and their weights were determined using the training data. Of the data, 30% was used to test the model's accuracy and evaluate the performance of the best model. The two stepwise methods (FSSM and BSEM) were employed to determine the optimum combination of variables from a total of 27 candidate variables. The performance of the LR models was evaluated based on their discrimination, calibration, and optimization rates. The results indicated that the FSSM with fewer variables had an absolute advantage in model optimization. Although the BSEM selected more variables than the FSSM, there was only a slight improvement in other indicators.

From the 25 potential variables without multicollinearity, eight crucial variables were chosen by the final LR model. Three variables were the distances to various types of buildings. The proximity to residential buildings was more important than that to commercial buildings. The most crucial factor was the proximity to residential quarters, whose importance was twice that of land price and proximity to a bus stop. The result was consistent with the preferences of customers for PPLs being located near their home addresses [52]. This study further supported the idea that the residential quarter was the most important among the four types of residential buildings, while the dormitory and CC types were relatively unimportant. The final model identified the boundaries of areas suitable for PPLs, accounting for 8% of the total area of GZ. The site-selection ranges for PPLs could be focused on these areas, which significantly reduced the difficulty of analysis and time costs. There were several limitations in this study due to the insufficient data. Future research should divide metropolitan areas into multiple zones for modeling and analyze the differences in the variables chosen by the model in the various zones.

Author Contributions: Conceptualization, Zilai Zheng, Takehiro Morimoto and Yuji Murayama; methodology, Zilai Zheng, Takehiro Morimoto and Yuji Murayama; software, Zilai Zheng; validation, Zilai Zheng; formal analysis, Zilai Zheng; investigation, Zilai Zheng; resources, Zilai Zheng; data curation, Zilai Zheng; writing—original draft preparation, Zilai Zheng; writing—review and editing, Takehiro Morimoto and Yuji Murayama; visualization, Zilai Zheng; supervision, Takehiro Morimoto and Yuji Murayama. All authors have read and agreed to the published version of the manuscript.

Funding: This research was partly supported by the JSPS grants of 21K01027 and 18H00763.

Institutional Review Board Statement: Not applicable.

Informed Consent Statement: Not applicable.

Data Availability Statement: The data presented in this study are available from the author upon reasonable request.

Acknowledgments: The comments and suggestions of the anonymous reviewers are gratefully acknowledged.

Conflicts of Interest: The authors declare no conflict of interest.

References

1. Gevaers, R.; van de Voorde, E.; Vanelslander, T. Characteristics and typology of last-mile logistics from an innovation perspective in an urban context. In *City Distribution and Urban Freight Transport: Multiple Perspectives*; Macharis, C., Melo, S., Eds.; Edward Elgar Publishing: Cheltenham, UK, 2011; pp. 56–71. [CrossRef]
2. Punakivi, M.; Yrjölä, H.; Holmström, J. Solving the last mile issue: Reception box or delivery box. *Int. J. Phys. Distrib. Logist. Manag.* **2001**, *31*, 427–439. [CrossRef]
3. Xiao, Z.; Wang, J.J.; Lenzer, J.; Sun, Y. Understanding the diversity of final delivery solutions for online retailing: A case of Shenzhen, China. *Transp. Res. Procedia* **2017**, *25*, 985–998. [CrossRef]
4. Slabinac, M. Innovative solutions for a "Last-Mile" delivery—A European experience. In Proceedings of the 15th International Scientific Conference Business Logistics in Modern Management, Osijek, Croatia, 15 October 2015; pp. 111–130.
5. Edwards, J.; McKinnon, A.; Cherrett, T.; McLeod, F.; Song, L. Carbon dioxide benefits of using collection–delivery points for failed home deliveries in the United Kingdom. *Transp. Res. Rec.* **2010**, *2191*, 136–143. [CrossRef]
6. Gevaers, R.; van de Voorde, E.; Vanelslander, T. Cost modelling and simulation of last-mile characteristics in an innovative B2C supply chain environment with implications on metropolitan areas and cities. *Procedia Soc. Behav. Sci.* **2014**, *125*, 398–411. [CrossRef]
7. Kedia, A.; Kusumastuti, D.; Nicholson, A. Acceptability of collection and delivery points from consumers' perspective: A qualitative case study of Christchurch city. *Case Stud. Transp. Policy* **2017**, *5*, 587–595. [CrossRef]
8. Rautela, H.; Janjevic, M.; Winkenbach, M. Investigating the financial impact of collection-and-delivery points in last-mile E-commerce distribution. *Res. Transp. Bus. Manag.* **2021**, 100681, in press. [CrossRef]
9. Van Duin, J.H.R.; Wiegmans, B.W.; van Arem, B.; van Amstel, Y. From home delivery to parcel lockers: A case study in Amsterdam. *Transp. Res. Procedia* **2020**, *46*, 37–44. [CrossRef]
10. Weltevreden, J.W. B2c e-commerce logistics: The rise of collection-and-delivery points in The Netherlands. *Int. J. Ret. Distrib. Manag.* **2008**, *36*, 638–660. [CrossRef]
11. State Post Bureau of The People's Republic of China. 2017. Available online: http://www.spb.gov.cn/zc/ghjbz_1/201702/t20170213_991162.html (accessed on 11 September 2021).
12. Collins, M.G.; Steiner, F.R.; Rushman, M.J. Land-use suitability analysis in the United States: Historical development and promising technological achievements. *Environ. Manag.* **2001**, *28*, 611–621. [CrossRef]
13. Cova, T.J.; Church, R.L. Exploratory spatial optimization in site search: A neighborhood operator approach. *Comput. Environ. Urban Syst.* **2000**, *24*, 401–419. [CrossRef]
14. Hopkins, L. Methods for generating land suitability maps: A comparative evaluation. *J. Am. Inst. Plann.* **1997**, *34*, 19–29. [CrossRef]
15. Malczewski, J. GIS-based land-use suitability analysis: A critical overview. *Prog. Plann.* **2004**, *62*, 3–65. [CrossRef]
16. Yang, G.; Huang, Y.; Fu, Y.; Huang, B.; Sheng, S.; Mao, L.; Huang, S.; Xu, Y.; Le, J.; Ouyang, Y.; et al. Parcel locker location based on a Bilevel programming model. *Math. Probl. Eng.* **2020**, *2020*. [CrossRef]
17. Zheng, Z.; Morimoto, T.; Murayama, Y. Optimal location analysis of delivery parcel-pickup points using AHP and Network Huff Model: A case study of Shiweitang Sub-District in Guangzhou City, China. *ISPRS Int. J. Geoinf.* **2020**, *9*, 193. [CrossRef]
18. Lachapelle, U.; Burke, M.; Brotherton, A.; Leung, A. Parcel locker systems in a car dominant city: Location, characterisation and potential impacts on city planning and consumer travel access. *J. Transp. Geogr.* **2018**, *71*, 1–14. [CrossRef]
19. Liu, S.; Liu, Y.; Zhang, R.; Cao, Y.; Li, M.; Zikirya, B.; Zhou, C. Heterogeneity of Spatial Distribution and Factors Influencing Unattended Locker Points in Guangzhou, China: The Case of Hive Box. *ISPRS Int. J. Geoinf.* **2021**, *10*, 409. [CrossRef]
20. Morganti, E.; Dablanc, L.; Fortin, F. Final deliveries for online shopping: The deployment of pickup point networks in metropolitan and suburban areas. *Res. Transp. Bus. Manag.* **2014**, *11*, 23–31. [CrossRef]
21. Xue, S.; Li, G.; Yang, L.; Liu, L.; Nie, Q.; Mehmood, M.S. Spatial Pattern and Influencing Factor Analysis of Attended Collection and Delivery Points in Changsha City, China. *Chin. Geogr. Sci.* **2019**, *29*, 1078–1094. [CrossRef]
22. Lin, L.; Han, H.; Yan, W.; Nakayama, S.; Shu, X. Measuring Spatial Accessibility to Pick-Up Service Considering Differentiated Supply and Demand: A Case in Hangzhou, China. *Sustainability* **2019**, *11*, 3448. [CrossRef]
23. Li, G.; Chen, W.; Yang, L. Spatial pattern and agglomeration mode of parcel collection and delivery points in Wuhan City. *Prog. Geogr.* **2019**, *38*, 407–416. (In Chinese)
24. Li, G.; Yang, L.; He, J. The spatial pattern and organization relation of the pickup points based on POI data in Xi'an: Focus on Cainiao stations. *Sci. Geogr. Sin.* **2018**, *38*, 2024–2030. (In Chinese)
25. Derdouri, A.; Murayama, Y. Onshore Wind Farm Suitability Analysis Using GIS-based Analytic Hierarchy Process: A Case Study of Fukushima Prefecture, Japan. *Geoinf. Geostat. Overv.* **2018**. [CrossRef]
26. Estoque, R.C.; Murayama, Y. Suitability analysis for beekeeping sites in La Union, Philippines, using GIS and multi-criteria evaluation techniques. *Res. J. Appl. Sci.* **2010**, *5*, 242–253. [CrossRef]

27. Kumar, M.; Shaikh, V.R. Site suitability analysis for urban development using GIS based multicriteria evaluation technique. *J. Indian Soc. Remote Sens.* **2013**, *41*, 417–424. [CrossRef]
28. Saha, S.; Sarkar, D.; Mondal, P.; Goswami, S. GIS and multi-criteria decision-making assessment of sites suitability for agriculture in an anabranching site of sooin river, India. *J. Adv. Model. Earth Syst.* **2021**, *7*, 571–588. [CrossRef]
29. Store, R.; Kangas, J. Integrating spatial multi-criteria evaluation and expert knowledge for GIS-based habitat suitability modelling. *Landsc. Urban Plann.* **2001**, *55*, 79–93. [CrossRef]
30. Pereira, J.M.; Duckstein, L. A multiple criteria decision-making approach to GIS-based land suitability evaluation. *Int. J. Geogr. Inf. Sci.* **1993**, *7*, 407–424. [CrossRef]
31. Lodwick, W.A.; Monson, W.; Svoboda, L. Attribute error and sensitivity analysis of map operations in geographical information systems: Suitability analysis. *J. Geogr. Inf. Sci.* **1990**, *4*, 413–428. [CrossRef]
32. Dreiseitl, S.; Ohno-Machado, L. Logistic regression and artificial neural network classification models: A methodology review. *J. Biomed. Inf.* **2002**, *35*, 352–359. [CrossRef]
33. State Post Bureau of The People's Republic of China. Statistical Communique on the Development of Postal Industry in 2014–2020. Available online: http://www.spb.gov.cn/sj/tjxx_1/ (accessed on 11 September 2021).
34. Oommen, T.; Baise, L.G.; Vogel, R.M. Sampling bias and class imbalance in maximum-likelihood logistic regression. *Math. Geosci.* **2011**, *43*, 99–120. [CrossRef]
35. Gholamy, A.; Kreinovich, V.; Kosheleva, O. Why 70/30 or 80/20 relation between training and testing sets: A pedagogical explanation. *Int. J. Intell. Syst.* **2018**, *11*, 105–111.
36. Hosmer, D.W.; Lemeshow, S. *Applied Logistic Regression*, 2nd ed.; John Wiley and Sons: New York, NY, USA, 2000.
37. Bland, J.M.; Altman, D.G. Multiple significance tests: The Bonferroni method. *Br. Med. J.* **1995**, *310*, 170. [CrossRef]
38. Midi, H.; Sarkar, S.K.; Rana, S. Collinearity diagnostics of binary logistic regression model. *J. Interdiscip. Math.* **2010**, *13*, 253–267. [CrossRef]
39. Belsley, D.; Kuh, E.; Welsch, R. *Regression Diagnostics: Identifying Influential Data and Sources of Collinearity*, 2nd ed.; John Wiley and Sons: New York, NY, USA, 2013.
40. Schroeder, M.A.; Lander, J.; Levine-Silverman, S. Diagnosing and dealing with multicollinearity. *West. J. Nurs. Res.* **1990**, *12*, 175–187. [CrossRef] [PubMed]
41. Mansfield, E.R.; Helms, B.P. Detecting multicollinearity. *Am. Stat.* **1982**, *36*, 158–160. [CrossRef]
42. Kroll, C.N.; Song, P. Impact of multicollinearity on small sample hydrologic regression models. *Water Resour. Res.* **2013**, *49*, 3756–3769. [CrossRef]
43. Zellner, D.; Keller, F.; Zellner, G.E. Variable selection in logistic regression models. *Commun. Stat. Simul. Comput.* **2004**, *33*, 787–805. [CrossRef]
44. Soroush, A.; Bahreininejad, A.; van den Berg, J. A hybrid customer prediction system based on multiple forward stepwise logistic regression model. *Intell. Data Anal.* **2012**, *16*, 265–278. [CrossRef]
45. Flack, V.F.; Chang, P.C. Frequency of selecting noise variables in subset regression analysis: A simulation study. *Am. Stat.* **1987**, *41*, 84–86. [CrossRef]
46. Lee, K.I.; Koval, J.J. Determination of the best significance level in forward stepwise logistic regression. *Commun. Stat. Simul. Comput.* **1997**, *26*, 559–575. [CrossRef]
47. Pearce, J.; Ferrier, S. Evaluating the predictive performance of habitat models developed using logistic regression. *Ecol. Modell.* **2000**, *133*, 225–245. [CrossRef]
48. Brier, G.W. Verification of forecasts expressed in terms of probability. *Mon. Weather Rev.* **1950**, *78*, 1–3. [CrossRef]
49. Swets, J.; Pickett, R.; Whitehead, S.; Getty, D.; Schnur, J.; Swets, J.; Freeman, B. Assessment of diagnostic technologies. *Science* **1979**, *205*, 753–759. [CrossRef] [PubMed]
50. Hanley, J.A.; McNeil, B.J. The meaning and use of the area under a receiver operating characteristic (ROC) curve. *Radiology* **1982**, *143*, 29–36. [CrossRef]
51. Maxim, L.D.; Niebo, R.; Utell, M.J. Screening tests: A review with examples. *Inhal. Toxicol.* **2014**, *26*, 811–828. [CrossRef]
52. Iwan, S.; Kijewska, K.; Lemke, J. Analysis of parcel lockers' efficiency as the last mile delivery solution—The results of the research in Poland. *Transp. Res. Procedia* **2016**, *12*, 644–655. [CrossRef]
53. Gaughan, A.E.; Stevens, F.R.; Huang, Z.; Nieves, J.J.; Sorichetta, A.; Lai, S.; Ye, X.; Linard, C.; Hornby, G.M.; Hay, S.I.; et al. Spatiotemporal patterns of population in mainland China, 1990 to 2010. *Sci. Data* **2016**, *3*, 1–11. [CrossRef]
54. Squires, G.D. Urban sprawl and the uneven development of metropolitan America. In *Urban Sprawl: Causes, Consequences, and Policy Responses*; Urban Institute Press: Washington, DC, USA, 2002; pp. 1–22.

International Journal of
Geo-Information

MDPI

Article

An Economic Development Evaluation Based on the OpenStreetMap Road Network Density: The Case Study of 85 Cities in China

Bo Liu [1], Yu Shi [1], Da-Jun Li [1,*], Yan-Dong Wang [2], Gabriela Fernandez [3] and Ming-Hsiang Tsou [3]

1 Faculty of Geomatics, East China University of Technology, 418# Guanglan Road, Nanchang 330013, China; liubo@ecut.edu.cn (B.L.); yushi19930807@outlook.com (Y.S.)
2 State Key Laboratory of Information Engineering in Surveying, Mapping and Remote Sensing, Wuhan University, 129# Luoyu Road, Wuhan 430079, China; ydwang@whu.edu.cn
3 Department of Geography, Center for Human Dynamics in the Mobile Age (HDMA), San Diego State University, 5500 Campanile Drive, San Diego, CA 92182-4493, USA; gfernandez2@sdsu.edu (G.F.); mtsou@mail.sdsu.edu (M.-H.T.)
* Correspondence: djli@ecut.edu.cn

Received: 14 July 2020; Accepted: 17 August 2020; Published: 28 August 2020

Abstract: The evaluation of urban economies has been one key concern identified by scholars. In the past, most research methods on urban development assessments have been based on statistical data, and the analysis results have been presented in the form of statistical tables. Moreover, the development of urban road networks reflects the status of urban development and spatial metrics, which are obtained from the urban road network which can be used to evaluate the growth of the urban economy. The OpenStreetMap (OSM) is collected through crowdsourcing, and the OSM road network has the characteristics of a simplified and efficient approach to collect data, update data, free available data, etc. Therefore, in this paper, the OSM road network density is used as a spatial metric which is taken as the main study subject, to evaluate the economic development of Chinese cities. In our experiment, results show that there is a significant regression correlation between the OSM road network density and municipal gross domestic product (GDP). For the 85 selected Chinese cities, a total of 71 cities with residuals between −0.1 and 0.1 account for 83.53%, and a total of 79 cities with residuals between −0.2 and 0.2 account for 92.94%. Therefore, it is apparent that the OSM road network density can be used as a spatial metric to evaluate the municipal GDP, and as a result, can be used by local governments and scholars to estimate, evaluate, and forecast the urban economic development of China.

Keywords: OpenStreetMap (OSM); road network density; urban economy; regression analysis; spatial metric

1. Introduction

In recent years, the rapid development of remote sensing, volunteered geographic information (VGI) and other technologies, spatial data acquisition have become easier to employ. As a result, there is more and more research on land-cover, land-use, and urban development using spatial metrics [1–8]. A few different approaches used to represent spatial concepts have resulted in the development of various spatial metrics [1–4]. More commonly applied metrics used by scholars have included patch size, dominance, number of patches, and density, edge length and density, nearest neighbor distance, fractal dimension, contagion, etc. [4]. Herold et al. [5] used spatial metrics and texture measures to describe the spatial characteristics of land-cover objects within each land-use region as derived from interpreted aerial photographs. These spatial metrics include percentage of landscape, patch density, mean patch size, area standard deviation, edge density, largest patch index, Euclidean mean nearest neighbor distance, Euclidean nearest neighbor distance standard deviation, area weighted mean patch fractal dimension, fractal dimension standard deviation, etc. Herold et al. [6] argued that remote sensing and spatial metrics lead to an improved understanding and representation of urban dynamics while helping to develop alternative conceptions of urban spatial structure and change. In order to cover as many metrics as possible, Reis et al. [7] presented an extended, updated, and more thorough portfolio of spatial metrics to measure the urban growth and urban shrinkage patterns. In their paper, spatial metrics were defined as the quantitative measures used to assess the spatial characteristics of urban settlements. Reis et al. [8] assembled spatial metrics into three groups: landscape metrics; geo-spatial metrics, and spatial statistics. They indicated that landscape metrics such as fractal dimension, shape, mean perimeter-area ratio mean shape, etc., have been traditionally used to quantify several aspects of landscape configuration and composition. Geospatial metrics has mostly been used to measure urban spatial patterns. However, there are metrics that are similar between geospatial metrics and landscape metrics. Moreover, an important difference between the metrics from landscape metrics is that the latter include a set of metrics that evolved in a "top-down" approach. Spatial statistics are metrics based on statistical tools, and these spatial metrics are often used in combination with regression and spatial econometric models. In this paper, the main purpose of this research is to discuss whether the OpenStreetMap (OSM) road network density can be used to evaluate the level of urban economic development in cities. Firstly, using a regression model to establish the relationship between the OSM road network density and municipal gross domestic product (GDP) from 2014 to 2017. Then, the OSM road network density and municipal GDP in 2018 were used to verify the regression model, to discuss whether the OSM road network density can be used as a spatial metric to evaluate the level of urban economic development in cities. Because the OSM road network has the advantage to be an easier approach to collect data, in a more efficient way, update data, and provide free available data, etc. If we can use the OSM road network density to evaluate the level of urban economic development in cities, then this research can help policy makers in China monitor and evaluate their cities towards more transparent and efficient cities. Moreover, the OSM road network density is based on statistics and mainly uses a regression model, therefore, we take the OSM road network density as a spatial metric to evaluate the level of urban development and transparency in cites.

In general, there is a significant correlation between urbanization and the level of economic development, and it seems that each country or region conforms to this rule to a certain degree [8–10], and urbanization has been a defining global phenomenon and a key driving force for social and economic development during the past century [11,12]. Despite China being the world's largest developed country, its urbanization has progressed at an unprecedented rate [13], as urbanization has shifted due to the rise of industries and population in and around cities, facilitating the development of economies of scale [14]. Cai et al. [15] indicated that the improvement of urban infrastructure attracted corporate investments, created new jobs, and led to an influx of labor. Therefore, the proportion of China's population living in cities increased from 17.9% in 1978 to 58.5% in 2017 [16]. Heshmati et al. [17] calculated a multi-dimensional composite index of urban infrastructure analyzing 31 provinces and six regions in China during 2005 to 2014, and indicated that the economy, employment, human development, utilities, and technology components of urban infrastructure had positive and significant effects on China's urbanization, and suggested that the government should guide investments to more efficient transportation systems that improve the development of a city.

As described in the above literature, the efficient transportation systems have a positive impact on urbanization, and the higher the urbanization level of a city, the greater the density of its road network, the higher economic level of a city. In order to discuss the relationship between the road network density and urban economic development, based on geospatial metrics and spatial statistics, many researchers have studied the relationship between spatial metrics and urban economic development. Yu et al. [18] explored the role of the motorway in the evolution of spatial economic agglomerations, they indicated that an improvement in the motorway network lead to a higher degree of geographic concentration of economic activities. Jiao et al. [19] studied the relation between road accessibility and economic growth in China from 1990 to 2010. They studied a total of 337 cities in China and explored the bivariate analysis framework of accessibility, economic growth, and increased rates. The analysis showed that there is a significant positive relationship between the accessibility and economic growth of a city, and that the economic increased rates are largely influenced by the change of accessibility. Worku [20] studied the trends, stock of achievements, and impact of the road network on the economic growth of Ethiopia, Africa. He indicated that the impact of the road network was seen to be less strong on the agricultural GDP growth, but had pronounced impact of the industrial and service sector of GDP. Ivanova and Masarova [21] used the series and correlation method to analyze the effects of road infrastructure development on the economic growth and competitiveness of Slovak's economy, they indicated that the road infrastructure was a prerequisite for economic growth. Beyzatlar et al. [22] investigated the Granger causality relationship between income and transportation of EU-15 countries. They indicated that there was an endogenous relationship between income and transportation. Gao et al. [23] studied the relationship between the comprehensive transportation freight index and GDP in China. They indicated that the volume of freight traffic and freight turnover in China are positively correlated with GDP. Fan and Chan-Kang [24] studied the impact of road investment on the overall economic growth, rural and urban growth, and rural and urban poverty reduction. They indicated that road investments yielded the highest economic returns in the eastern and central regions of China, while the contributions to poverty reduction were the greatest in western China. It can be seen from the above research that the road network can be correlated with GDP, and most researchers have used regression analysis to investigate the relationship between spatial metrics and urban economic development in cities.

In the last decade, OSM has achieved tremendous development with the popularity and development of the Internet. OSM is a user-generated street map, most of the data sources are provided by the public or volunteers [25], and until now, there are more than six million registered members in the world, and more than six billion nodes in the OSM database [26]. Today, the OSM has become available to be applied in many ways, such as 3D city modeling, road updating, etc. [27–37]. For example, based on the free geographic data provided by the OSM project and the public domain height information provided by the Shuttle Radar Topography Mission, Over et al. [27] studied the prospects of using an interactive 3D city model in Germany and pointed out that the point of interest (POI) in the OSM data provides new opportunities for 3D city modeling. Fonte et al. [28,29] used the OSM data and GlobeLand30 image data to process, resulting in a more accurate and detailed land use coverage map, and more details can be shown by this method than any other methods. Mobasheri et al. [34] explored the feasibility of using the OSM data as geo-navigation data in several German cities by using factors such as the number of features and integrity, and the results showed that the sidewalk data in the OSM can be used to route navigation. Zhang et al. [37] explored the relationship between road density and road type diversity based on data obtained from China's OSM road network in May 2014, taking 340 prefecture-level cities in China as its study area, it is concluded that the OSM road diversity reflects the demand and value of road-related geographic information and it also reflects the interests of users towards employing the OSM geographical information; Goetz [38] used the points of interest data in the OSM to focus on the detailed display in 3D city modeling. Wang and Zipf [39] used an algorithm to extract the building information in the OSM data for modeling, and the building interior details can be displayed by using the proposed method. In the study of path navigation, Bergman and Oksanen [40] took the OSM data and mobile sport tracking data as research objects, the hidden Markov model (HMM) based as a research method, pointed out that the OSM data has feasibility in bicycle path navigation. In terms of geographic mapping, Rosina et al. [41] took Slovenia and Austria as research objects, and added the OSM data to the Copernicus imperviousness layer to improve the population distribution map, drawing methods of the two countries. The experimental result showed that the total error is reduced after adding the OSM data for auxiliary processing, and the addition of the OSM data has certain improvement effects; Zhao et al. [42] studied the evolution of the OSM road network in Beijing from four aspects. Through the experimental analysis, they believed that the development of the OSM road network in Beijing was significantly related to the number of volunteers, and the growth of the OSM road network was very similar to the development process of the real road network. Dingil et al. [43] used the OSM to estimate and analyze the passenger transport energy per person per year of 57 cities, distributed over 33 countries, the results indicating that high private car mode share is a main cause for the high transport energy usage of such cities.

From the above application research on OSM, it is apparent that using OSM data to do research has become more widespread. In this paper, we aim to focus on developing an economic development evaluation for 85 Chinese cities by using OSM spatial metric road network density. The main purpose is to explore the application of the OSM to evaluate the urban economic development of Chinese cities, and take the OSM road network density as a spatial metric to evaluate the level of urban economic development which is measured by the municipal GDP. In our experiment, we selected 85 cities to verify the proposed method, the results show that the correlation between the OSM road network density and municipal GDP are significant. As a result, it is feasible to predict the level of urban economic development by using the OSM road network density.

This paper is organized as follows: the data sources and basic methods are introduced in Sections 2 and 3, respectively. Experimental results and analysis are reported in Section 4. Conclusions are drawn in Section 5.

2. Study Areas and Data Source

2.1. Study Areas

Since the reform and reopening in 1978, China's urbanization level has continuously improved. As the pioneer area of China's economy development, coastal cities and provincial cities have a higher urbanization rate, which has exceeded 60% in 2017 [43]. This is because eastern coastal and provincial cities have comparative advantages in resource adsorption, innovation, transportation, and so on, making them leaders in China's overall economic and social development [43]. Since eastern coastal and provincial cities are the pioneers of China's economic and urbanization development, the open-up policy, new technical innovation, and adjustment of input structure methods has improved the urbanization efficiency of China [44,45].

In this paper, we selected a total of 85 cities in China. Among the 85 cities, there were a total of 62 eastern cities, 12 central cities, and 11 western cities. In addition, among the 85 cities, the study included a total of 27 provincial cities and 4 municipalities.

2.2. Data Collection

There are three main types of data: OSM road network, municipal GDP, and urban area in each selected city. More details about data sources and data formats are as follows.

2.2.1. OSM Road Network

Currently, OSM is one of the most successful and popular VGI projects, and has achieved tremendous development. Until now, there are more than six million registered members in the world, and more than six billion nodes in the OSM database [26], and a lot of research on OSM. Because OSM data are collected through crowdsourcing, the quality of OSM has been often discussed, and usually evaluated based on its quality with authority data. Haklay [46] was the first researcher who analyzed and investigated the data quality of the OSM road network for England, UK. Since then, many researchers have analyzed the data quality of OSM for Germany [47,48], France [49], and China [50]. Luo et al. [50] selected three large, medium, and small cities in China, and compared the length integrity of the OSM road network with the Baidu road network, and Google road network. They indicated that the length integrity of the OSM road network is basically consistent with Baidu's road network and the Google road network, and the length integrity of the OSM road network is better than Baidu's road network, and the Google road network in some regions. Hecht et al. [51] measured the completeness of the OSM data on buildings by comparing them with official survey data. It is clear from their research that the OSM building data in urban areas, particularly near the town center, achieve a much higher level of completeness. Singh Sehra et al. [52] introduced a comprehensive review of the assessment of OSM data. Some researchers indicated that OSM has a higher location accuracy, completeness, etc., data quality characteristics in urban areas [48,53]. Because, most urban areas with a higher population density inherit larger numbers of contributors, who influence the quantity and quality of the collaboratively crowdsourced OSM objects [45–53]. Therefore, in this paper, we use the urban road network of the OSM to calculate the road network density for 85 Chinese cities.

The OSM road network is downloaded from the OSM website (http://download.geofabrik.de/), which is the ESRI Shapefile. In addition, the projected coordinate system used is the Universal Transverse Mercator (UTM) coordinate system. In this paper, data were collected and downloaded from 2014 to 2018. The OSM road network has about 20 road classes such as cycle way, footway, motorway, residential, primary, and secondary, etc.

2.2.2. Municipal Gross Domestic Product

Over the past 40 years, China's economic reform has been successful, becoming one of the most important economic power engines around the world, and the second largest economy measured by GDP. China's rapid economic growth has largely depended on abundant use of natural resources, low-cost investment, and labor with support of a high saving rate, and government policies have also played an important role in promoting infrastructure construction [54]. In recent years, China's economic growth rate has fallen from the double-digit rate from 5% to 7%. China's economy has entered "The New Normal Economy". There are three main characteristics of The New Normal of China's Economy: (a) a shift from high growth rates to medium-high growth rates; (b) an on-going process of optimizing and upgrading the economic structure, and narrowing the urban-rural gap, with higher personal income as a share of GDP, and an increasing number of people benefiting from economic development; and (c) a transition from growth driven by input and investment to one driven by innovation. These characteristics and measures can promote the steady growth of the Chinese economy, enhance development potential, and further unleash market vitality [55]. No matter if China's economy is in the double-digit growth rate or the current "new normal economy" medium-high growth rate, the GDP is an index used to describe the economic development level of a city, and the efficient transportation system, population, technology, etc., have a positive impact on GDP. Among the 85 selected cities, a total of 5 first-grade cities, 31 second-grade cities, and 49 third-grade cities were identified (https://www.yicai.com/news/5293378.html). Overall, the GDP and urbanization rate of the first-grade cities are higher than the second-grade cities, and the GDP and urbanization rate of the second-grade cities are higher than the third-grade cities.

In our research, municipal GDP data are collected from the National Bureau of Statistics (http://www.stats.gov.cn/), data on GDP are from 2014 to 2018, and municipal GDP unit is in trillion CNY.

2.2.3. Exploring the Urban Area of Each Selected City

The study scope analyzed a total of 85 main urban areas. The study applied the unit of area square kilometers. Take Shanghai for example, the scope of the main urban area is extracted from the Shanghai Bureau of Planning and Natural Resources Department (http://ghzyj.sh.gov.cn/). Figure 1 shows the location and the municipal GDP in 2018 of the selected 85 cities.

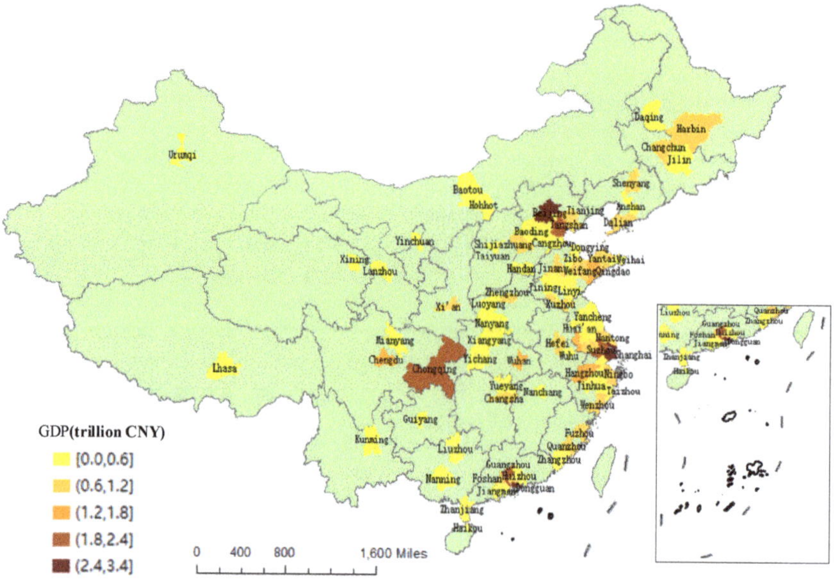

Figure 1. Study research area and gross domestic product (GDP) in 2018.

3. Methodology

Regression analysis is one of the most commonly statistical analysis methods used to describe the correlation between independent variables and dependent variables [56]. In our paper, we used a regression model to study the correlation between the OSM road network density and the municipal GDP, and employed the OSM road network density as a spatial metric to evaluate the level of urban economic development of cities. In this section, we do not describe the details of the regression model, but only introduce the calculation method of the OSM road network density.

Calculating the OSM Road Network Density of a City

The road network density is an important index for the evaluation of regional road traffic [57–59]. The OSM road network density comprises of geographical information that reflect a real-world road network, and an index for assessing the quality of OSM geographic data. In this study, we calculated the OSM road network density of 85 Chinese cities [60,61]:

$$D_i = L_i/A_i \; i \in [1, 2, 3, \ldots, 85], \tag{1}$$

where one main urban area is i, the OSM road network density in the main urban area i is D_i, the OSM road network length for the main urban area i is L_i, and the area of the main urban area i is A_i.

Figure 2 shows the OSM road network in three different grades based on city. The blue polylines are the OSM road network in 2018. The red polylines are the OSM road networks in 2014. Figure 1 shows the cities of Beijing, Shanghai, and Guangzhou as first-grade cities. Compared to the cities of Nanjing, Wuhan, and Chengdu with second-grade cities, and the cities of Guiyang, Haikou, and Lanzhou with third-grade cities, these cities have different levels of economic development and OSM road network density.

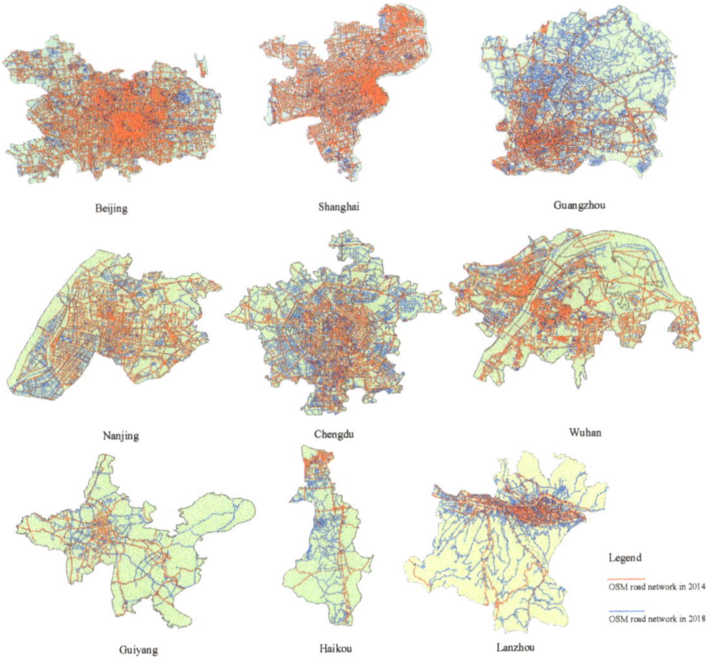

Figure 2. The OpenStreetMap (OSM) road network with three different grades in Chinese cities.

4. Results and Analysis

4.1. Fit Analysis

In this paper, we used Equation (1) to calculate the OSM road network density from 2014 to 2018, then applied the OSM road network density data from 2014 to 2017 as the independent variable, and the municipal GDP from 2014 to 2017 as the dependent variable. Moreover, the regression models of the 85 Chinese cities were obtained using the unary linear regression model, and then applied the OSM road network density and the municipal GDP data in 2018 to validate the regression model.

The OSM road network density, the municipal GDP, the regression models, and the coefficient of determination (R^2) of the 85 Chinese cities from 2014 to 2017 are shown in Table A1 (Appendix A), the statistics of R^2 are shown in Figure 3. Among all the cities, the maximum value of R^2 is Guangzhou, and its R^2 is 0.999; the minimum value of R^2 is Hohhot, and its R^2 is 0.0005. The distribution of coefficient of determination is shown in Figure 4.

In this paper, we also used the OSM road network density and population data from 2014 to 2017 as the independent variables, the municipal GDP from 2014 to 2017 as the dependent variable, the regression models of 85 Chinese cities were obtained using a binary linear regression model, and then used the OSM road network density and the municipal GDP data in 2018 to validate the same regression model.

The statistics of R^2 are shown in Figure 5, and the distribution of coefficient of determination is shown in Figure 6. Among all the selected cities, the maximum value of R^2 is Taizhou, its R^2 is 0.9999. The minimum value of R^2 is Baotou, its R^2 is 0.3353. In this regression model, the R^2 of Guangzhou is 0.9996, and the R^2 of Hohhot is 0.4842. The distribution of coefficient of determination is shown in Figure 6. In addition, we can see that the R^2, which is calculated by the OSM road network density and population, is higher than the R^2, which is calculated by the OSM road network density and is consistent with reference [24,62]. Fan and Chan-Kang [24], Savaş [62] indicated that there is a high correlation in cities with road investments, population, and economic growth.

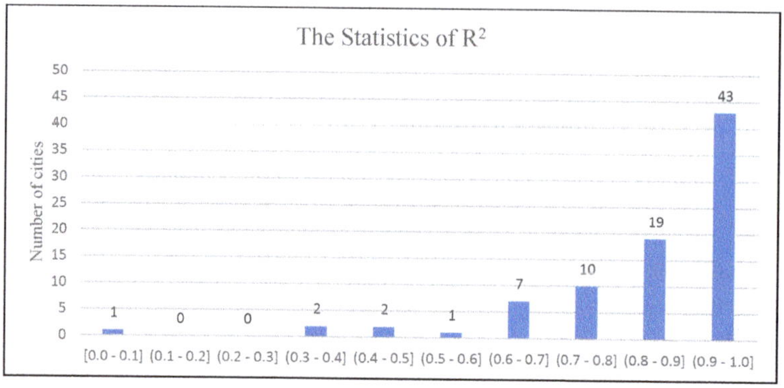

Figure 3. Statistics of R^2 calculated by the OSM road network density.

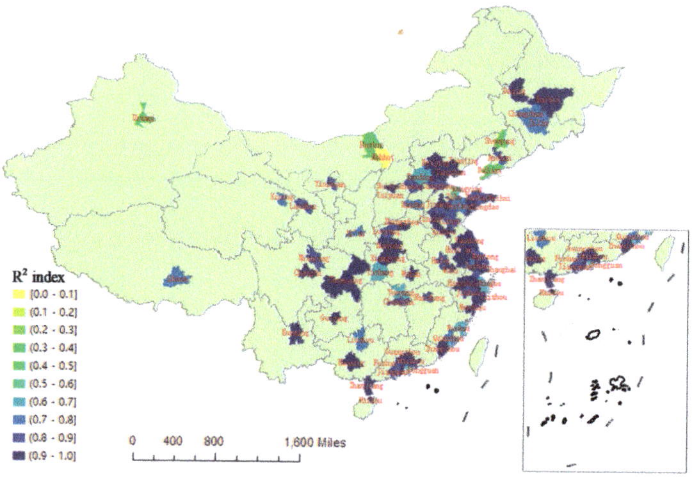

Figure 4. Distribution of R^2 calculated by the OSM road network density.

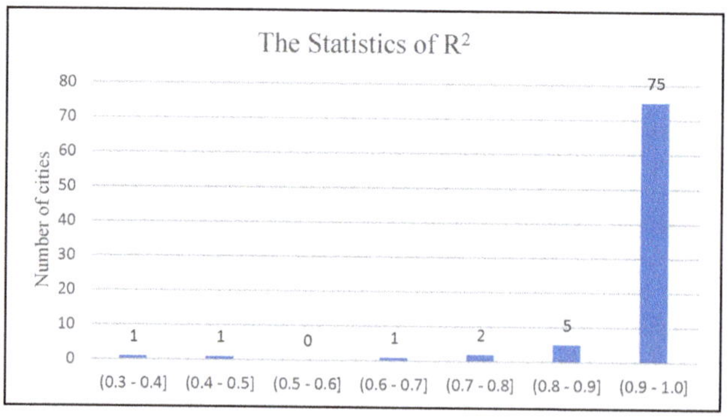

Figure 5. Statistics of R^2 calculated by the OSM road network density and population.

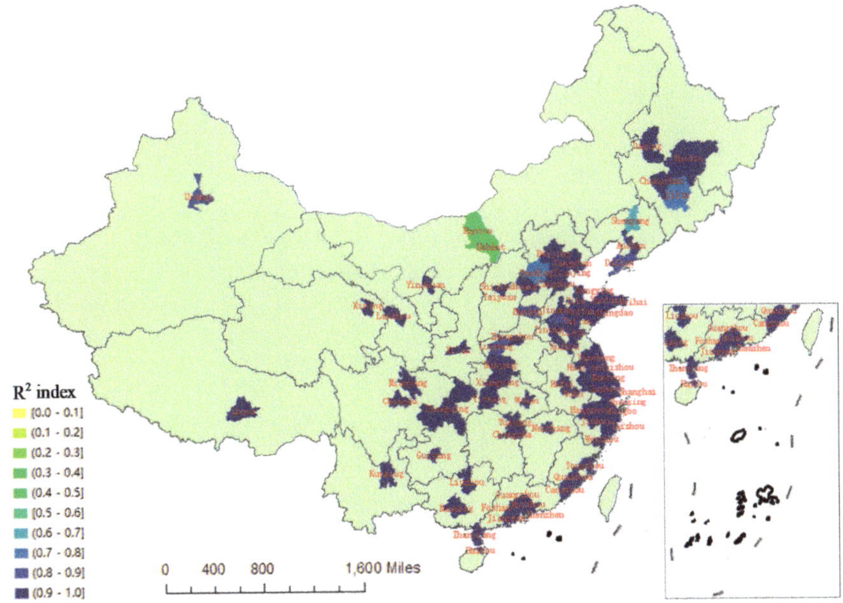

Figure 6. Distribution of R^2 which is calculated by the OSM road network density and population.

Figure 3 shows a total of 72 cities with significant correlation with coefficient of determination R^2 above 0.7, accounting for 84.71%. These 72 cities experienced rapid economic development during the four years from 2014 to 2017, and with their municipal GDP increasing steadily year by year; 5 cities with coefficient of determination R^2 below 0.5, they are Shenyang, Urumqi, Dalian, Baotou, and Hohhot, respectively, and their R^2 are 0.499, 0.483, 0.342, 0.303, and 0.0005, respectively. The municipal GDP statistical data and the OSM road network density for these five cities are shown in Tables 1 and 2. It can be seen that the municipal GDP for these five cities does not increase year by year, but the density of the OSM road network increases year by year.

Table 1. The Chinese municipal GDP (trillion CNY) statistical results of five cities.

City	GDP (2014)	GDP (2015)	GDP (2016)	GDP (2017)	GDP (2018)
Shenyang	0.709871	0.728000	0.546001	0.586497	0.62924
Urumqi	0.246147	0.263164	0.245898	0.274382	0.309962
Dalian	0.765558	0.773164	0.68102	0.73639	0.76685
Baotou	0.363631	0.378193	0.386763	0.275303	0.29518
Hohhot	0.289405	0.309052	0.317359	0.274372	0.29035

Table 2. The OSM road network density of the five cities.

City	OSM RND (2014)	OSM RND (2015)	OSM RND (2016)	OSM RND (2017)	OSM RND (2018)
Shenyang	2.014	2.150	2.203	2.292	2.449
Urumqi	1.472	1.489	1.549	1.676	1.851
Dalian	1.750	1.838	1.917	1.981	2.441
Baotou	0.601	0.629	0.852	0.900	1.399
Hohhot	0.567	0.656	0.860	0.870	0.962

Note: OSM RND represents the OSM road network density.

Hohhot shows the minimum coefficient of determination, and the coefficient of determination is 0.0005. We found that the GDP of Hohhot decreased from 317.359 billion CNY in 2016 to 274.372 billion CNY in 2017. However, the density of the OSM road network in the main urban area gradually increased from 2014 to 2017. More specifically, in 2016, with an increase of 0.204 compared with 2015, which is much higher than other years. Baotou and Hohhot, which are located in the Inner Mongolia Autonomous Region of China, have similar patterns. In 2017, the GDP of Baotou was 275.303 billion CNY, a decrease of 111.46 billion CNY from 386.763 billion CNY in 2016. The density of the OSM road network in the main urban area increased from 0.629 in 2015 to 0.852 in 2016, an increase of 0.22, which is much higher than changes in other years.

4.2. Validation of the Model

In order to validate the regression model obtained by the OSM road network density in the above section, we used the real municipal GDP of 85 Chinese cities in 2018 to validate the regression model. The statistical results of the absolute residuals and relative residuals when using the OSM road network density are shown in Table A2, Figure 7, and Figure 8, respectively. The distribution of absolute residual and relative residual results when using the OSM road network density is shown in Figures 9 and 10, respectively.

The calculation method of absolute residual ($R_{absolute}$) and relative residual ($R_{relative}$) is as follows:

$$R_{absolute} = Predictive_i - Real_i \; i \in [1,2,3,\ldots,85], \tag{2}$$

$$R_{relative} = \frac{(Predictive_i - Real_i)}{Real_i} \times 100 \; i \in [1,2,3,\ldots,85], \tag{3}$$

where one main urban area is i, the predictive GDP of the main urban area i in 2018 is $Predictive_i$, the real GDP of the main urban area i in 2018 is $Real_i$.

At the same time, we validated the regression model obtained by using the OSM road network density and population, we also used the real municipal GDP of 85 cities in 2018 to validate the regression model, and the statistical results of the absolute residuals are shown in Figures 11 and 12, respectively. The distribution of residual results is shown in Figures 13 and 14, respectively.

We calculated the difference between the residual, which is obtained by using the OSM road network density, and population. The statistical results of the differences are shown in Figure 15, and the distribution of the differences are shown in Figure 16. The results show 69 cities with residual differences between −0.1 and 0.1, accounting for 81.18%; a total of 76 cities with residual differences between −0.2 and 0.2, accounting for 89.41%. Overall, the forecasting results by using the OSM road network density are found to have similar characteristics in the forecasting results when using the OSM road network density and population. This shows that the OSM road network density can be used as a spatial metric to evaluate the level of urban economic development in cities.

Considering that the main purpose of this paper is to study whether the OSM road network density can be used as a spatial metric to evaluate the level of urban economic development, the following discussion will focus on forecasting results obtained by using the OSM road network density.

As shown in Figure 7 and Table A2, there are a total of 50 Chinese cities that have negative absolute residuals, and 35 cities with positive absolute residuals. The largest positive absolute residual is Chongqing, with a value of the absolute residual of 0.444; the smallest negative absolute residual is Shenzhen, with a value of the absolute residual of −0.2817. The absolute residuals of 44 cities are between −0.1 and 0.0, accounting for 51.76%, and the absolute residuals of 27 cities are between 0.0 and 0.1, accounting for 31.76%. The absolute residuals of 71 cities are between −0.1 and 0.1, accounting for 85.53%. The absolute residuals of 4 cities are between −0.2 and −0.1, accounting for 4.71%. The absolute residuals of 4 cities are between 0.1 and 0.2, accounting for 4.71%. The absolute residuals of 79 cities are between −0.2 and 0.2, accounting for 92.94%.

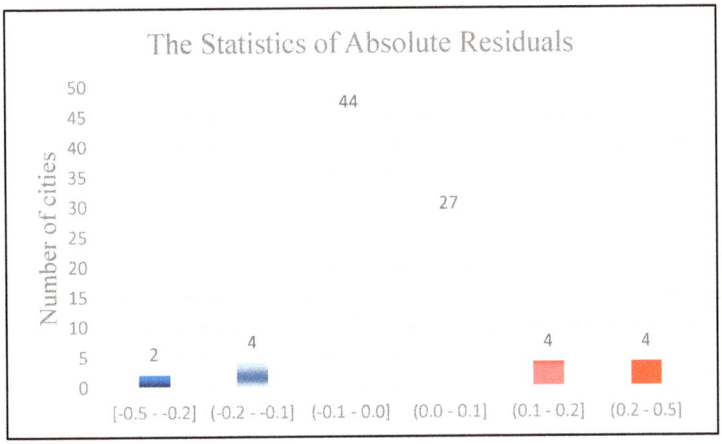

Figure 7. The statistical result of the absolute residuals by using the OSM road network density.

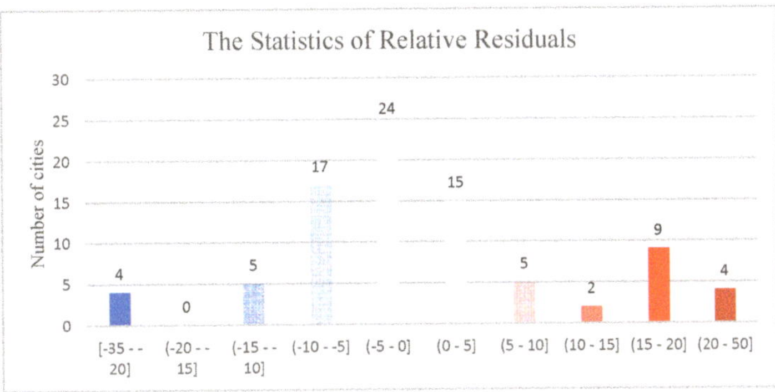

Figure 8. The statistical result of the relative residuals by using the OSM road network density.

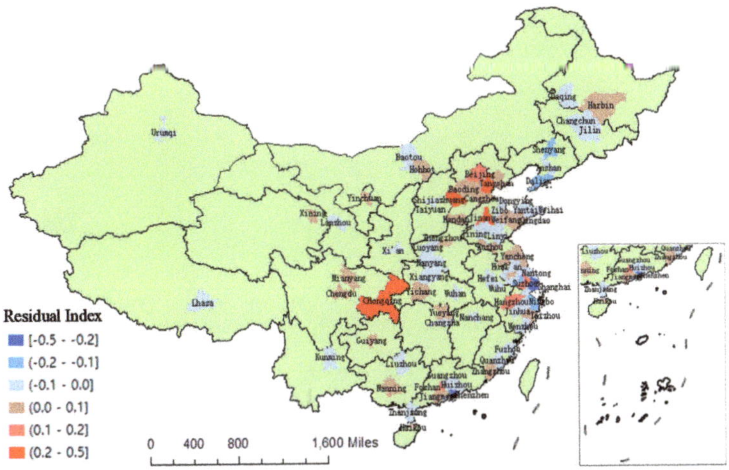

Figure 9. Distribution of the absolute residuals of 85 Chinese cities by using OSM road network density.

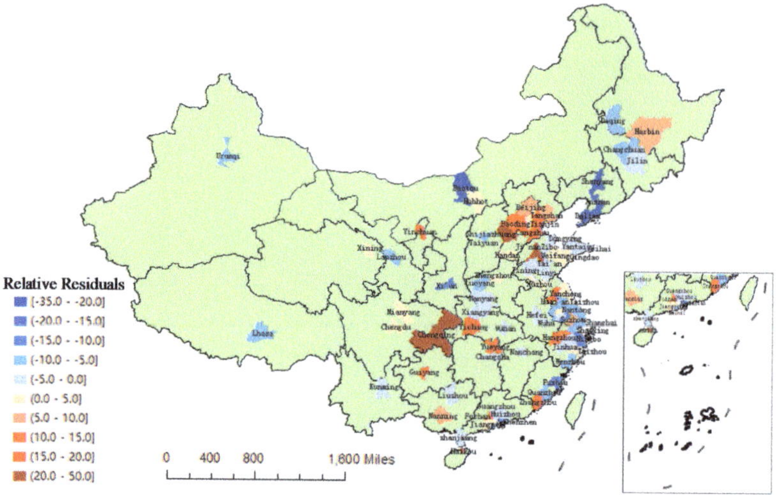

Figure 10. Distribution of the relative residuals of 85 Chinese cities by using OSM road network density.

As shown in Figure 8 and Table A2, there are a total of 50 Chinese cities that have negative relative residuals, and 35 cities with positive relative residuals. The smallest negative relative residual is Anshan, with a relative residual of −34.6088, and the largest positive relative residual is Shijiazhuang, with a value of the relative residual of 47.8711. The relative residuals of 41 cities are between −10 and 0.0, accounting for 48.24%, and the relative residuals of 20 cities are between 0.0 and 10, accounting for 23.53%. The relative residuals of 61 cities are between −10.0 and 10.0, accounting for 71.76%. The relative residuals of 77 cities are between −20.0 and 20.0, accounting for 90.59%.

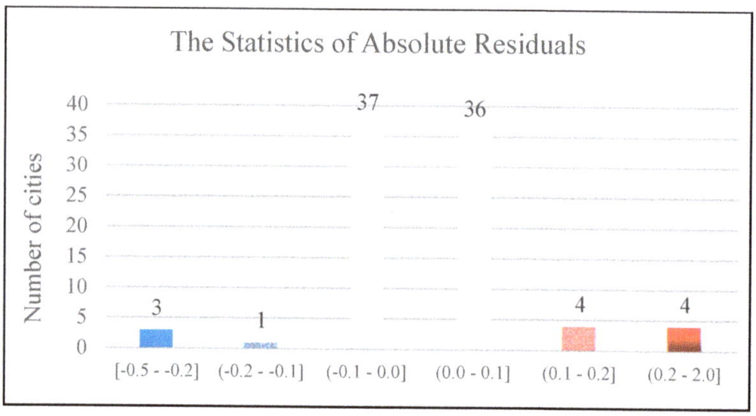

Figure 11. The statistical result of the absolute residuals by using OSM road network density and population.

Results indicate that the regression models obtained from the above section have a high prediction accuracy, and the OSM road network density can be used as a spatial metric to forecast the municipal GDP. In order to directly discuss the difference between the predicted GDP in 2018 and GDP in 2018, the following discussion is based on the absolute residuals in Figure 7 and Table A2.

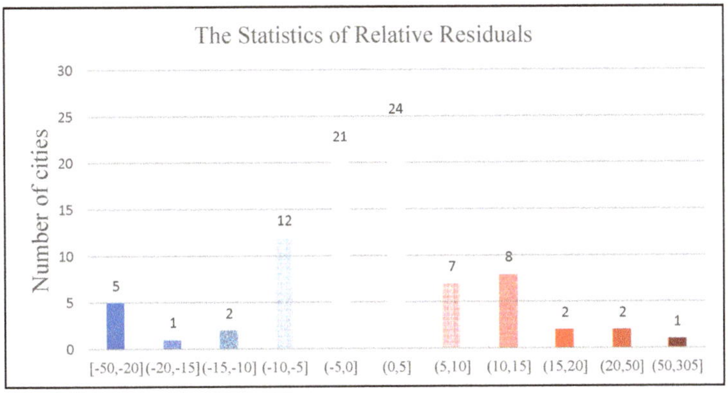

Figure 12. The statistical result of the relative residuals by using OSM road network density and population.

The largest positive value of residual is Chongqing, the value of residual is 0.444. Chongqing is located in the southwest of China, and is the only municipality in the southwest of China, its total area is 82,400 square kilometers, and its main urban area is 7220 square kilometers. The municipal GDP and OSM road network density from 2014 to 2018 are shown in Tables A1 and A2. From 2014 to 2017, the municipal GDP of Chongqing increased year by year, the average annual municipal GDP growth of Chongqing is 172.084 billion CNY. However, the municipal GDP growth in 2018 is 93.824 billion CNY, far lower than the average annual municipal GDP growth in the previous 3 years. The OSM road network density growth in 2018 is 0.38, and the average OSM road network density growth is 0.13, different from the municipal GDP growth change in 2018, the OSM road network density growth in 2018 is much higher than the average OSM road network density growth during the previous 3 years, and the inconsistent trends in municipal GDP growth and OSM road network growth have resulted in excessive forecast residual.

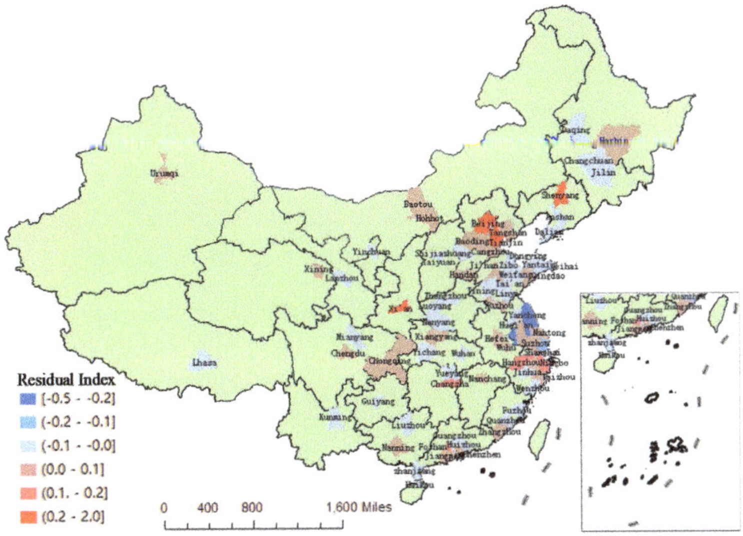

Figure 13. Distribution of 85 Chinese cities' absolute residuals using OSM road network density and population.

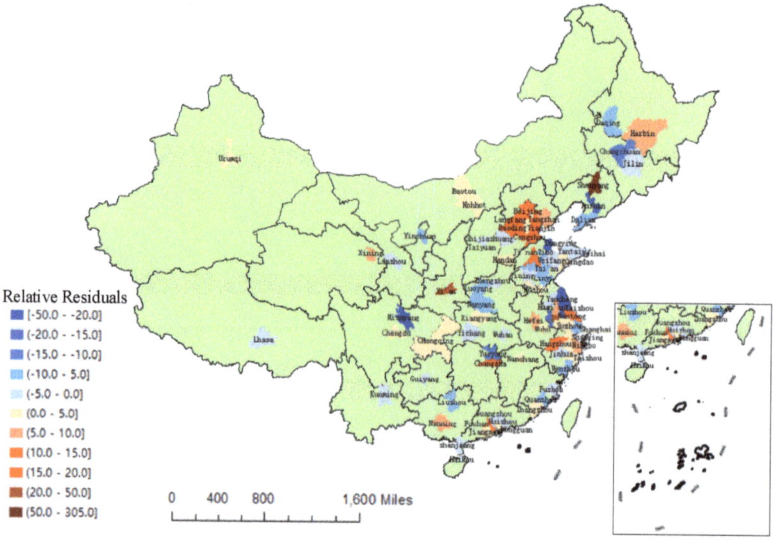

Figure 14. Distribution of 85 Chinese cities' relative residuals using OSM road network density and population.

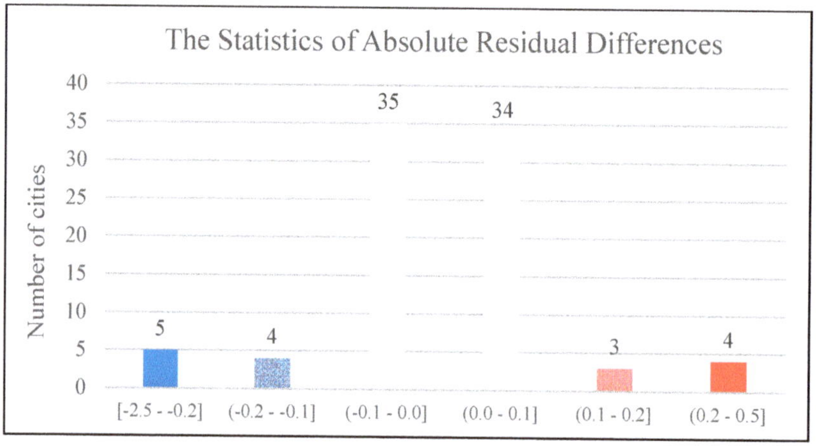

Figure 15. Statistical result of the difference between two residuals.

Tianjin has the second largest residual, the value of the residual is 0.3149. Tianjin is a municipality located in the north of China, total area is 11,900 square kilometers, and the main urban area is 1007 square kilometers. From 2014 to 2018, the total municipal GDP of Tianjin increased year by year, but the municipal GDP growth varied greatly. The municipal GDP growth in 2018 is 260.45 billion CNY, compared to the average GDP growth of 94.075 billion CNY. As a result, the municipal GDP growth in 2018 is much lower than the average GDP growth from the previous three years. However, the OSM road network density of Tianjin is increasing quickly, more specifically in 2018, the OSM road network density growth is 0.89, and the average OSM road network density growth is 0.26, the OSM road network density growth in 2018 is much higher than the average OSM road network density growth during the previous 3 years, and these factors finally caused the excessive forecast residuals.

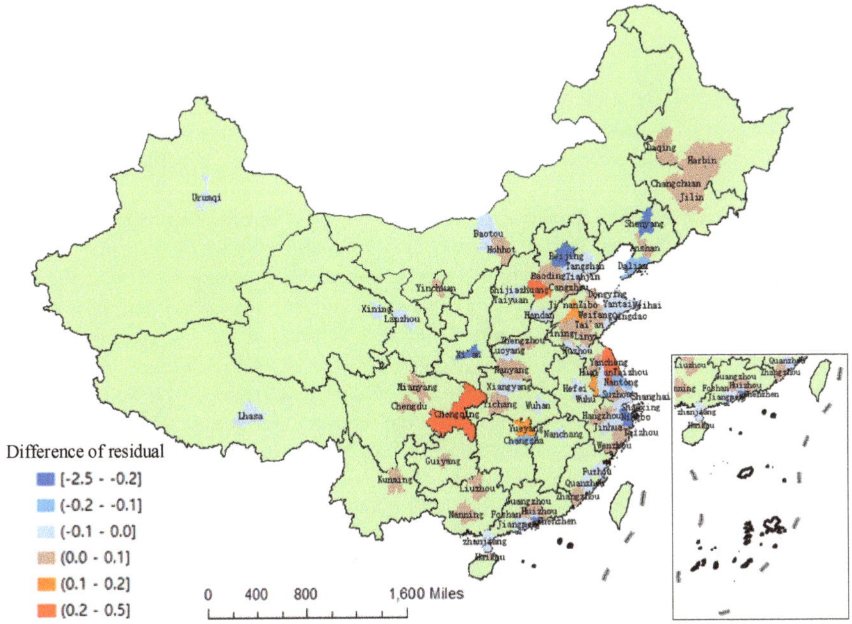

Figure 16. Distribution of absolute residual differences.

On the contrary, Shenzhen has the smallest negative residual, the value of the residual is −0.2817. Shenzhen is an important special economic development zone in the south of China, its total area is 1997 square kilometers, and the main urban area is 927 square kilometers. Shenzhen's economy is developed with a steady growth from 1.600182 billion CNY in 2014 to 2.4222 billion CNY in 2018. However, during 2014 to 2017, the annual growth of road density in Shenzhen showed a downward trend, but an annual growth in 2018. Thus, the difference between the variation tendency of the municipal GDP and the variation tendency of the OSM road network density makes the forecast value of the municipal GDP smaller than the real data, which leads to Shenzhen's residual to have the smallest negative residual.

Overall, between the 85 cities, a total of 8 cities were found with absolute residual larger than 0.1, and 6 cities with absolute residual lower than −0.1, these 14 cities include Beijing, Shanghai, Shenzhen, Tianjin, Chongqing, Hangzhou, Dalian, Ningbo, Ji'nan, Suzhou, Shenyang, Wuxi, Foshan, and Shijiazhuang. These 14 cities all have characteristic of developed cities. In recent years, the pace of urban construction of these developed cities has increased at a rapid pace, having large populations, which can contribute a lot to the municipal GDP growth. From 2014 to 2018, the changes of municipal GDP are not consistent with the OSM road network density. More specifically, the municipal GDP annual growth for the above 14 cities in 2018 is much lower than that in 2017, but at the same time, the OSM road network density is increasing, which caused a big absolute residual value.

However, when considering the other 71 cities, the absolute residuals are found to be between −0.1 and 0.1. From 2014 to 2018, the municipal GDP growth of these cities did not change drastically, the change trend of the municipal GDP and OSM road network density has similarities, resulting in a smaller absolute residual. Among these 71 cities, most cities are second and third grade cities, these cities were found to have smaller populations. Therefore, to some extent, the analysis of the absolute residual in the paper also confirms that the current economic development strategy of China with a high-speed economic stage to a medium-high-speed economic stage has a relatively predicted impact in developed cities, when compared to the second and third grade cities, where the impact is not significant.

5. Conclusions

This study focused on using the urban OSM road network density as a spatial metric to evaluate the urban economic development of 85 Chinese cities. The following conclusions are identified in this paper:

(1) The OSM road network density can be used as a spatial metric to estimate and predict the level of urban economic development of cities which is commonly measured by the municipal GDP.

(2) It is feasible to analyze the level of urban economic development of cities by using the OSM road network density. There is a significant correlation between the OSM road network density and the municipal GDP. Results demonstrated a total of 71 Chinese cities with absolute residuals between −0.1 and 0.1, accounting for 83.53%; and 79 cities with absolute residuals between −0.2 and 0.2, accounting for 92.94%. There are 61 cities with relative residuals between −10 and 10, accounting for 71.76%; and there are 77 cities with relative residuals between −20 and 20, accounting for 90.59%.

(3) In our experiment, the R^2 is calculated with the OSM road network density and population is higher than the $R^{2\cdot}$ which is calculated by the OSM road network density, but for the residuals, the absolute and relative residuals by using the OSM road network density are found to have similar characteristics as the absolute and relative residuals by using the OSM road network density and population. Among the 85 cities, there are 73 cities with absolute residuals between −0.1 and 0.1, accounting for 85.88%; and 78 cities with absolute residuals between −0.2 and 0.2, accounting for 91.76%.There are a total of 64 cities with relative residuals between −10 and 10, accounting for 75.29%; and there are 77 cities with relative residuals between −20 and 20, accounting for 90.59%.

(4) The development strategy of the Chinese economy is changing from high-speed development to medium-high speed development, these findings are more apparent in economically developed cities, compared to ordinary cities. Moreover, economically developed cities pay more attention to the influence of high technology.

The research in this paper has identified the imperfections found in the method when using the OSM road network density to evaluate and predict the level of urban economic development in Chinese cities. It is evident that the OSM road network density can be used as a spatial metric for predictive analysis and policy making to gain transparency. In the future, we will use the proposed spatial metric to do some related research.

Author Contributions: Methodology, Bo Liu, Yu Shi, Da-Jun Li and Ming-Hsiang Tsou; Software, Yu Shi; Visualization, Yu Shi and Bo Liu; Writing—original draft, Bo Liu, Yu Shi and Yan-Dong Wang; Writing—review and editing, Yan-Dong Wang, Gabriela Fernandez and Ming-Hsiang Tsou. All authors have read and agreed to the published version of the manuscript.

Funding: This work was funded by the National Key Research and Development Program of China (No.2016YFB0501403), The National Natural Science Foundation of China (Nos. 41201395, 416014160), The Key Laboratory of Earth Observation and Geospatial Information Science of NASG (201811), and the China Scholarship Council Foundation of China (No. 201808360267).

Acknowledgments: Thank you to the Center for Human Dynamics in the Mobile Age, Department of Geography, San Diego State University, San Diego, California, USA.

Conflicts of Interest: The authors declare no conflict of interest.

Appendix A

Table A1. The statistical results of the municipal GDP, coefficient of determination R^2, and regression equation of 85 Chinese cities from 2014 to 2017.

No.	Cities	Location of the Cities	Municipal GDP (trillion CNY)				OSM RND Represent the OSM Road Network Density				Regression Equation	Coefficient of Determination (R^2)
			2014	2015	2016	2017	2014	2015	2016	2017		
1	Beijing	Eastern	2.13308	2.2[illegible]686	2.48993	2.80004	4.958224	5.141546	5.357883	5.499061	y = 1.174x − 3.721	0.9515
2	Shanghai	Eastern	2.356094	2.49[illegible]499	2.746615	3.013386	3.155228	3.566878	4.213839	4.775253	y = 0.4047x + 1.0635	0.9954
3	Guangzhou	Eastern	1.670687	1.81[illegible]041	1.954744	2.150315	2.676991	3.534283	4.42808	5.49531	y = 0.1697x + 1.212	0.999
4	Shenzhen	Eastern	1.600182	1.75[illegible]286	1.94926	2.249006	6.827965	8.959036	9.394507	9.633126	y = 0.1792x + 0.3274	0.6705
5	Tianjin	Eastern	1.572693	1.65[illegible]819	1.788539	1.854919	5.901592	6.248347	6.441155	6.67073	y = 0.3837x − 0.706	0.956
6	Chongqing	Central	1.42622	1.57[illegible]727	1.774059	1.942473	0.662438	0.81481	0.900102	1.039407	y = 1.4165x + 0.4687	0.9746
7	Hangzhou	Eastern	0.920616	1.00[illegible]021	1.131372	1.260336	2.782885	3.022515	3.985984	4.35391	y = 0.1934x + 0.3956	0.9616
8	Nanjing	Eastern	0.882075	0.97[illegible]077	1.050302	1.17151	2.497384	2.930633	3.550194	4.445725	y = 0.1445x + 0.5339	0.9909
9	Qingdao	Eastern	0.86921	0.93[illegible]007	1.001129	1.103728	1.760078	1.887759	2.111393	2.364463	y = 0.3787x + 0.2068	0.9962
10	Dalian	Eastern	0.765558	0.77[illegible]164	0.68102	0.73639	1.75048	1.837818	1.917477	1.981108	y = −0.2446x + 1.1969	0.3415
11	Ningbo	Eastern	0.761028	0.80[illegible]361	0.868649	0.98421	1.560726	1.969064	2.056487	2.125499	y = 0.31x + 0.256	0.6433
12	Xiamen	Eastern	0.327358	0.34[illegible]603	0.378427	0.43517	6.883743	7.503913	7.690256	7.821306	y = 0.0947x − 0.3358	0.6945
13	Ji'nan	Eastern	0.57706	0.61[illegible]023	0.653612	0.720196	1.30219	1.353739	1.427773	1.692866	y = 0.3468x + 0.1394	0.9479
14	Suzhou	Eastern	1.376089	1.45[illegible]407	1.54751	1.731951	3.440061	3.859497	4.561366	6.548412	y = 0.1107x + 1.0168	0.9822
15	Wuhan	Central	1.006948	1.0[illegible]056	1.191261	1.341034	2.750418	2.951396	3.20614	3.695856	y = 0.3509x + 0.0517	0.9939
16	Chengdu	Western	1.005683	1.08[illegible]116	1.217023	1.388939	3.358756	4.236498	4.594147	5.561516	y = 0.1801x + 0.3738	0.9487
17	Changsha	Central	0.782481	0.85[illegible]013	0.945536	1.021013	1.533576	1.756367	2.052184	3.297249	y = 0.1215x + 0.6375	0.8346
18	Xi'an	Western	0.549264	0.5[illegible]012	0.625718	0.746985	3.196316	3.865494	4.135966	4.471107	y = 0.1421x + 0.0688	0.7828
19	Shenyang	Eastern	0.709871	0.[illegible]28	0.546001	0.586497	2.014167	2.150415	2.203039	2.292415	y = −0.5466x + 1.8259	0.4997
20	Zhengzhou	Eastern	0.667699	0.73[illegible]152	0.802531	0.91302	1.740456	2.115113	2.676731	2.880727	y = 0.1937x + 0.3227	0.9216
21	Dongguan	Eastern	0.588118	0.62[illegible]506	0.682767	0.758209	1.030598	1.501366	2.197582	2.854381	y = 0.092x + 0.4898	0.992
22	Fuzhou	Eastern	0.516916	0.56[illegible]808	0.619764	0.71034	0.902237	2.082021	2.257173	2.395907	y = 0.0982x + 0.4147	0.6464
23	Wuxi	Eastern	0.820531	0.85[illegible]826	0.921002	1.05118	2.153178	3.162821	3.301254	3.867276	y = 0.1253x + 0.5199	0.7635
24	Harbin	Eastern	0.534007	0.57[illegible]121	0.610161	0.635505	0.483118	0.51934	0.584794	0.615797	y = 0.7215x + 0.1913	0.9782
25	Foshan	Eastern	0.760328	0.80[illegible]392	0.863	0.95496	4.658889	5.297152	5.517081	5.916113	y = 0.1522x + 0.0309	0.8897
26	Changchun	Eastern	0.534243	0.55[illegible]003	0.591794	0.653003	0.937574	1.229054	1.432111	1.486014	y = 0.1836x + 0.3497	0.7557
27	Shijiazhuang	Eastern	0.517027	0.5[illegible]406	0.592773	0.646088	2.650926	2.756092	2.828802	2.846626	y = 0.5834x − 1.0413	0.8326
28	Taiyuan	Central	0.253109	0.27[illegible]3534	0.29556	0.338218	0.925942	0.96981	1.230886	1.398845	y = 0.1582x + 0.1111	0.9397
29	Yantai	Eastern	0.600208	0.64[illegible]608	0.69257	0.733895	0.571782	0.624688	0.692235	0.923017	y = 0.3473x + 0.4237	0.8592
30	Hefei	Central	0.515797	0.56[illegible]027	0.627438	0.721345	1.07847	1.916291	2.42456	2.662393	y = 0.1166x + 0.372	0.8532
31	Kunming	Western	0.371299	0.39[illegible]801	0.430008	0.485764	0.764001	0.918156	1.01496	1.177625	y = 0.2808x + 0.1489	0.9709
32	Wenzhou	Eastern	0.430281	0.46[illegible]984	0.50454	0.545317	0.987931	1.447027	1.584584	2.000244	y = 0.1169x + 0.3096	0.9462

Table A1. *Cont.*

No.	Cities	Location of the Cities	Municipal GDP (trillion CNY)				OSM RND Represent the OSM Road Network Density				Regression Equation	Coefficient of Determination (R^2)
			2014	2015	2016	2017	2014	2015	2016	2017		
33	Nanning	Western	0.31483	0.341009	0.370339	0.411883	0.304956	0.405285	0.406003	0.589919	y = 0.3344x + 0.2169	0.9107
34	Nanchang	Central	0.366796	0.400001	0.435499	0.500319	4.445648	5.080118	5.207499	5.528507	y = 0.1164x − 0.1642	0.8569
35	Tangshan	Eastern	0.62253	0.61	0.63062	0.71061	0.692859	0.858997	0.905533	1.472886	y = 0.1277x + 0.518	0.9037
36	Zibo	Eastern	0.402977	0.41302	0.441201	0.478132	0.953503	0.988958	1.8103	2.013233	y = 0.058x + 0.3503	0.8951
37	Changzhou	Eastern	0.490187	0.52732	0.577386	0.662228	1.340996	1.801608	1.848598	3.019328	y = 0.1003x + 0.3634	0.9294
38	Quanzhou	Eastern	0.573336	0.613774	0.664663	0.754801	0.634438	0.920326	1.384851	1.342822	y = 0.1867x + 0.4517	0.7318
39	Guiyang	Eastern	0.249727	0.289116	0.31577	0.353796	1.4045	1.561081	1.73026	2.215922	y = 0.1197x + 0.0952	0.9203
40	Jiaxing	Eastern	0.33528	0.351706	0.37601	0.435524	1.127291	1.266025	1.598774	1.725241	y = 0.1447x + 0.1678	0.8476
41	Nantong	Eastern	0.565279	0.61484	0.67682	0.77346	0.587234	1.243394	2.189651	2.424369	y = 0.0989x + 0.4982	0.8832
42	Jinhua	Eastern	0.320664	0.34065	0.363501	0.387022	0.314445	0.663407	0.705874	0.910499	y = 0.1096x + 0.2819	0.8947
43	Zhuhai	Eastern	0.185732	0.202498	0.222637	0.256473	2.257383	2.892806	3.178339	3.573405	y = 0.0525x + 0.0605	0.9154
44	Huizhou	Eastern	0.30007	0.314003	0.341217	0.383058	0.54594	0.681892	0.733018	0.937024	y = 0.2204x + 0.1749	0.9566
45	Xuzhou	Eastern	0.496391	0.531988	0.580852	0.660595	0.839667	0.833549	1.604102	3.131913	y = 0.0638x + 0.4653	0.9418
46	Haikou	Eastern	0.10917	0.116196	0.125767	0.139058	1.215575	1.39047	1.391586	1.593147	y = 0.0796x + 0.0113	0.9019
47	Urumqi	Western	0.246147	0.263164	0.245898	0.274382	1.471562	1.488615	1.548517	1.676137	y = 0.1043x + 0.0962	0.4828
48	Shaoxing	Eastern	0.426583	0.446665	0.471	0.510804	0.758167	0.803163	1.000226	1.245672	y = 0.1613x + 0.3103	0.9789
49	Zhongshan	Eastern	0.28233	0.301003	0.320278	0.345031	1.636848	1.759837	2.144733	2.393204	y = 0.0758x + 0.1617	0.9693
50	Taizhou (Zhejiang)	Eastern	0.338751	0.355813	0.384281	0.438822	0.531115	0.576276	0.832387	1.119472	y = 0.1603x + 0.2568	0.9833
51	Lanzhou	Western	0.200094	0.209599	0.226423	0.252354	0.623852	0.644902	0.723011	1.678987	y = 0.0412x + 0.1843	0.8392
52	Weifang	Eastern	0.478674	0.517053	0.55227	0.585863	0.808287	1.067039	1.251496	2.839208	y = 0.0438x + 0.4681	0.7604
53	Baoding	Eastern	0.30352	0.300034	0.32273	0.35809	1.08439	1.307091	1.321359	1.482171	y = 0.1299x + 0.1524	0.637
54	Zhenjiang	Eastern	0.325238	0.350248	0.383384	0.401036	0.877717	1.211188	1.385888	1.712882	y = 0.0949x + 0.2419	0.955
55	Yangzhou	Eastern	0.369789	0.401684	0.444938	0.506492	1.440231	1.801123	2.298118	2.526604	y = 0.117x + 0.1947	0.9365
56	Hohhot	Western	0.289405	0.309052	0.317359	0.274372	0.567368	0.655892	0.859838	0.869946	y = −0.0029x + 0.2997	0.0005
57	Langfang	Eastern	0.217596	0.24019	0.27063	0.28806	0.437621	0.726791	0.791253	1.032201	y = 0.1226x + 0.1626	0.9129
58	Luoyang	Central	0.328457	0.350875	0.37829	0.429019	0.908334	1.110387	1.912928	2.084188	y = 0.069x + 0.2679	0.856
59	Weihai	Eastern	0.279034	0.300157	0.32122	0.34801	1.158146	1.249603	1.297057	1.523613	y = 0.1835x + 0.0723	0.935
60	Yancheng	Eastern	0.383562	0.42125	0.457608	0.508269	0.657349	0.692985	0.894144	1.233145	y = 0.194x + 0.274	0.9275
61	Linyi	Eastern	0.35698	0.37632	0.402675	0.434539	1.480754	2.084111	2.317847	2.54728	y = 0.0686x + 0.2481	0.8741
62	Jiangmen	Eastern	0.208276	0.224002	0.241878	0.269025	1.3465	1.558839	1.660016	1.74101	y = 0.1434x + 0.0097	0.8806
63	Taizhou (Jiangsu)	Eastern	0.337089	0.365553	0.410178	0.474453	0.822494	1.257907	1.596425	2.136174	y = 0.107x + 0.2414	0.9821
64	Zhangzhou	Eastern	0.250636	0.276745	0.312534	0.352853	0.916913	1.151741	1.200081	1.38604	y = 0.2206x + 0.0414	0.9201
65	Handan	Eastern	0.308001	0.31454	0.33371	0.36663	0.675282	0.813087	0.914555	0.949443	y = 0.1836x + 0.1769	0.7366

Table A1. *Cont.*

No.	Cities	Location of the Cities	Municipal GDP (trillion CNY)				OSM RND Represent the OSM Road Network Density				Regression Equation	Coefficient of Determination (R^2)
			2014	2015	2016	2017	2014	2015	2016	2017		
66	Jining	Western	0.380006	0.4[illegible]1312	0.430182	0.465057	0.471934	0.526896	0.776055	0.91366	y = 0.1739x + 0.3022	0.966
67	Wuhu	Eastern	0.23079	0.2[illegible]5732	0.269944	0.306552	0.672768	0.965252	1.227687	1.957186	y = 0.0598x + 0.1912	0.9873
68	Yinchuan	Central	0.139567	0.1[illegible]8073	0.161728	0.180317	0.497675	0.59603	0.693312	0.716513	y = 0.164x + 0.0548	0.8525
69	Liuzhou	Eastern	0.220851	0.2[illegible]9862	0.247694	0.275564	0.455206	0.485649	0.604588	0.599445	y = 0.2719x + 0.0977	0.7542
70	Mianyang	Western	0.157989	0.1[illegible]0033	0.183042	0.207475	0.445944	0.554054	0.707322	0.734055	y = 0.1434x + 0.0921	0.838
71	Zhanjiang	Eastern	0.22587	0.2[illegible]8002	0.258478	0.282403	1.405769	1.659609	2.261492	2.575143	y = 0.0455x + 0.1613	0.9729
72	Anshan	Eastern	0.2349	0.[illegible]326	0.14408	0.16021	0.727266	0.768087	0.816542	0.828872	y = −0.9284x + 0.9219	0.8297
73	Daqing	Eastern	0.407	0.[illegible]9835	0.261	0.26805	0.499983	1.191177	1.2228	1.222723	y = −0.1858x + 0.5007	0.9601
74	Yichang	Central	0.313221	0.[illegible]3848	0.370936	0.385717	1.996736	2.126815	2.172778	2.900025	y = 0.0641x + 0.2048	0.6429
75	Baotou	Eastern	0.363631	0.3[illegible]8193	0.386763	0.275303	0.601152	0.628997	0.851639	0.900335	y = −0.1854x + 0.4892	0.3029
76	Jilin	Eastern	0.27302	0.[illegible]4552	0.253135	0.23028	0.601762	0.695863	0.781348	0.880384	y = −0.1324x + 0.3484	0.7854
77	Huai'an	Eastern	0.245539	0.2[illegible]4509	0.3048	0.338743	0.549558	0.656979	0.695744	0.776926	y = 0.4165x + 0.0119	0.9658
78	Cangzhou	Eastern	0.313338	0.[illegible]2406	0.35334	0.38169	0.746255	0.809976	0.888409	1.36473	y = 0.1017x + 0.2462	0.8629
79	Xiangyang	Central	0.31293	0.3[illegible]8212	0.369451	0.40649	0.347927	0.424744	0.563634	0.594033	y = 0.3347x + 0.1952	0.9255
80	Yueyang	Central	0.266939	0.2[illegible]8628	0.310087	0.325803	0.78718	0.829211	0.870221	1.291411	y = 0.0898x + 0.2131	0.67
81	Taian	Eastern	0.300219	0.[illegible]1584	0.33168	0.358528	0.935667	1.041736	1.769295	1.918874	y = 0.0462x + 0.2612	0.8588
82	Dongying	Eastern	0.343049	0.3[illegible]5064	0.34796	0.380178	0.455815	0.496966	0.737227	0.774951	y = 0.0783x + 0.3058	0.5307
83	Nanyang	Central	0.267688	0.2[illegible]7502	0.311877	0.33777	0.32572	0.387065	0.39798	0.513531	y = 0.3688x + 0.1515	0.9074
84	Xining	Western	0.106578	0.1[illegible]3162	0.124817	0.12849	2.148855	2.336419	2.553052	2.433314	y = 0.0526x − 0.0062	0.7804
85	Lhasa	Western	0.034745	0.0[illegible]8946	0.042495	0.047916	0.497234	0.788384	0.857679	0.901169	y = 0.0273x + 0.0203	0.7902

Table A2. The statistical results for predictive GDP of 85 Chinese cities in 2018.

No.	Cities	Location of the Cities	GDP in 2018 (trillion CNY)	Predictive GDP in 2018 by Using OSM Road Network Density (trillion CNY)	Predictive GDP in 2018 by Using OSM Road Network Density and Population (trillion CNY)	Absolute Residuals by Using OSM Road Network Density	Relative Residuals by Using OSM Road Network Density	Absolute Residuals by Using OSM Road Network Density and Population	Relative Residuals by Using OSM Road Network Density and Population
1	Beijing	Eastern	3.0320	3.1938	3.4422	0.1618	5.3364	0.4102	13.5290
2	Shanghai	Eastern	3.2680	3.0637	3.0811	−0.2043	−6.2515	−0.1869	−5.7191
3	Guangzhou	Eastern	2.2859	2.2386	2.2264	−0.0473	−2.0692	−0.0595	−2.6029
4	Shenzhen	Eastern	2.4222	2.1405	2.4473	−0.2817	−11.6299	0.0251	1.0362
5	Tianjin	Eastern	1.8810	2.1958	2.2128	0.3149	16.7358	0.3318	17.6396
6	Chongqing	Central	2.0363	2.4803	2.0682	0.4440	21.8043	0.0319	1.5666
7	Hangzhou	Eastern	1.3509	1.5446	1.5063	0.1937	14.3386	0.1554	11.5034

Table A2. *Cont.*

No.	Cities	Location of the Cities	GDP in 2018 (trillion CNY)	Predictive GDP in 2018 by Using OSM Road Network Density (trillion CNY)	Predictive GDP in 2018 by Using OSM Road Network Density and Population (trillion CNY)	Absolute Residuals by Using OSM Road Network Density	Relative Residuals by Using OSM Road Network Density	Absolute Residuals by Using OSM Road Network Density and Population	Relative Residuals by Using OSM Road Network Density and Population
8	Nanjing	Eastern	1.2820	1.2024	1.0471	−0.0797	−6.2090	−0.2349	−18.3229
9	Qingdao	Eastern	1.2002	1.2330	1.2537	0.0328	2.7329	0.0535	4.4576
10	Dalian	Eastern	0.7669	0.5997	0.7002	−0.1672	−21.8021	−0.0667	−8.6974
11	Ningbo	Eastern	1.0746	0.9531	1.1785	−0.1215	−11.3065	0.1039	9.6687
12	Xiamen	Eastern	0.4791	0.4474	0.4887	−0.0317	−6.6166	0.0096	2.0038
13	Ji'nan	Eastern	0.7857	1.0287	0.8826	0.2430	30.9278	0.0969	12.3330
14	Suzhou	Eastern	1.8565	1.9590	1.9216	0.1025	5.5211	0.0651	3.5066
15	Wuhan	Central	1.4847	1.4228	1.4250	−0.0619	−4.1692	−0.0597	−4.0210
16	Chengdu	Western	1.5254	1.5929	1.5392	0.0675	4.4251	0.0138	0.9047
17	Changsha	Central	1.1003	1.0682	1.2354	−0.0322	−2.9174	0.1351	12.2785
18	Xi'an	Western	0.8350	0.7505	1.2377	−0.0845	−10.1198	0.4027	48.2275
19	Shenyang	Eastern	0.6292	0.4871	2.5457	−0.1421	−22.5842	1.9165	304.5931
20	Zhengzhou	Eastern	1.0143	1.0272	1.0107	0.0128	1.2718	−0.0036	−0.3549
21	Dongguan	Eastern	0.8279	0.8223	0.8294	−0.0056	−0.6764	0.0015	0.1812
22	Fuzhou	Eastern	0.7857	0.6961	0.7764	−0.0895	−11.4038	−0.0093	−1.1837
23	Wuxi	Eastern	1.1439	1.0292	1.1492	−0.1146	−10.0271	0.0053	0.4633
24	Harbin	Eastern	0.6301	0.6759	0.6757	0.0458	7.2687	0.0456	7.2369
25	Foshan	Eastern	0.9936	1.1602	1.1390	0.1666	16.7673	0.1454	14.6337
26	Changchun	Eastern	0.7176	0.6533	0.6331	−0.0643	−8.9604	−0.0845	−11.7754
27	Shijiazhuang	Eastern	0.6083	0.8995	0.5967	0.2913	47.8711	−0.0116	−1.9070
28	Taiyuan	Central	0.3884	0.3584	0.3837	−0.0301	−7.7240	−0.0047	−1.2101
29	Yantai	Eastern	0.7833	0.7658	0.7777	−0.0174	−2.2341	−0.0056	−0.7149
30	Hefei	Central	0.7823	0.7452	0.8277	−0.0371	−4.7424	0.0454	5.8034
31	Kunming	Western	0.5207	0.5186	0.4968	−0.0021	−0.4033	−0.0239	−4.5900
32	Wenzhou	Eastern	0.6006	0.5519	0.5490	−0.0487	−8.1086	−0.0516	−8.5914
33	Nanning	Western	0.4147	0.4548	0.4532	0.0401	9.6696	0.0385	9.2838
34	Nanchang	Central	0.5275	0.5144	0.5443	−0.0131	−2.4834	0.0168	3.1848
35	Tangshan	Eastern	0.6955	0.7423	0.7444	0.0468	6.7290	0.0489	7.0309
36	Zibo	Eastern	0.5068	0.4839	0.4693	−0.0230	−4.5185	−0.0375	−7.3994
37	Changzhou	Eastern	0.7050	0.6780	0.7372	−0.0270	−3.8298	0.0322	4.5674
38	Quanzhou	Eastern	0.8468	0.7687	0.7780	−0.0781	−9.2230	−0.0688	−8.1247

Table A2. *Cont.*

No.	Cities	Location of the Cities	GDP in 2018 (trillion CNY)	Predictive GDP in 2018 by Using OSM Road Network Density (trillion CNY)	Predictive GDP in 2018 by Using OSM Road Network Density and Population (trillion CNY)	Absolute Residuals by Using OSM Road Network Density	Relative Residuals by Using OSM Road Network Density	Absolute Residuals by Using OSM Road Network Density and Population	Relative Residuals by Using OSM Road Network Density and Population
39	Guiyang	Eastern	0.3798	0.4347	0.3671	0.0549	14.4550	−0.0127	−3.3439
40	Jiaxing	Eastern	0.4872	0.4513	0.4875	−0.0359	−7.3686	0.0003	0.0616
41	Nantong	Eastern	0.8427	0.7887	0.9330	−0.0540	−6.4080	0.0903	10.7156
42	Jinhua	Eastern	0.4100	0.3952	0.4049	−0.0148	−3.6098	−0.0051	−1.2439
43	Zhuhai	Eastern	0.2915	0.2626	0.3010	−0.0288	−9.9142	0.0095	3.2590
44	Huizhou	Eastern	0.4103	0.3910	0.3898	−0.0193	−4.7039	−0.0205	−4.9963
45	Xuzhou	Eastern	0.6755	0.6919	0.7017	0.0164	2.4278	0.0262	3.8786
46	Haikou	Eastern	0.1511	0.2083	0.1459	0.0573	37.8557	−0.0052	−3.4414
47	Urumqi	Western	0.3100	0.2892	0.3181	−0.0207	−6.7097	0.0081	2.6129
48	Shaoxing	Eastern	0.5417	0.5172	0.5427	−0.0244	−4.5228	0.001	0.1846
49	Zhongshan	Eastern	0.3633	0.3738	0.4142	0.0105	2.8902	0.0509	14.0105
50	Taizhou (Zhejiang)	Eastern	0.4875	0.4979	0.4914	0.0105	2.1333	0.0039	0.8000
51	Lanzhou	Western	0.2733	0.2572	0.2673	−0.0161	−5.8910	−0.006	−2.1954
52	Weifang	Eastern	0.6157	0.6289	0.6077	0.0132	2.1439	−0.008	−1.2993
53	Baoding	Eastern	0.3590	0.4259	0.4253	0.0669	18.6351	0.0663	18.4680
54	Zhenjiang	Eastern	0.4050	0.4095	0.5013	0.0045	1.1111	0.0963	23.7778
55	Yangzhou	Eastern	0.5466	0.5026	0.5821	−0.0440	−8.0498	0.0355	6.4947
56	Hohhot	Western	0.2904	0.2969	0.2957	0.0066	2.2383	0.0053	1.8251
57	Langfang	Eastern	0.3108	0.3112	0.3174	0.0004	0.1287	0.0066	2.1236
58	Luoyang	Central	0.4641	0.4307	0.4405	−0.0334	−7.1967	−0.0236	−5.0851
59	Weihai	Eastern	0.3641	0.3568	0.3589	−0.0073	−2.0049	−0.0052	−1.4282
60	Yancheng	Eastern	0.5487	0.5544	0.3228	0.0057	1.0388	−0.2259	−41.1700
61	Linyi	Eastern	0.4718	0.4571	0.4416	−0.0147	−3.1157	−0.0302	−6.4010
62	Jiangmen	Eastern	0.2900	0.2910	0.3001	0.0010	0.3448	0.0101	3.4828
63	Taizhou (Jiangsu)	Eastern	0.5108	0.4790	0.2763	−0.0317	−6.2255	−0.2345	−45.9084
64	Zhangzhou	Eastern	0.3948	0.4722	0.4006	0.0774	19.6049	0.0058	1.4691
65	Handan	Eastern	0.3455	0.3577	0.3600	0.0122	3.5311	0.0145	4.1968
66	Jining	Western	0.4931	0.4905	0.4776	−0.0026	−0.5273	−0.0155	−3.1434
67	Wuhu	Eastern	0.3279	0.3260	0.3294	−0.0019	−0.5794	0.0015	0.4575
68	Yinchuan	Central	0.1901	0.2211	0.1720	0.0309	16.3072	−0.0181	−9.5213

Table A2. *Cont.*

No.	Cities	Location of the Cities	GDP in 2018 (trillion CNY)	Predictive GDP in 2018 by Using OSM Road Network Density (trillion CNY)	Predictive GDP in 2018 by Using OSM Road Network Density and Population (trillion CNY)	Absolute Residuals by Using OSM Road Network Density	Relative Residuals by Using OSM Road Network Density	Absolute Residuals by Using OSM Road Network Density and Population	Relative Residuals by Using OSM Road Network Density and Population
69	Liuzhou	Eastern	0.3084	0.3025	0.2880	−0.0059	−1.9131	−0.0204	−6.6148
70	Mianyang	Western	0.2304	0.2376	0.1645	0.0072	3.1250	−0.0659	−28.6024
71	Zhanjiang	Eastern	0.3008	0.2947	0.2966	−0.0061	−2.0279	−0.0042	−1.3963
72	Anshan	Eastern	0.1751	0.1145	0.1012	−0.0606	−34.6088	−0.0739	−42.2045
73	Daqing	Eastern	0.2801	0.2620	0.2582	−0.0181	−6.4620	−0.0219	−7.8186
74	Yichang	Central	0.4064	0.4784	0.3904	0.0720	17.7165	−0.016	−3.9370
75	Baotou	Eastern	0.2952	0.2298	0.3095	−0.0654	−22.1545	0.0143	4.8442
76	Jilin	Eastern	0.2210	0.2180	0.2167	−0.0030	−1.3575	−0.0043	−1.9457
77	Huai'an	Eastern	0.3601	0.4267	0.3499	0.0666	18.4949	−0.0102	−2.8325
78	Cangzhou	Eastern	0.3676	0.4266	0.4065	0.0590	16.0501	0.0389	10.5822
79	Xiangyang	Central	0.4310	0.4096	0.4381	−0.0214	−4.9652	0.0071	1.6473
80	Yueyang	Central	0.3411	0.4007	0.3001	0.0596	17.4729	−0.041	−12.0199
81	Taian	Eastern	0.3652	0.3579	0.3465	−0.0072	−1.9989	−0.0187	−5.1205
82	Dongying	Eastern	0.4152	0.4105	0.3266	−0.0047	−1.1320	−0.0886	−21.3391
83	Nanyang	Central	0.3567	0.3501	0.3379	−0.0066	−1.8503	−0.0188	−5.2705
84	Xining	Western	0.1286	0.1322	0.1360	0.0035	2.7994	0.0074	5.7543
85	Lhasa	Western	0.0528	0.0486	0.0515	−0.0042	−7.9545	−0.0013	−2.4621

References

1. Gustafson, E.J. Quantifying landscape spatial pattern: What is the state of the art? *Ecosystems* **1998**, *1*, 143–156. [CrossRef]
2. Hargis, C.D.; Bissonette, J.A.; David, J.L. The behavior of landscape metrics commonly used in the study of habitat fragmentation. *Landsc. Ecol.* **1998**, *13*, 167–186. [CrossRef]
3. O'Neill, R.V.; Krummel, J.R.; Gardner, R.H.; Sugihara, G.; Jackson, B.; DeAngelis, D.L.; Milne, B.T.; Turner, M.G.; Zygmunt, B.; Christensen, S.W.; et al. Indices of landscape pattern. *Landsc. Ecol.* **1988**, *1*, 153–162. [CrossRef]
4. McGarigal, K.; Cushman, S.A.; Neel, M.C.; Ene, E. FRAGSTATS: Spatial Pattern Analysis Program for Categorical Maps. 2002. Available online: www.umass.edu/landeco/research/fragstats/fragstats.html (accessed on 15 May 2020).
5. Herold, M.; Liu, X.H.; Clarke, K.C. Spatial Metrics and Image Texture for Mapping Urban Land use. *Photogramm. Eng. Remote Sens.* **2003**, *69*, 991–1001. [CrossRef]
6. Herold, M.; Couclelis, H.; Clarke, K.C. The role of spatial metrics in the analysis and modeling of urban land use change. *Comput. Environ. Urban Syst.* **2005**, *29*, 369–399. [CrossRef]
7. Reis, J.P.; Silva, E.A.; Pinbo, P. Measure space: A review of spatial metrics for urban growth and shrinkage. In *The Routledge Handbook of Planning Research Methods*; Silva, E.A., Healey, P., Harris, N., Van den Broeck, P., Eds.; Routledge: New York, NY, USA, 2014; pp. 279–292.
8. Reis, J.P.; Silva, E.A.; Pinbo, P. Spatial metrics to study urban patterns in growing and shrinking cities. *Urban Geogr.* **2016**, *37*, 246–271. [CrossRef]
9. Chen, M.; Huang, Y.; Tang, Z.; Lu, D.; Liu, H.; Ma, L. The provincial pattern of the relationship between urbanization and economic development in China. *J. Geogr. Sci.* **2014**, *24*, 33–45. [CrossRef]
10. Henderson, V. The Urbanization Process and Economic Growth: The So-What Question. *J. Econ. Growth* **2003**, *8*, 47–71. [CrossRef]
11. Njoh, A.J. Urbanization and development in sub-Saharan Africa. *Cities* **2003**, *20*, 167–174. [CrossRef]
12. Liu, Y.S.; Fang, F.; Li, Y.H. Key issues of land use in China and implications for policy making. *Land Use Policy* **2014**, *40*, 6–12. [CrossRef]
13. Xiang, W.N.; Stuber, R.M.B.; Meng, X.C. Meeting critical challengers and striving for urban sustainability in China. *Landsc. Urban Plan.* **2011**, *100*, 418–420. [CrossRef] [PubMed]
14. Li, Y.Z. Urbanization and economic growth in China: An empirical research based on VAR model. *Int. J. Econ. Financ.* **2017**, *9*, 210–219.
15. Cai, Z.Y.; Liu, Q.; Cao, S.X. Real estate supports rapid development of China's urbanization. *Land Use Policy* **2020**, *95*, 104582. [CrossRef]
16. National Bureau of Statistics (NBS). *China Statistical Yearbook*; China statistics Press: Beijing, China, 2018.
17. Heshmati, A.; Rashidghalam, M. Measurement and Analysis of Urban Infrastructure and Its Effects on Urbanization in China. *J. Infrastruct. Syst.* **2020**, *26*, 04019030. [CrossRef]
18. Yu, N.N.; de Roo, G.; de Jong, M.; Storm, S. Does the expansion of a motorway network lead to economic agglomeration Evidence from China. *Transp. Policy* **2016**, *45*, 218–227. [CrossRef]
19. Jiao, J.; Wang, J.; Jin, F.; Du, C. Understanding Relationship between Accessibility and Economic Growth: A Case Study from China (1990–2010). *Chin. Geogr. Sci.* **2016**, *26*, 803–816. [CrossRef]
20. Worku, I. Road Sector Development and Economic Growth in Ethiopia. *Ethiop. J. Econ.* **2010**, *19*, 101–146.
21. Ivanova, E.; Masarova, J. Importance of road infrastructure in the economic development and competitiveness. *Compet. Nations Global Econ.* **2013**, *18*, 263–274. [CrossRef]
22. Beyzatlar, M.A.; Karacal, M.; Yetkiner, H. Granger-causality between transportation and GDP: A panel data approach. *Transp. Res. Part A Policy Pract.* **2014**, *63*, 43–55. [CrossRef]
23. Gao, Y.; Zhang, Y.P.; Li, H.J.; Peng, T.; Hao, S.Q. Study on the Relationship Between Comprehensive Transportation Freight Index and GDP in China. *Procedia Eng.* **2016**, *137*, 571–580. [CrossRef]
24. Fan, S.G.; Chan-Kang, C. Regional road development, rural and urban poverty: Evidence from China. *Transp. Policy* **2008**, *15*, 305–314. [CrossRef]
25. Haklay, M.; Weber, P. OpenStreetMap: User-Generated Street Maps. *IEEE Pervasive Comput.* **2008**, *7*, 12–18. [CrossRef]

26. OpenStreetMap. Stats-OpenStreetMap Wiki [Online]. Available online: https://wiki.openstreetmap/wiki/.orgStats (accessed on 20 June 2020).
27. Over, M.; Schilling, A.; Neubauer, S.; Zipf, A. Generating web-based 3D City Models from OpenStreetMap: The current situation in Germany. *Comput. Environ. Urban Syst.* **2010**, *34*, 496–507. [CrossRef]
28. Fonte, C.C.; Martinho, N. Assessing the applicability of OpenStreetMap data to assist the validation of land use/land cover maps. *Int. J. Geogr. Inf. Sci.* **2017**, *31*, 2382–2440. [CrossRef]
29. Fonte, C.; Minghini, M.; Patriarca, J.; Antoniou, V.; See, L.; Skopeliti, A. Generating Up-to-Date and Detailed Land Use and Land Cover Maps Using OpenStreetMap and GlobeLand30. *ISPRS Int. J. Geo-Inf.* **2017**, *6*, 125. [CrossRef]
30. Bittner, C. Diversity in volunteered geographic information: Comparing OpenStreetMap and Wikimapia in Jerusalem. *GeoJournal* **2016**, *82*, 887–906. [CrossRef]
31. Mobasheri, A. A Rule-Based Spatial Reasoning Approach for OpenStreetMap Data Quality Enrichment; Case Study of Routing and Navigation. *Sensors* **2017**, *17*, 2498. [CrossRef]
32. Zhou, Q. Exploring the relationship between density and completeness of urban building data in OpenStreetMap for quality estimation. *Int. J. Geogr. Inf. Sci.* **2018**, *32*, 257–281. [CrossRef]
33. Goetz, M.; Zipf, A. Towards defining a framework for the automatic derivation of 3D CityGML models from volunteered geographic information. *Int. J. 3-D Inf. Model.* **2012**, *1*, 1–16. [CrossRef]
34. Hennig, S. OpenStreetMap used in protected area management. The example of the recreational infrastructure in Berchtesgaden National Park. *J. Prot. Mt. Areas Res. Manag.* **2017**, *1*, 30–41. [CrossRef]
35. Mobasheri, A.; Sun, Y.; Loos, L.; Ali, A. Are Crowdsourced Datasets Suitable for Specialized Routing Services? Case Study of OpenStreetMap for Routing of People with Limited Mobility. *Sustainability* **2017**, *9*, 997. [CrossRef]
36. Juhász, L.; Hochmair, H.H. How do volunteer mappers use crowdsourced Mapillary street level images to enrich OpenStreetMap? In Proceedings of the 20th AGILE Conference on Geo-Information Science, Wageningen, The Netherlands, 11 May 2017; pp. 18–21.
37. Zhang, Y.J.; Li, X.M.; Wang, A.M.; Bao, T.L.; Tian, S.Z. Density and diversity of OpenStreetMap road networks in China. *J. Urban Manag.* **2015**, *4*, 135–146. [CrossRef]
38. Goetz, M. Towards generating highly detailed 3D CityGML models from OpenStreetMap. *Int. J. Geogr. Inf. Sci.* **2013**, *27*, 845–865. [CrossRef]
39. Wang, Z.Y.; Zipf, A. Using Openstreetmap Data to Generate Building Models with Their Inner Structures for 3D Maps. *ISPRS Ann. Photogramm. Remote Sens. Spat. Inf. Sci.* **2017**. [CrossRef]
40. Bergman, C.; Oksanen, J. Conflation of OpenStreetMap and Mobile Sports Tracking Data for Automatic Bicycle Routing. *Trans. GIS* **2016**, *20*, 848–868. [CrossRef]
41. Rosina, K.; Hurbã¡Nek, P.; Cebecauer, M. Using OpenStreetMap to improve population grids in Europe. *Am. Cartogr.* **2016**, *44*, 139–151. [CrossRef]
42. Zhao, P.X.; Jia, T.; Qin, K.; Shan, J.; Jiao, C.J. Statistical analysis on the evolution of OpenStreetMap road networks in Beijing. *Physica A* **2015**, *420*, 59–72. [CrossRef]
43. Dingil, A.E.; Schweizer, J.; Rupi, F.; Stasiskiene, Z. Updated Models of Passenger Transport Related Energy Consumption of Urban Areas. *Sustainability* **2019**, *11*, 4060. [CrossRef]
44. Shang, Y.S.; Liu, S.G.; Liu, C.Y.; Yin, P. Spatial-temporal characteristics of urbanization efficiency in coastal cities of China. In Proceedings of the 7th Annual International Conference on Geo-Spatial Knowledge and Intelligence, Guangzhou, China, 20–21 December 2019. [CrossRef]
45. Zou, Y.F.; Deng, M.; Li, Y.J.; Rong, Y. Evolution characteristics and policy implications of new urbanization in provincial capital cities in western China. *PLoS ONE* **2020**, *15*, e0233555. [CrossRef]
46. Haklay, M. How good is volunteered geographical information? A comparative study of OpenStreetMap and Ordnance Survey datasets. *Environ. Plan. B Plan. Des.* **2010**, *37*, 682–703. [CrossRef]
47. Zielstra, D.; Zipf, A. A comparative study of proprietary geodata and volunteered geographic information for Germany. In Proceedings of the 13th AGILE International Conference on Geographic Information Science, Guimarães, Portugal, 10–14 May 2010.
48. Neis, P.; Zielstra, D.; Zipf, A. The street network evolution of crowdsourced maps: OpenStreetMap in Germany 2007–2011. *Future Internet* **2012**, *4*, 1–21. [CrossRef]
49. Girres, J.-F.; Touya, G. Quality assessment of the French OpenStreetMap dataset. *Trans. GIS* **2010**, *14*, 435–459. [CrossRef]

50. Luo, L.C.; Liu, B.; Liu, X.C. Data Quality Assessment and Application Analysis for OpenStreetMap Road Network. *Jiangxi Sci.* **2017**, *35*, 151–157.
51. Hecht, R.; Kunze, C.; Hahmann, S. Measuring Completeness of Building Footprints in OpenStreetMap over Space and Time. *ISPRS Int. J. Geo-Inf.* **2013**, *2*, 1066–1091. [CrossRef]
52. Singh Sehra, A.; Singh, J.; Singh Rai, H. Assessment of OpenStreetMap Data—A Review. *Int. J. Comput. Appl.* **2013**, *76*, 17–20.
53. Ludwig, I.; Voss, A.; Krause-Traudes, M. A comparison of the street networks of Navteq and OSM in Germany. In *Advancing Geoinformation Science for a Changing World*; Geertman, S., Reinhardt, W., Toppen, F., Eds.; Springer: Berlin, Germany, 2011; pp. 65–84.
54. Zhang, J.; Chen, J. Introduction to China's new normal economy. *J. Chin. Econ. Bus. Stud.* **2017**, *15*, 1–4. [CrossRef]
55. Li, C.; Zhang, X.J. Renminbi Internationalization in the New Normal: Progress, Determinants and Policy Discussions. *China World Econ.* **2017**, *25*, 22–44. [CrossRef]
56. Montgomery, D.C.; Peck, E.A. *Introduction to Linear Regression Analysis*; Wiley: Hoboken, NJ, USA, 1982.
57. Hawbaker, T.J.; Radeloff, V.C.; Hammer, R.B.; Clayton, M.K. Road density and land scape pattern in relation to housing density, and ownership, land cover, and soils. *Landsc. Ecol.* **2005**, *20*, 609–625. [CrossRef]
58. Shen, J.; Wu, R. *Urban Road and Transportation*; Wuhan University Press: Wuhan, China, 2006. (In Chinese)
59. Zhang, Q.; Wang, J.; Peng, X.; Gong, P.; Shi, P. Urban built-up land change detection with road density and spectral information from multi-temporal Landsat TM data. *Int. J. Remote Sens.* **2002**, *23*, 3057–3078. [CrossRef]
60. Feng, Z.; Liu, D.; Yang, Y. Evaluation of transportation ability of China: From county to province level. *Geogr. Res.* **2009**, *28*, 419–429.
61. Jin, F.J.; Wang, C.J.; Li, X.W. Discrimination method and its application analysis of regional transport superiority. *Acta Geogr. Sin.* **2008**, *63*, 787–798.
62. Savaş, B. The Relationship between Population and Economic Growth: Empirical Evidence from the Central Asian Economies. *Orta Asya Kafkasya Araştırmaları* **2008**, *6*, 135–153.

International Journal of
Geo-Information

MDPI

Article

Measuring Impacts of Urban Environmental Elements on Housing Prices Based on Multisource Data—A Case Study of Shanghai, China

Liujia Chen [1,2,3], Xiaojing Yao [1,3,*], Yalan Liu [1,3], Yujiao Zhu [4], Wei Chen [4], Xizhi Zhao [5] and Tianhe Chi [1,3]

1 Airspace Information Research Institute, Chinese Academy of Sciences, Beijing 100094, China; chenliujia2013@gmail.com (L.C.); liuyl@radi.ac.cn (Y.L.); chith@126.com (T.C.)
2 University of Chinese Academy of Science, Beijing 100049, China
3 Lab of Spatial Information Integration, Institute of Remote Sensing and Digital Earth, Chinese Academy of Sciences, Beijing 100101, China
4 School of Geosciences & Surveying Engineering, China University of Mining & Technology, Beijing 100083, China; 15621341509@163.com (Y.Z.); chenw@cumtb.edu.cn (W.C.)
5 Research Center of Government Geographic Information System, Chinese Academy of Surveying and Mapping, Beijing 100830, China
* Correspondence: yaoxj@aircas.ac.cn; Tel.: +86-1860-043-0682

Received: 26 December 2019; Accepted: 7 February 2020; Published: 10 February 2020

Abstract: Diverse urban environmental elements provide health and amenity value for residents. People are willing to pay a premium for a better environment. Thus, it is essential to assess the benefits and values of these environmental elements. However, limited by the interpretability of the machine learning model, existing studies cannot fully excavate the complex nonlinear relationships between housing prices and environmental elements, as well as the spatial variations of impacts of urban environmental elements on housing prices. This study explored the impacts of urban environmental elements on residential housing prices based on multisource data in Shanghai. A SHapley Additive exPlanations (SHAP) method was introduced to explain the impacts of urban environmental elements on housing prices. By combining the ensemble learning model and SHAP, the contributions of environmental characteristics derived from street view data and remote sensing data were computed and mapped. The experimental results show that all the urban environmental characteristics account for 16 percent of housing prices in Shanghai. The relationships between housing prices and two green characteristics (green view index from street view data and urban green coverage rate from remote sensing) are both nonlinear. Shanghai's homebuyers are willing to pay a premium for green only when the green view index or urban green coverage rate are of higher value. However, there are significant differences between the impacts of the green view index and urban green coverage rate on housing prices. The sky view index has a negative influence on housing prices, which is probably because the high-density and high-rise residential area often has better living facilities. Residents in Shanghai are willing to pay a premium for high urban water coverage. The case of Shanghai shows that the proposed framework is practical and efficient. This framework is believed to provide a tool to inform the decisions of housing buyers, property developers and policies concerning land-selling and buying, property development and urban environment improvement.

Keywords: street view; remote sensing; urban environmental elements; ensemble learning; green view; sky view; building view; SHAP

1. Introduction

Urban green spaces, sky and other urban environmental elements can significantly affect the quality of urban life [1,2]. Various studies have shown that urban environmental elements have a significant influence on people's physical and mental health. For instance, urban green spaces have multiple ecological benefits, including air purification [3,4], climate regulation [5], carbon storage [6] and noise reduction [7]. In addition, green spaces provide plenty of spaces for pressure releasing and, consequently, positively affect mental health [8–10]. Higher levels of sky view visibility were associated with lower psychological distress [11]. Contrary to green and sky, high-rise buildings make people feel stressed [12]. With rapid urbanization and improvement of living standards, increasing concern about the quality and quantity of urban environmental elements has grown all over the world. Many people display a marked preference for natural over built environmental elements [13]. This preference is often shown by the housing choices of consumers in the residential housing market. People are willing to pay extra for a home with more natural environmental elements [14].

The explanatory variables of housing prices have been widely discussed in the housing literature. Bangura, Lee and Al-Masum discussed the ability of market fundamentals in explaining housing prices from the macroeconomic perspective [15–17], while Trojanek and Yamagata examined the importance of housing attributes in explaining housing prices from the microeconomic perspective [18,19]. In recent years, a great deal of research has studied the impacts of environmental elements on housing price. For instance, a house with a water view could attract a premium of 8%–10% in the Netherlands [20]. In Guangzhou, the view of green spaces and proximity to water bodies can lead to a considerable increase in house price, contributing at 7.1% and 13.2%, respectively [1]. An additional street tree increases a house's monthly rental price by $21.00 in Portland, Oregon, USA [21]. In Singapore, vegetation had positive effects on housing prices, accounting for 3% of a property's value [22]. On the contrary, both street and building views would depress housing price, with the influence of street view more significant than building view in Hong Kong [23]. However, most of the existing studies analyze the impacts of urban environmental elements on housing prices by using field survey data [1,24] and satellite remote sensing data [25,26]. Field survey data is time-consuming and hard to be applied in large-scale studies. Satellite remote sensing data is limited by an overhead view perspective and spatiotemporal resolution. Street view images bring a new opportunity to obtain urban environmental elements. This type of data has the advantages of easy obtaining, wide coverage and high spatial resolution. More importantly, street view images represent a horizontal view perspective, which is closer to the general population's perception of urban environmental elements. The rapid development of computer vision provides an efficient method for the information extraction of street view images. In this context, a great number of studies have been conducted to measure street-level green [27], estimate the spatiotemporal patterns of urban mobility [28], examine the relationship between street view and perceived safety [29] and assess the visual quality of urban environment [30] Therefore, in this study, street view data is used to evaluate the relationship between urban environmental elements and housing prices.

Most of the existing studies conducted on the impacts of urban environmental elements on housing prices used the hedonic pricing model (HPM) as the research method. This method assumes that real estate is heterogeneous and three types of characteristics have significant impacts on housing price, namely structure, neighborhood and location characteristics [31,32]. In empirical research, HPM mainly has three forms, including linear models [24,30], semi-log models [1] and double-log models [33]. However, most studies combine linear regression with HPM to interpret the impact of different independent variables [34,35]. No matter which form HPM is, only the log transformation of independent variables or dependent variables is performed for reducing the heteroscedasticity of the model. Therefore, the hedonic model is limited to revealing the complicated nonlinear relationships between housing prices and a variety of potential determinants [36]. In addition, the combination of linear regression and HPM explains the impact of a housing characteristic on housing prices by the value of this characteristic and the same corresponding regression coefficients of the regression equation.

Thus, this method could not reveal the spatial variations of the contribution of each characteristic. To address these problems, we propose an analytical framework which combines ensemble learning and SHapley Additive exPlanations (SHAP). By combing the individual machine learning methods to form a new classifier, ensemble learning algorithms such as Random Forest Regression (RFR) and XGBoost Regression (XGBoost) achieve better performance than any of the individual ones [37]. Compared to traditional methods, these ensemble learning algorithms show obvious advantages in three aspects: (1) capability to capture nonlinear relationships, (2) high prediction accuracy and (3) capability to capture high-order interactions between inputs. Recent urban housing prices studies have shown the advantage of ensemble learning algorithms over traditional methods [28,38]. Hu compared the performance of six machine learning algorithms in monitoring housing rental prices and found that ExtraTrees and RFR get better results [39]. However, because the nature of ensemble learning models are not interpretable models, almost all of these studies only range the importance when measuring the impacts of a housing characteristic on housing prices. It is hard to analyze the contribution of each characteristic to the housing price. SHAP, which is based on the game theoretically optimal Shapley values, falls into this specific scope and provides a new opportunity for solving this problem. Unlike methods that provide a specific global predictor, the SHAP framework provides an explanation of the model overall behavior in the form of particular feature contributions. Thus, this method can be used to explain the spatial variations of the contribution of each characteristic and the complex nonlinear relationships between each characteristic and housing prices. SHAP is becoming an increasingly popular tool to interpret natural and social phenomena [40,41].

In brief, the main contributions of this study are as follows. (1) Considering the perception of the urban environment from the horizontal view perspective, which could be easier for ordinary people to understand, street view data is used to calculate the environmental characteristics. (2) Tree-based ensemble learning regression algorithms are employed to model the housing prices and a method for explanting these ensemble learning models—SHAP is introduced to interpret the relationships between urban environmental elements and housing prices. By combining tree-based ensemble learning regression algorithms and the SHAP model, the complex and nonlinear relationships between most of the environmental elements and housing prices are revealed, which is more elaborate than the results of previous studies. (3) SHAP models are employed for the geospatial analysis of housing prices. The spatial distribution of SHAP for five environmental characteristics were mapped to improve the understanding of the spatial variations of each urban environmental characteristic's contribution. (4) The impacts of the green view index from street view data and green coverage rate from remote sensing data are compared in this study. The difference impacts of the same urban environmental elements from different observation perspectives provide new insights into urban environment research.

The remainder of this paper is organized as follows. Section 2 introduces the study area, data and methods used in this study. Section 3 presents the research results and discusses the reasons behind these results and suggestions for future work. Section 4 provides a conclusion of our study.

2. Data and Methods

2.1. Study Area

Shanghai, one of the financial, trade, economic and shipping hubs in the world, is located on China's east coast. Since the implementation of housing reforms that transformed the housing system from an administrative allocation model to a market mechanisms model in 1980, housing prices in Shanghai have ballooned over the years [42]. At present, Shanghai has become one of the most expensive housing markets, with a large number of housing transactions. The area within the outer ring road, which has a population density of 17,070 per square kilometer, is regarded as the central city of Shanghai [43]. With such a high-density population, a large number of housing transactions occur in this area. Therefore, an empirical analysis in the area within the outer ring road can supply essential references for relevant studies. The study areas in this paper are shown in Figure 1a.

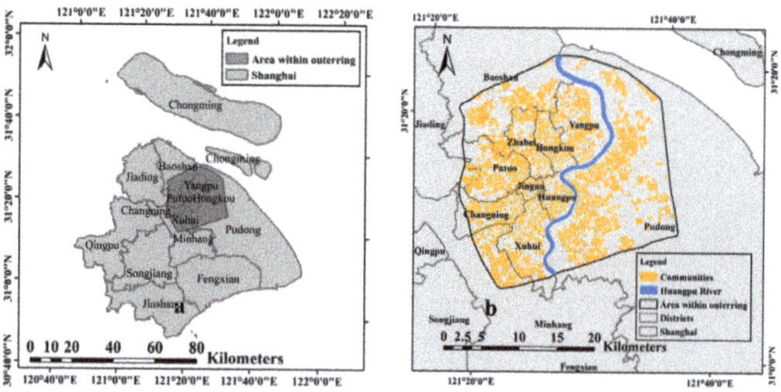

Figure 1. Location map of the study area: the area within the outer ring road of Shanghai (**a**) and the distribution of communities (**b**).

2.2. Overall Methodological Framework

Figure 2 presents the overall methodological framework, which follows three major steps to complete the analysis. First, multisource data were gathered and cleaned to extract the housing prices and corresponding characteristics at the community level. Second, we used these housing prices and characteristics to select the most appropriate machine learning model. Finally, by inputting the selected machine learning model and the characteristics of the communities into the SHapley Additive exPlanations model (SHAP), the SHAP value of these characteristics were computed to analyze the global importance of the characteristics and the contribution of urban environmental characteristics.

2.3. Characteristics Extraction

In China, taking the form of a gated residential area, a community is regarded as a basic management unit of urban planning [44]. In addition, houses located in the same community share a similar urban environment. Therefore, we chose communities as the basic analytical units in this paper. By crawling Baidu Maps, we obtained 7043 community boundaries in the study area (Figure 1b). All the housing characteristics involved in this study were transformed to the same community units for further study.

2.3.1. Housing Price

In this study, based on a web crawler, we collected the historical transactional data of preowned houses from Lianjia.com in 2018. There were four steps in the processing of preowned houses transaction data. First, a web crawler was used to download the historical transaction data of preowned houses, which occurred in 2018 from Lianjia.com. The transaction data recorded a number of housing attributes, including address, community name, total price, total area, price per square meter, elevator and construction time of building. Then, the collected data were cleared for (1) records whose spatial position are outside the area within the outer ring road; (2) records with missing important attributes, such as "elevator" and "construction time of building" and (3) repeated records. Finally, the price per square meter was averaged for each community. As a result of housing transactional data processing, we obtained 2547 study units with observed historical transactional data. Figure 3 presents the spatial distribution of the community-level housing prices.

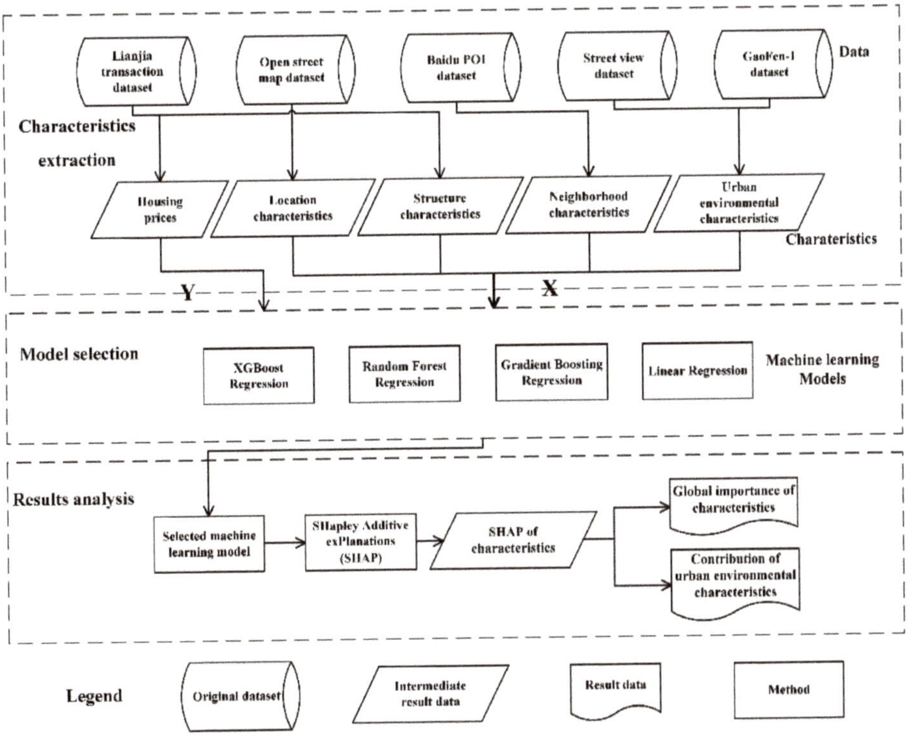

Figure 2. The overall methodological framework.

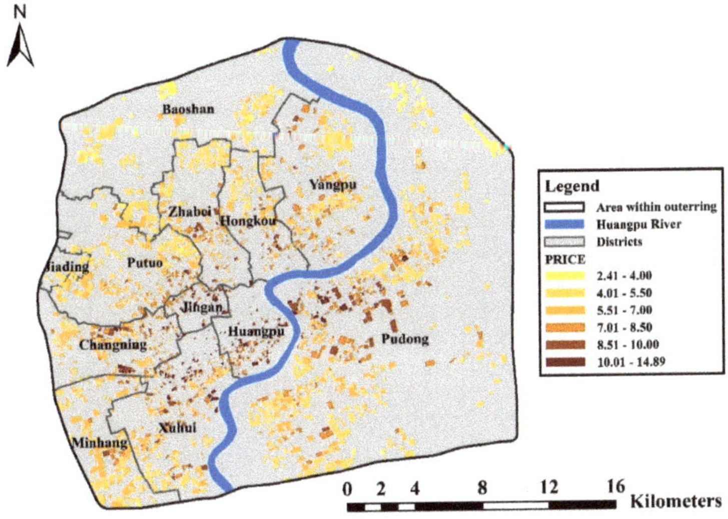

Figure 3. Spatial distribution of community-level housing prices.

2.3.2. Urban Environmental Characteristics from Street View Data

Street view data represents the urban environmental elements from a horizontal view perspective, which is closer to the general population's perception and could be easier for ordinary people to understand. Therefore, street view data was employed to measure the impacts of urban environmental elements on housing prices in this study.

The process for computing urban environmental characteristics from street view data involves three steps: street view data crawling, environmental elements extraction and characteristic calculation.

First, we selected main roads within the area of the outer ring road based on Shanghai's OpenStreetMap dataset. After that, the centerlines of these main roads were extracted, and then, we got street view sample sites along the centerlines at 50-m intervals. Each sample site was represented by a panoramic street view image. Finally, by inputting the spatial coordinate of sample sites into a Baidu static picture API, we crawled 84,520 panoramic street view images, which were acquired on August and September, 2017. Each of them has a size of 1024 by 290 pixels.

In this study, we mainly focused on three horizontal view environmental elements, including green, sky and building. Each of the elements was defined as the ratio of pixels associated with the specific element to the total pixels in a street view image. Specifically, the values of the green view index (GVI), the sky view index (SVI) and the building view index (BVI) were calculate by following equations:

$$\text{GVI} = \frac{\text{Pixels}_{\text{green}}}{\text{Pixels}_{\text{total}}} \tag{1}$$

$$\text{SVI} = \frac{\text{Pixels}_{\text{sky}}}{\text{Pixels}_{\text{total}}} \tag{2}$$

$$\text{BVI} = \frac{\text{Pixels}_{building}}{\text{Pixels}_{\text{total}}} \tag{3}$$

The rapid development of computer vision, especially the deep convolutional neural network (DCNN), provides a new method for the information extraction of images. The state-of-the-art DCNNs such as SegNet [45], PSPNet [46] and DeepLabv3 [47] were employed for image semantic segmentation and exhibited an outstanding performance in image interpretation [27]. In this study, DeepLabv3, one of the most popular image semantic segmentation models, was applied to extract street-level environmental elements at the pixel level. Figure 4 shows the flow charts of the street view images' semantic segmentation. DeepLabv3 was first pretrained using the Cityscapes dataset and was then used to segment the street view data for extracting green space, sky and building. DeepLabv3 combines an atrous convolution with upsampled filters to solve the problem of segmenting objects at multiple scales. The performance of this model outperformed the state-of-the-art models on the PASCAL VOC 2012 semantic image segmentation benchmark [47]. The Cityscapes dataset was employed to pretrain the DeepLabv3 model. Cityscapes is a large-scale dataset containing a variety of stereo video sequences at street level from 50 different cities. Five-thousand of these images have high-quality pixel-level labeling [48]. DeepLabv3 achieved 81.3% accuracy on the Cityscapes dataset. The configuration of the hardware devices used in this study were an Intel i7-8700k CPU, a NVIDIA 1080ti graphics card with 12GB video memory and 32 GB physical memory. The operation system of the computer is 64-bit Windows 10 Professional.

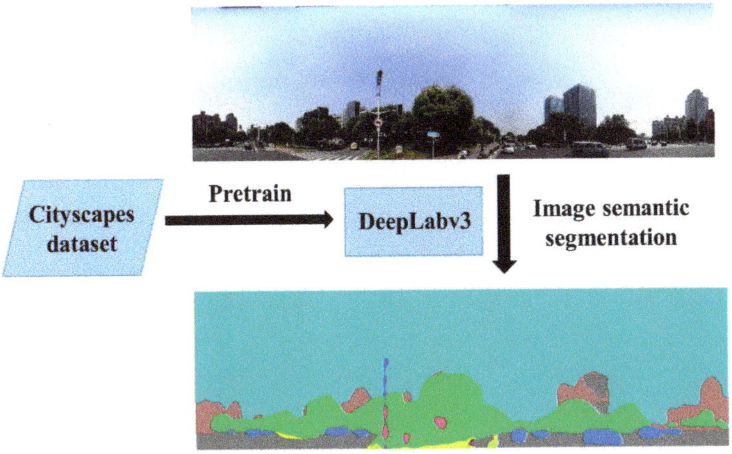

Figure 4. The flow chart of the street view images' semantic segmentation.

For the characteristics calculations, the GVI, SVI and BVI for each community with a 400 m radius buffer were averaged to obtain environmental characteristics at the community level. The reason why we chose 400 m is that the square root of the average area of Shanghai's communities is about 400 m, and the scope of citizens' public lives has been well-covered by this buffer. The willingness to buy a house are influenced not only by the view from their apartment but also by the view from their public life. After the calculation, there were 115 sample sites per community.

2.3.3. Urban Environmental Characteristics from Remote Sensing Data

To compare the urban environmental characteristics derived from street view data with remotely sensed characteristics, GaoFen-1 data were used to calculate the urban green coverage rate (UG) and urban water coverage rate (UW). Four GaoFen-1 images used in this paper were acquired on April and May, 2015, all of which consisted of four multispectral bands at an 8 m spatial resolution and one panchromatic band at a 2 m spatial resolution. The supervised classification was conducted to extract green and water by the support vector machine (SVM) tool in ENVI 5.3. Specifically, 80 green water samples and 80 water samples were randomly selected by visual interpretation. For each type of land cover, 50 samples were chosen for the training classification model and 30 samples for testing. The classification performance was assessed by a confusion matrix of test samples. The total precision was 96.75%, and the Kappa coefficient was 0.9578. The classification results are shown in Figure 5. For the characteristics calculations, the UG and UW for each community with a 400 m radius buffer were averaged to obtain the environmental characteristics from remote sensing data at the community level.

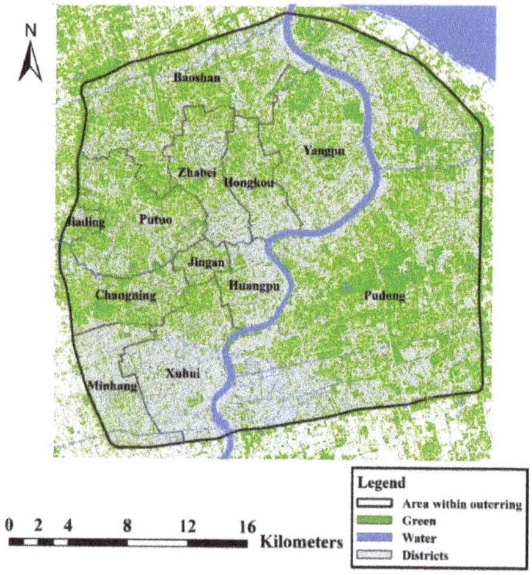

Figure 5. The classification results of green and water within the study area based on GaoFen-1 images.

2.3.4. Other Characteristics

In light of the attributes of preowned house transaction data and the spatial scale of study units, the year of construction (YEAR), average construction area of the apartment (AREA), plot ratio (PR) and whether the elevator is available (EL) were selected as structure characteristics. The variable of AREA should be introduced, because that area significantly affects the housing prices in Chinese megacities. Specifically, small houses often have a higher price per square meter because of lower total prices. Big houses (AREA > 200 m^2) also have a higher price per square meter due to better facilities and management. EL in original transaction data is a dummy variable. If the elevator is available in the apartment building, the value is 1; otherwise, the value is 0. For PR, the plot ratio of a community was obtained by dividing the gross floor area of the building by the area of the total community area on which the building was erected. In this study, this variable was calculated by the building footprint and Baidu community data.

For location characteristics, the distance to the city center (C_DIS), the city employment center (EC_DIS), river (R_DIS) and the Huangpu River (HPR_DIS) were chosen. In detail, the Bund was selected as the city center of Shanghai, and the employment center identified by Sun was used in this study [49]. The reason why the HPR_DIS was chosen is that the distance from each neighborhood centroid to the Huangpu River notably affects residential housing prices. The housing prices decrease with the increase of the distance [30].

For neighborhood characteristics, the variables which measured the accessibility to bus stations, subway stations, primary schools and first-class hospitals at grade 3 (hospitals with high-quality facilities and services) were included in our study. Using the points of interest (POI) data collected from the Baidu Map, the distance from each community to its nearest facility and the number of facilities within a specified distance were calculated. Specifically, 500 m and 1000 m were selected as the distance threshold in the density calculation, considering the 15-min community life circle proposed by the Chinese government.

General descriptive statistics of the selected housing characteristics are shown in Table 1.

Table 1. General descriptive statistics of the housing characteristics.

Variables	Description	Mean	Standard Deviation	Range
Dependent variable				
PRICE	Transaction price (10,000 RMB/m^2)	6.347	1.678	2.413–14.894
Location characteristics				
C_DIS	Distance to the city center (10 km)	0.792	0.331	0.046–1.650
EC_DIS	Distance to the city employment centers (10 km)	0.295	0.188	0–1.092
R_DIS	Distance to the river (10 km)	0.278	0.198	0.02–1.138
HPR_DIS	Distance to the Huangpu River (10 km)	0.420	0.283	0.003–1.311
Structure characteristics				
YEAR	2019 minus the construction time of building	21.622	9.121	2–106
AREA	Average construction area in the apartment (m^2)	78.788	38.743	22–346
PR	Plot ratio	2.600	1.234	0–13.703
EL	Dummy variable, 1 if elevator is available.	0.398	0.470	0–1
Neighborhood characteristics				
BUS_NEAR	Distance to the nearest bus station (km)	0.083	0.091	0–0.996
BUS_500M	Number of bus stations within 500 m	9.894	3.862	0–25
BUS_1000M	Number of bus stations within 1000 m	30.740	8.887	4–73
SUB_NEAR	Distance to the nearest subway station (km)	0.704	0.548	0–4.167
SUB_500M	Number of subway stations within 500 m	0.577	0.641	0–3
SUB_1000M	Number of subway stations within 1000 m	1.940	1.297	0–7
PRI_NEAR	Distance to the nearest primary school (km)	0.365	0.298	0–2.259
PRI_500M	Number of primary schools within 500 m	1.611	1.280	0–7
PRI_1000M	Number of primary schools within 1000 m	4.847	2.769	0–18
FH3_NEAR	Distance to the nearest first-class hospital at grade 3 (km)	2.221	1.641	0.026–7.614
FH3_500M	Number of first-class hospitals at grade 3 within 500 m	0.154	0.435	0–3
FH3_1000M	Number of first-class hospitals at grade 3 within 1000 m	0.547	0.976	0–6
Urban Environmental characteristics				
GVI	Mean green view index within 400 m distance	0.315	0.123	0–0.828
SVI	Mean sky view index within 400 m distance	0.470	0.124	0–0.798
BVI	Mean building view index within 400 m distance	0.117	0.071	0–0.403
UG	Urban green coverage rate	0.381	0.154	0.020–0.755
UW	Urban water coverage rate	0.025	0.032	0–0.380

2.4. Ensemble Learning Algorithms

The relationships between housing prices and housing characteristics is complex and nonlinear. By combing a bunch of individual models and averaging the individual result, ensemble learning algorithms are more flexible and less data-sensitive. Thus, ensemble learning algorithms are suitable for modeling housing prices. The most commonly used ensemble learning methods are bagging and boosting. The difference between these two methods is that bagging methods train a number of individual models by a random subset of train data in a parallel way while boosting methods train models in a sequential way for learning mistakes made by the previous model. In this study, three tree-based ensemble learning algorithms and linear regressions were employed to model housing prices for selecting the algorithm. Random forest regression (RFR) uses bagging as the ensemble method and decision tree as the individual model. Since RFR trains each tree independently and uses random subsets from the training set, this method is less likely to overfit [50]. Gradient boosting regression (GBR), a boosting model, builds trees one at a time, where each new tree aims to correct errors in the predictions made by all previous trees [51]. Achieving high accuracy in a wide range of practical applications, XGBoost is an optimized distributed gradient boosting method based on ensembles of classification and regression trees (CARTs) [52]. This method provides a parallel tree-boosting to solve problems in a fast and accurate way.

Different algorithms have their own strengths and weaknesses. Therefore, to choose the optimal ensemble learning algorithms, we compared their performances in the explanation of housing prices. In detail, the regression performances of the four algorithms were measured by five common metrics, including explained variance score, mean absolute error (MAE), mean squared error (MSE), median absolute error (MedAE) and the coefficient of determination (R^2):

$$\text{explained variance}(y, \hat{y}) = 1 - \frac{Var\{y - \hat{y}\}}{Var\{y\}} \tag{4}$$

$$MAE(y, \hat{y}) = \frac{1}{n}\sum_{i=0}^{n-1} |y_i - \hat{y}_i| \tag{5}$$

$$MSE(y, \hat{y}) = \frac{1}{n}\sum_{i=0}^{n-1} (y_i - \hat{y}_i)^2 \tag{6}$$

$$MedAE(y, \hat{y}) = \text{median}(|y_1 - \hat{y_1}|, \ldots, |y_n - \hat{y_n}|) \tag{7}$$

$$R^2(y, \hat{y}) = 1 - \frac{\sum_{i=1}^{n} (y_i - \hat{y})^2}{\sum_{i=1}^{n} (y_i - \bar{y})^2} \tag{8}$$

where y and $\hat{y}$ are the true housing price and the estimated housing price, Var is Variance, n denotes the total number of communities, y_i and $\hat{y}_i$ represent the predicted housing price of the i-th community and the corresponding true value, $\hat{y_n}$ means the predicted housing price of the n-th community and $\bar{y}$ is the mean true housing price.

All the experiments in this study were performed by using a scikit-learn and XGBoost Python package. For the hyperparameter tuning and the accuracy evaluation, we chose a 10-fold cross-validation, which is a common method for performance validation.

2.5. Shapley Additive Explanations

Proposed by Lundberg and Lee, SHapley Additive exPlanations (SHAP) is a method to explain the prediction of a specific instance by calculating the contribution of each feature to the prediction [53]. The SHAP method computes Shapley values from coalitional game theory. The Shapley value of a feature value is its contribution to the output value, weighted and summed over all possible feature value combinations. The value of the j-th feature contributed ϕ_j was calculated as follow:

$$\phi_j(val) = \sum_{S \subseteq \{x_1,\ldots,x_p\} \setminus \{x_j\}} \frac{|S|!(p - |S| - 1)!}{p!}(val(S \cup \{x_j\}) - val(S)) \tag{9}$$

where p is the number of features, S represents a subset of the features used in the model, x denotes the vector of feature values of an instance to be explained and val(S) means the prediction for feature values in set S.

The advantages of SHAP include: (1) global interpretability—the collective SHAP value is able to identify the positive or negative relationship for each variable with the target and (2) local interpretability—each feature of an instance gets its own corresponding SHAP values. Traditional variable importance algorithms are limited to obtain the results across the entire population but not on each individual instance. Meanwhile, we can also measure the global importance of characteristics by computing the absolute Shapley values per characteristic:

$$I_j = \sum_{i=1}^{n} |\phi_j^{(i)}| \tag{10}$$

where $\phi_j^{(i)}$ represents the SHAP value of the j-th feature for instance i.

In this paper, we employed SHAP feature attributions, SHAP explanation force plots, SHAP summary plots and SHAP partial dependence plots and interaction plots to explore the relationships between housing prices and urban environmental elements. The XGBoost and shap Python packages were used for implementing SHAP.

3. Results and Discussion

3.1. Spatial Dstribution of Urban Environmental Characteristics

To enhance the understanding of the environmental elements of study area, we plotted the spatial distribution of five urban environmental characteristics in Figure 6. Each characteristic was mapped using seven value intervals by the natural breaks method. The average value of the green view index (GVI), sky view index (SVI), building view index (BVI), urban green coverage rate (UG) and urban water coverage rate (UW) at the community level were 0.315, 0.473, 0.117, 0.381 and 0.025, respectively. Figure 6a shows that the communities with high GVI were mainly located in the Yangpu District, Hongku District, Changning District, the northeast of Putuo District and the south of Baoshan District. Figure 6b indicates the value of SVI were the lowest in the central area and increase to the outskirts gradually, while the BVI values show the opposite pattern in Figure 6c. Figure 6d demonstrates that the UG values also increased from the central area to the outskirts gradually. From Figure 6e, we can find that the communities with high UW were mainly concentrated along the Huangpu River and the Suzhou Creek.

3.2. Model Selection

The multicollinearity between variables, which were measured by the variance inflation factor (VIF), and the results of the hedonic model, which was built by the linear regression model, are shown in Table 2. The VIFs of all the characteristics were lower than four, which indicated that these characteristics did not have serious multicollinearity. The performances of the ensemble learning regression algorithms and linear regression algorithms are compared in Table 3. Table 3 shows that the explained variance score ranged from 0.5023 to 0.6820, the MAE ranged from 0.6554 to 0.8509, the MSE ranged from 0.8556 to 1.3784, the MedAE ranged from 0.4848 to 0.6549 and the R^2 ranged from 0.4847 to 0.7045. The performances of the three ensemble methods were much better than linear regression. Among the three ensemble methods, XGBoost regression presented the best performance and was selected to be trained for interpreting the impact of urban environmental elements on housing prices.

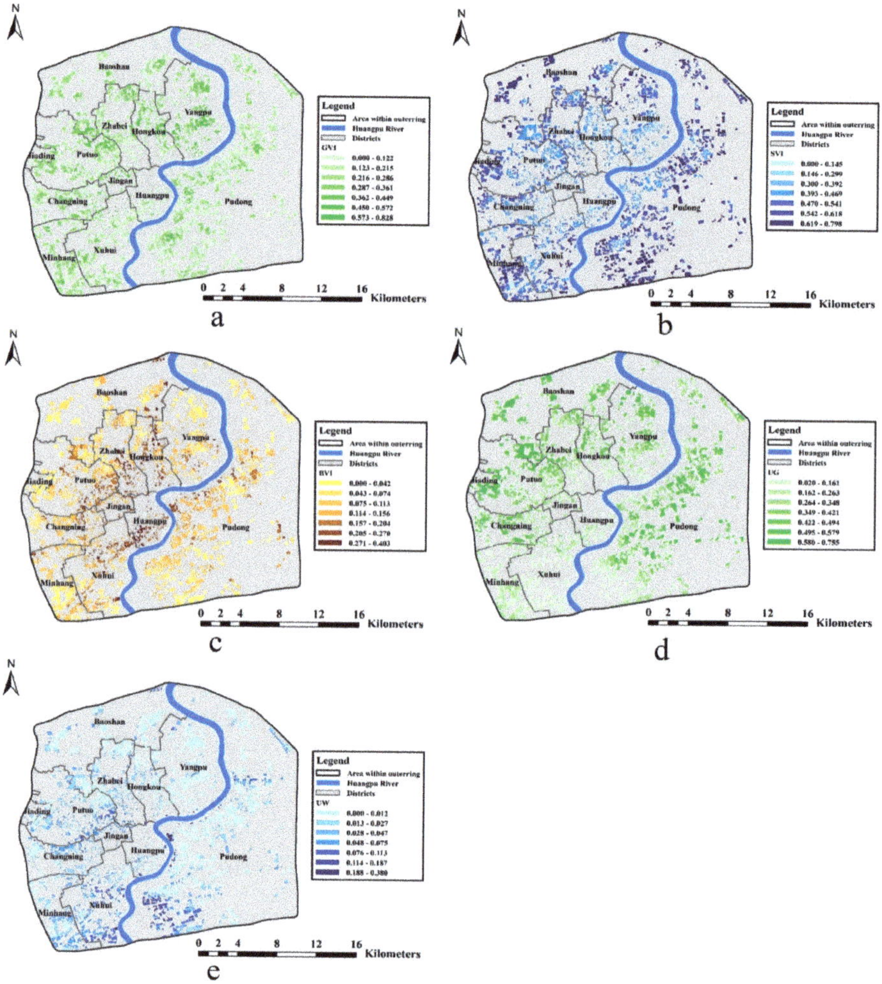

Figure 6. The spatial distributions of urban environmental characteristics: (**a**) green view index (GVI), (**b**) sky view index (SVI), (**c**) building view index (BVI), (**d**) urban green coverage (UG) and (**e**) urban water coverage (UW).

Table 2. The unstandardized coefficients, standard error and variance inflation factor (VIF) values of variables.

Variables	Unstandardized Coefficients	Standard Error	VIF
Constant	8.342	0.368	
Location characteristics			
C_DIS	−1.196 ***	0.123	3.191
EC_DIS	−1.771 ***	0.168	1.932
R_DIS	0.021	0.154	1.781
HPR_DIS	−0.236 **	0.102	1.602
Structure characteristics			
YEAR	−0.042 ***	0.004	2.343
AREA	0.003 ***	0.001	1.987
PR	−0.161 ***	0.022	1.453
EL	0.467 ***	0.073	2.277
Neighborhood characteristics			
BUS_NEAR	−0.075	0.282	1.255
BUS_500M	-9.972×10^{-5}	0.009	2.577
BUS_1000M	0.002	0.004	2.788
SUB_NEAR	−0.094 **	0.060	2.115
SUB_500M	0.122	0.048	1.856
SUB_1000M	0.156 ***	0.025	2.100
PRI_NEAR	0.356 ***	0.098	1.634
PRI_500M	0.069 ***	0.026	2.061
PRI_1000M	0.025 *	0.013	2.497
FH3_NEAR	−0.077 ***	0.023	2.824
FH3_500M	−0.005	0.067	1.658
FH3_1000M	0.180 ***	0.034	2.153
Urban Environmental characteristics			
GVI	0.710 **	0.329	3.143
SVI	−1.235 ***	0.317	2.964
BVI	0.088	0.539	2.838
UG	−0.053	0.191	1.652
UW	6.494 ***	0.856	1.475

* Indicates significance at the 10% level, ** indicates significance at the 5% level and *** indicates significance at the 1% level.

Table 3. Performance of linear regression algorithms and three ensemble learning regression algorithms. MAE: mean absolute error, MSE: mean squared error, MedAE: median absolute error and R^2: coefficient of determination.

	Linear Regression	XGBoost Regression	Random Forest Regression	Gradient Boosting Regression
Explained variance score	0.5023	**0.6820**	0.6398	0.5887
MAE	0.8509	**0.6554**	0.6918	0.7697
MSE	1.3784	**0.8556**	0.9703	1.1340
MedAE	0.6549	**0.4848**	0.4891	0.5876
R^2	0.4847	**0.7045**	0.6306	0.5747

In order to investigate whether urban environmental characteristics from the horizontal view and from the overhead view will affect the housing prices, we estimated the R^2 of four additional models: model 1 only with location, structure and neighborhood characteristics; three horizontal view urban environmental characteristics (GVI, SVI and BVI) were added to model 2 based on model 1; two overhead view urban environmental characteristics (UG and UW) were added to model 3 based on model 1 and model 4 included all the characteristics. As shown in Table 4, adding either horizontal view urban environmental characteristics or overhead view ones led to a significant improvement of R^2. Specifically, horizontal view urban environmental characteristics increased R^2 by 0.0249 and overhead view ones increased R^2 by 0.0265. Adding all the urban environmental characteristics resulted in the highest R^2 of 0.7045. These results suggested that both urban environmental characteristics from the horizontal view and from the overhead view can affect housing prices. The following section further analyzed the impacts of urban environmental elements on housing prices based on model 4.

Table 4. Model performance with difference characteristics.

	Model 1	Model 2 (Model 1 + GVI + SVI + BVI)	Model 3 (Model 1 + UG + UW)	Model 4 (Model 1 + GVI + SVI + BVI + UG + UW)
R^2	0.6722	0.6971	0.6987	0.7045

3.3. Global Importance of Characteristics

In this section, we compared the global importance of all characteristics by calculating the SHAP feature importance. We run SHAP for communities based on the trained XGBoost models and got a matrix of Shapley values.

To facilitate the understanding, we took Aijian mansion as an example. Figure 7 shows characteristics each contributing to push the model output from the base value (the baseline for Shapley values is the average of all outputs) to the model output. Characteristics pushing the prices higher were shown in red; those pushing the prices lower were in blue. The baseline—the average predicted housing prices, was 6.373. The predicted price of Aijian mansion was 5.90. EC_DIS increased the price by 0.04922, while HPR_DIS decreased the price by 0.6428.

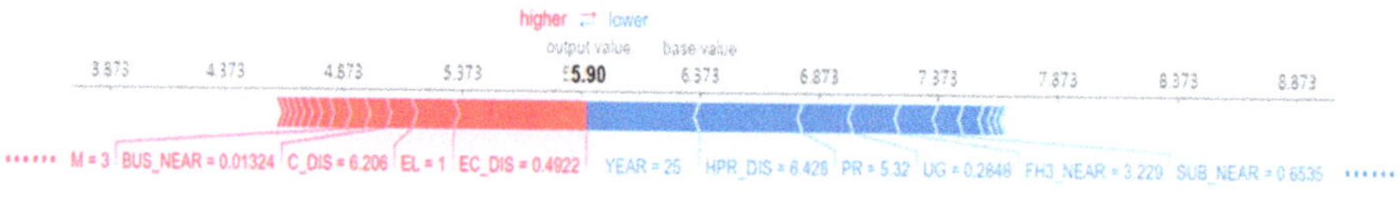

Figure 7. SHapley Additive exPlanations (SHAP) explanation force plots for Aijian mansion.

Based on the matrix of Shapley values, the absolute Shapley values per characteristic across the data were computed for measuring the global importance of characteristics by Formula (10). We sorted the characteristics by decreasing importance and plotted them in Figure 8. The top characteristics contributed more to the model than the bottom ones, and thus, had a greater impact on the housing prices. Overall, the four categories of the characteristics' SHAP importance could be ranked as follows: location characteristics (0.8491) > neighborhood characteristics (0.7055) > structure characteristics (0.6939) > urban environmental characteristics (0.4266). This result indicated that the location characteristics were the dominant determinants of housing prices in Shanghai. The importance of neighborhood characteristics and structure characteristics were roughly equivalent. Although urban environmental characteristics had relatively minimal impacts on housing prices, we cannot neglect the impacts of urban environmental characteristics, which accounted for 16 percent of the total importance. Specifically, the top five characteristics were YEAR (0.4259), EC_DIS (0.3720), C_DIS (0.2494), FH3_NEAR (0.1759) and HPR_DIS (0.1306). For five urban environmental characteristics, the SHAP importance was ranked as follows: UG (0.1145) > UW (0.1043) > SVI (0.0908) > GVI (0.0601) > BVI (0.0570). The SHAP importance of the overhead view environmental characteristics (0.2187) was slightly higher than those from the horizontal view (0.2079). The horizontal view environmental characteristics could account for 8 percent of total housing prices.

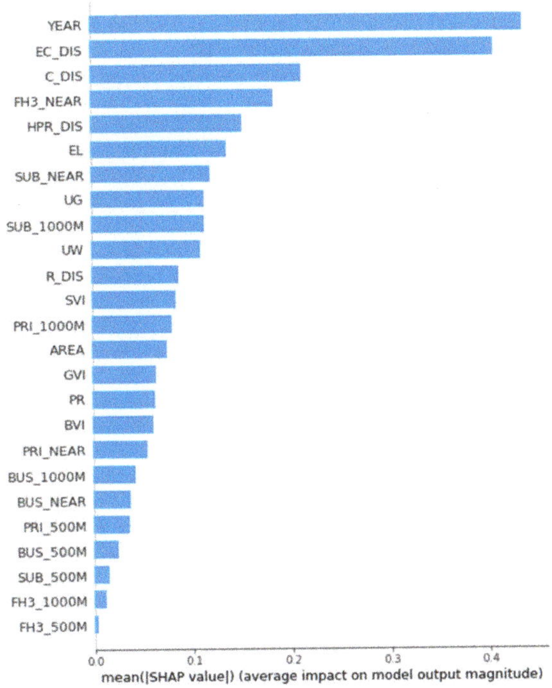

Figure 8. SHAP features importance for the determinants of housing prices.

Given that SHAP features importance only contains the absolute value of feature contributions, a density scatter plot of SHAP values for each characteristic was used to further analyze the relationships of the determinants with the housing prices. Characteristics were sorted by the values of SHAP importance. As shown in Figure 9, each point on the summary plot was the Shapley value for a characteristic of a community. The position on the x-axis was determined by the Shapley value, and the color denoted the value from low (blue) to high (red). The dispersion in the y-axis direction represented the number of points, which demonstrated the distribution of the Shapley values per characteristic. If the SHAP value of a characteristic increases with the increase of the corresponding feature value, this characteristic has a positive impact on housing prices, and vice versa. Figure 9 indicates that the four location characteristics all had strong negative relationships with housing prices. YEAR, FH3_NEAR and SUB_NEAR had apparent negative influences on housing prices. EL, SUB_1000M and PRI_1000M showed positive influences on housing prices. In terms of urban environmental characteristics, UW had a strong positive correlation with housing prices. The relationship between SVI and housing prices had a negative correlation. For UG and GVI, although the communities with high SHAP values had relatively high feature values, the SHAP values were not always increased with the increase of the feature values. This result showed that the relationships between these two characteristics and housing prices were complicated and nonlinear. In addition, it was difficult to identify the impacts of BVI on housing prices because of no obvious pattern.

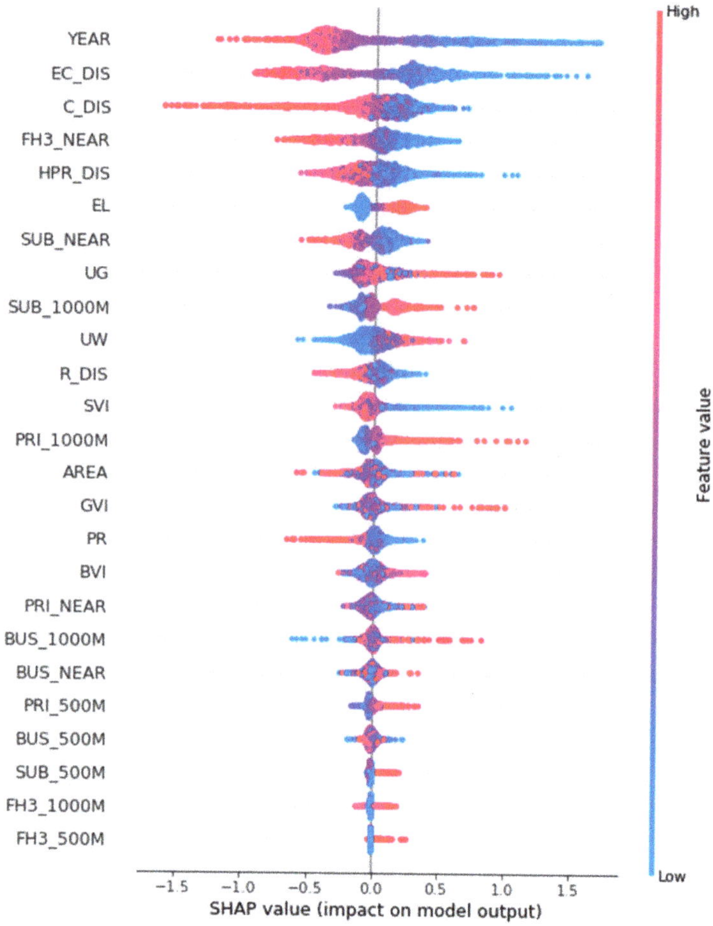

Figure 9. SHAP summary plots of housing prices.

3.4. Contribution of Urban Environmental Characteristics

Due to that SHAP summary plots couldn't fully reveal the complex and nonlinear relationships between most of the urban environmental characteristics and housing prices, we delved into the specific contributions of characteristics on housing prices by using the SHAP feature dependence plot. The SHAP feature dependence plot for five urban environmental characteristics were drawn in Figure 10 to describe their impacts on housing prices. The spatial distribution of SHAP for the five environmental characteristics were also mapped in Figure 11 to improve the understanding of the contribution of each urban environmental characteristic.

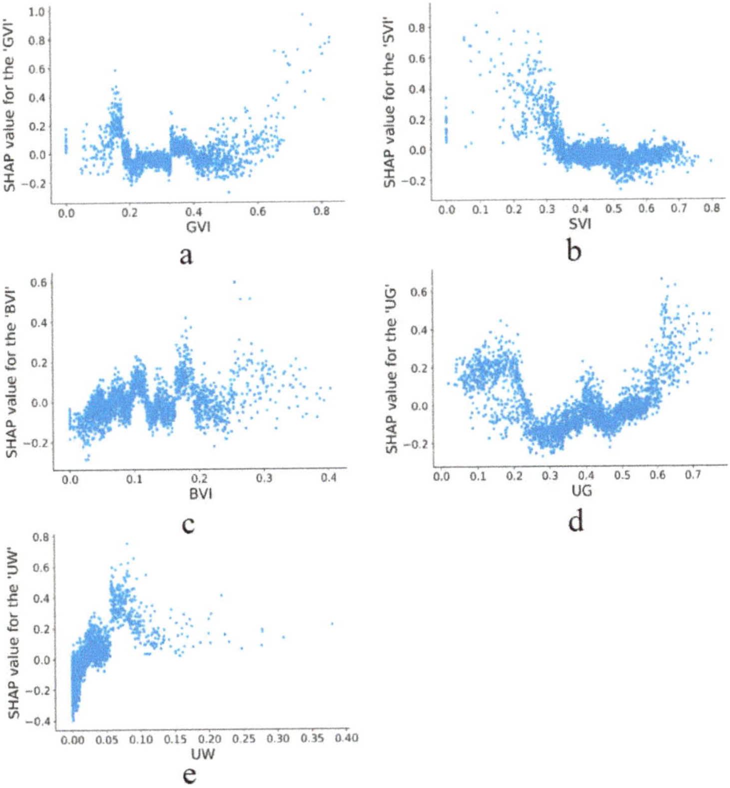

Figure 10. SHAP feature dependence plots for the five urban environmental characteristics: (**a**) GVI_SHAP, (**b**) SVI_SHAP, (**c**) BVI_SHAP, (**d**) UG_SHAP and (**e**) UW_SHAP.

3.4.1. Contribution of Green View Index (GVI) and Urban Green Coverage (UG)

The SHAP values of the GVI showed a decreasing, stable and increasing tendency, and the two inflection points were approximately 0.2 and 0.5 (Figure 10a). Most of the GVI SHAP values were positive when GVI was less than 0.2 or greater than 0.5. When the GVI exceeded 0.5, the GVI SHAP value increased as the GVI increased. The result of the traditional hedonic model, which was built by the linear regression model, showed that the GVI had a significant positive effect on housing prices (Table 2). Every one percent increase in the GVI can increase housing prices by 71 RMB/m^2. Our method indicated that the relationship between the GVI and housing prices was complex and nonlinear rather than linear positive. Shanghai's homebuyers were willing to pay a premium for a green view only when the GVI was of higher value, which was more elaborate than the results of previous studies. A study in the Netherlands showed that a green view can attract an extra price increase of 8% [20]. Another study in Hong Kong also suggested green space views have notably enhanced residential housing prices [23]. To better interpret the results, the spatial distribution of the GVI (Figure 6a) and GVI SHAP (Figure 11a) played an important role. From the distribution of the communities whose GVI and GVI SHAP were both high, we could find that most of these communities were near large parks, such as Changshou Park in the Putuo District, Xujiahui Park in the Xuhui District and Huashan Green Park in the Changning District. These parks could serve as recreational venues and provide pleasant views to residents [54]. The reason why the communities with low GVI values had positive effects on housing prices might be that most of these communities have been built for many years.

Although these older residential communities are lacking a horizontal green view, most of them have diverse public service facilities due to long-term developments.

Compared with the GVI SHAP, a similar trend was observed for the UG SHAP. Figure 10d showed that the SHAP value of the UG was positive when the UG was less than 0.23 and then fluctuated around zero. When the UG was greater than 0.5, the UG SHAP value presented a significant increase. The positive influence of the UG on housing prices when the UG was less than 0.23 or greater than 0.5 indicated that homebuyers were willing to pay more for higher UGs. The reasons for these results were also similar to those reasons for the GVI. Table 2 showed that the GVI was not significant in the traditional hedonic model, which was not consistent with our method. To investigate whether the impacts of the GVI and UG on housing prices show the same pattern, we carried out a comparison between the GVI and UG. The coefficient of determination for the GVI and UG was 0.0799. The spatial distribution of the GVI and UG were quite different. These results suggested that there were no obvious correlations between the GVI and UG. For the SHAP value of the GVI and UG, the coefficient of determination for them was 0.0098. The spatial distribution of the GVI SHAP and UG SHAP were also quite different. Thus, there were no obvious correlations between the GVI SHAP and UG SHAP. All of these results indicated that, although both higher GVI and higher UG had positive impacts on housing prices, there were significant differences between the patterns of their impacts on housing prices. These finding demonstrate that the impacts of the same urban environmental elements from different observation perspectives (horizontal view and overhead view) are different.

In general, the relationships between housing prices and two green characteristics (green view index from street view data and urban green coverage rate from remote sensing) are both nonlinear. Shanghai's homebuyers are willing to pay extra for green only when the green view index or urban green coverage rate are of higher value.

3.4.2. Contribution of Sky View Index (SVI)

The SVI of a community could reflect the amount of open spaces, as well as the height and density of buildings in and around this community. In this study, when the SVI value was less than 0.35, the SHAP value of most communities was positive and decreased from 0.8 to zero. For every one percent increase in the SVI, the housing prices decreased by 320 RMB/m^2. When the SVI value was greater than 0.35, the SVI SHAP value maintained stable at around zero. The result of the traditional hedonic model showed that the SVI had a significant negative effect on housing prices in Table 2. Every one percent increase in the SVI can decrease housing prices by 123.5 RMB/m2. The findings of our method indicated that the relationship between the SVI and housing prices was also nonlinear rather than linear. By comparing Figures 6b and 11b, we could find the values of the SVI SHAP were the highest in the central area and decreased to the outskirts gradually, which was opposite to the distribution of the SVI. Contrary to expectation, these results mean that the SVI has a strong and negative impact on housing prices in Shanghai when its value is less than 0.35. This finding contrasted with a previous study indicating both street and building views suppressed housing price in Hong Kong [23]. The opposite result in Shanghai could be explained as follows. The high housing prices in Shanghai has resulted in a vertical and compact city, with most residents living in high-density and high-rise residential buildings. The high-rise buildings mean enjoyment of wider views and less noise and air pollution in the higher floors, resulting in a better environmental quality.

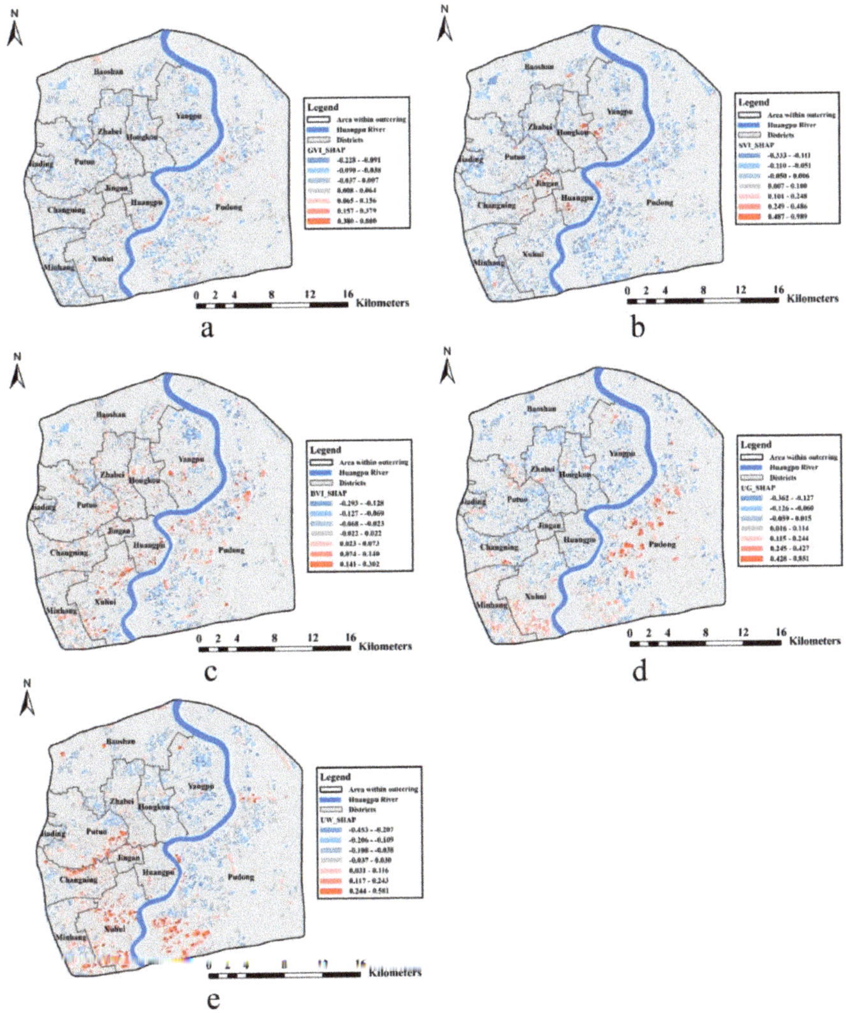

Figure 11. Spatial distribution of SHAP for the five urban environmental characteristics: (**a**) GVI_SHAP, (**b**) SVI_SHAP, (**c**) BVI_SHAP, (**d**) UG_SHAP and (**e**) UW_SHAP.

3.4.3. Contribution of Building View Index (BVI)

With regards to the BVI, the BVI SHAP value always fluctuated around zero, with a small variance between 0.2 and -0.2. This result demonstrated that the influence of the BVI on housing prices was not obvious. Table 2 showed that the BVI was not significant in the traditional hedonic model, which was consistent with our method. The reason for this result might be that many buildings are blocked by trees and cars in street view images. This leads to how the BVI couldn't depict the distribution of buildings accurately. In most cases, the SVI is the better choice than the BVI for the description of buildings from a horizontal view.

3.4.4. Contribution of Urban Water Coverage (UW)

The UW SHAP value increased sharply when the UW was lesser than 0.08, and a one percent increase in the UW SHAP could increase housing prices by 800 RMB/m^2. When the UW was greater than 0.08, the UW SHAP value maintained stable. This result indicated that Shanghai's homebuyers would be willing to pay a premium for houses in communities with a higher UW, which was consistent with studies in Hangzhou [55] and Hong Kong [23]. Table 2 showed that the UW was significant and positive in the traditional hedonic model, which is consistent with our method. In spatial distribution, Figures 6e and 11e show that the UW SHAP and the UW presented similar patterns. Communities with a high UW SHAP value were mainly concentrated along the Huangpu River and Suzhou Creek. These two main rivers provide a large amount of water coverage for the communities along them. In a compact city, water bodies have the effect of adjusting air temperature and humidity, which improves human comfort. The water also provides residents with precious spaces where air circulation and solar access are less impeded.

4. Conclusions

In this study, we proposed a new framework for measuring the impacts of urban environmental elements on housing prices in the area within Shanghai's outer ring. The green view index (GVI), the sky view index (SVI) and the building view index (BVI) were extracted as horizontal-view urban environmental characteristics based on the Baidu street view images using a deep convolutional neural network. The overhead view environmental characteristics were computed by remote sensing data. Comparing the results of three tree-based ensemble learning models and linear regression models, the XGBoost model showed the best performance. Thereafter, a SHapley Additive exPlanations (SHAP) method, which has the ability to explain the model's overall behavior in the form of particular feature contributions, was introduced to uncover the complex and nonlinear relationships between urban environmental characteristics and housing prices. The spatial distribution of SHAP for the five environmental characteristics were mapped to improve the understanding of the contribution of each urban environmental characteristic. In addition, the impacts of horizontal-view and overhead-view green characteristics on housing prices were compared to analyze the differences of the same urban environmental elements' impacts on housing prices from different observation perspectives. The experimental results are demonstrated as follows. Compared with location, neighborhood and structure characteristics, urban environmental characteristics have relatively minimal impacts that account for 16 percent of housing prices. The relationship between the GVI and housing prices is nonlinear rather than linear positive or linear negative. Similar to the GVI, the urban green coverage rate (UG) also has a nonlinear relation with housing prices. These findings indicated that Shanghai's homebuyers are willing to pay a premium for green only when the GVI or UG are of higher values. Although both a higher GVI and higher UG have positive impacts on housing prices, there are significant differences between their impacts on housing prices. Contrary to previous studies, when the SVI value is less than 0.35, every one percent increase in the SVI, decreases the housing prices by 320 RMB/m2. The potential reason is that high-density and high-rise residential areas often have better living facilities. Compared with the GVI and SVI, the influence of the BVI on housing prices is not obvious. A one percent increase in the urban water coverage rate (UW) can increase housing prices by 800 RMB/m2, which indicates residents in Shanghai are willing to pay a premium for water coverage. In summary, the case of Shanghai shows that the proposed framework is practical and efficient.

This study was limited in several ways. First, the applicability of the proposed framework was tested in Shanghai. Considering the geographical heterogeneity, the relationships between the urban environmental elements and housing prices may be different in a different city. Using this framework to quantify the differences among cities is expected to achieve a promising result. Second, the housing transaction data used in this study were only obtained in 2018. Thus, further studies could be conducted to integrate multi-year data to analyze the temporal dynamics of the impacts of the urban environmental elements on housing prices. Third, our housing model does not consider some housing characteristics, such as floor level and urban village, because these characteristics cannot be captured at present. It is worth discussing these characteristics of the Chinese housing market in future research. Last, the acquisition time of the data for extracting urban environmental characteristics was different.

The Baidu street view data were obtained in 2017, while the remote sensing data were obtained in 2015. Due to the rapid development of Shanghai and seasonal differences of nature environmental elements, differences in data acquisition time could have adverse effects on research findings. Therefore, street view data and remote sensing data with similar acquisition times could be used in future research to improve the results.

Author Contributions: Conceptualization, L.C. and X.Y.; methodology, L.C. and X.Z.; software, L.C.; validation, W.C. and T.C.; formal analysis, L.C. and Y.Z.; investigation, L.C.; resources, X.Y.; data curation, L.C. and Y.Z.; writing—original draft preparation, L.C.; writing—review and editing, X.Y., Y.L., W.C., X.Z. and T.C.; visualization, L.C. and supervision, X.Y., Y.L. and T.C. All authors have read and agree to the published version of the manuscript.

Funding: This research was funded by the National Natural Science Foundation of China (41701438).

Conflicts of Interest: The authors declare no conflicts of interest.

References

1. Jim, C.Y.; Chen, W.Y. Impacts of urban environmental elements on residential housing prices in Guangzhou (China). *Landsc. Urban Plan.* **2006**, *78*, 422–434. [CrossRef]
2. Chiesura, A. The role of urban parks for the sustainable city. *Landsc. Urban Plan.* **2004**, *68*, 129–138. [CrossRef]
3. Haaland, C.; von Den Bosch, C.K. Challenges and strategies for urban green-space planning in cities undergoing densification: A review. *Urban For. Urban Green.* **2015**, *14*, 347–354. [CrossRef]
4. Sæbø, A.; Popek, R.; Nawrot, B.; Hanslin, H.; Gawronska, H.; Gawronski, S. Plant species differences in particulate matter accumulation on leaf surfaces. *Sci. Total Environ.* **2012**, *427*, 347–354. [CrossRef]
5. Chen, X.-L.; Zhao, H.-M.; Li, P.-X.; Yin, Z.-Y. Remote sensing image-based analysis of the relationship between urban heat island and land use/cover changes. *Remote Sens. Environ.* **2006**, *104*, 133–146. [CrossRef]
6. Strohbach, M.W.; Arnold, E.; Haase, D. The carbon footprint of urban green space—A life cycle approach. *Landsc. Urban Plan.* **2012**, *104*, 220–229. [CrossRef]
7. Ridder, K.D.; Adamec, V.; Bañuelos, A.; Bruse, M.; Bürger, M.; Damsgaard, O.; Dufek, J.; Hirsch, J.; Lefebre, F.; Pérez-Lacorzana, J.M. An integrated methodology to assess the benefits of urban green space. *Sci. Total Environ.* **2004**, *334–335*, 489–497. [CrossRef]
8. Van den Berg, M.; van Poppel, M.; van Kamp, I.; Andrusaityte, S.; Balseviciene, B.; Cirach, M.; Danileviciute, A.; Ellis, N.; Hurst, G.; Masterson, D. Visiting green space is associated with mental health and vitality: A cross-sectional study in four european cities. *Health Place* **2016**, *38*, 8–15. [CrossRef]
9. Gubbels, J.S.; Kremers, S.P.; Droomers, M.; Hoefnagels, C.; Stronks, K.; Hosman, C.; de Vries, S. The impact of greenery on physical activity and mental health of adolescent and adult residents of deprived neighborhoods: A longitudinal study. *Health Place* **2016**, *40*, 153–160. [CrossRef]
10. De Vries, S.; van Dillen, S.M.E.; Groenewegen, P.P.; Spreeuwenberg, P. Streetscape greenery and health: Stress, social cohesion and physical activity as mediators. *Soc. Sci. Med.* **2013**, *94*, 26–33. [CrossRef]
11. Nutsford, D.; Pearson, A.L.; Kingham, S.; Reitsma, F. Residential exposure to visible blue space (but not green space) associated with lower psychological distress in a capital city. *Health Place* **2016**, *39*, 70–78. [CrossRef] [PubMed]
12. Asgarzadeh, M.; Koga, T.; Hirate, K.; Farvid, M.; Lusk, A. Investigating oppressiveness and spaciousness in relation to building, trees, sky and ground surface: A study in Tokyo. *Landsc. Urban Plan.* **2014**, *131*, 36–41. [CrossRef]
13. Hartig, T.; Evans, G.W.; Garling, T.; Golledge, R.G. Psychological Foundations of Nature Experience. *Adv. Psychol. Amst.* **1993**, *96*, 427. [CrossRef]
14. Benson, E.D.; Hansen, J.L.; Schwartz, A.L.; Smersh, G.T. Pricing residential amenities: The value of a view. *J. Real Estate Financ. Econ.* **1998**, *16*, 55–73. [CrossRef]
15. Lee, C.L. An examination of the risk-return relation in the Australian housing market. *Int. J. Hous. Mark. Anal.* **2017**. [CrossRef]
16. Al-Masum, M.A.; Lee, C.L. Modelling housing prices and market fundamentals: Evidence from the Sydney housing market. *Int. J. Hous. Mark. Anal.* **2019**. [CrossRef]
17. Bangura, M.; Lee, C.L. House price diffusion of housing submarkets in Greater Sydney. *Hous. Stud.* **2019**, 1–32. [CrossRef]

18. Trojanek, R.; Gluszak, M. Spatial and time effect of subway on property prices. *J. Hous. Built Environ.* **2018**, *33*, 359–384. [CrossRef]
19. Yamagata, Y.; Murakami, D.; Yoshida, T.; Seya, H.; Kuroda, S. Value of urban views in a bay city: Hedonic analysis with the spatial multilevel additive regression (SMAR) model. *Landsc. Urban Plan.* **2016**, *151*, 89–102. [CrossRef]
20. Luttik, J. The value of trees, water and open space as reflected by house prices in the Netherlands. *Landsc. Urban Plan.* **2000**, *48*, 161–167. [CrossRef]
21. Donovan, G.H.; Butry, D.T. The effect of urban trees on the rental price of single-family homes in Portland, Oregon. *Urban For. Urban Green.* **2011**, *10*, 163–168. [CrossRef]
22. Belcher, R.N.; Chisholm, R.A. Tropical vegetation and residential property value: A hedonic pricing analysis in Singapore. *Ecol. Econ.* **2018**, *149*, 149–159. [CrossRef]
23. Jim, C.Y.; Chen, W.Y. Value of scenic views: Hedonic assessment of private housing in Hong Kong. *Landsc. Urban Plan.* **2009**, *91*, 226–234. [CrossRef]
24. Chen, W.Y.; Jim, C.Y. Amenities and disamenities: A hedonic analysis of the heterogeneous urban landscape in Shenzhen (China). *Geogr. J.* **2010**, *176*, 227–240. [CrossRef]
25. Donovan, G.H.; Butry, D.T. Trees in the city: Valuing street trees in Portland, Oregon. *Landsc. Urban Plan.* **2010**, *94*, 77–83. [CrossRef]
26. McPherson, E.G.; Simpson, J.R.; Xiao, Q.F.; Wu, C.X. Million trees Los Angeles canopy cover and benefit assessment. *Landsc. Urban Plan.* **2011**, *99*, 40–50. [CrossRef]
27. Li, X.; Chuanrong, Z.; Weidong, L. Does the Visibility of Greenery Increase Perceived Safety in Urban Areas? Evidence from the Place Pulse 1.0 Dataset. *ISPRS Int. J. Geo-Inf.* **2015**, *4*, 1166–1183. [CrossRef]
28. Yoo, S.; Im, J.; Wagner, J.E. Variable selection for hedonic model using machine learning approaches: A case study in Onondaga County, NY. *Landsc. Urban Plan.* **2012**, *107*, 293–306. [CrossRef]
29. Zhang, F.; Zhou, B.; Liu, L.; Liu, Y.; Fung, H.H.; Lin, H.; Ratti, C. Measuring human perceptions of a large-scale urban region using machine learning. *Landsc. Urban Plan.* **2018**, *180*, 148–160. [CrossRef]
30. Ye, Y.; Xie, H.; Fang, J.; Jiang, H.; Wang, D. Daily accessed street greenery and housing price: Measuring economic performance of human-scale streetscapes via new urban data. *Sustainability* **2019**, *11*, 1741. [CrossRef]
31. Rosen, S. Hedonic prices and implicit markets: Product differentiation in pure competition. *J. Political Econ.* **1974**, *82*, 34–55. [CrossRef]
32. Lancaster, K.J. A new approach to consumer theory. *J. Political Econ.* **1966**, *74*, 132–157. [CrossRef]
33. Zhang, Y.; Dong, R. Impacts of street-visible greenery on housing prices: Evidence from a hedonic price model and a massive street view image dataset in Beijing. *Int. J. Geo Inf.* **2018**, *7*, 104. [CrossRef]
34. Wen, H.; Tao, Y. Polycentric urban structure and housing price in the transitional China: Evidence from Hangzhou. *Habitat Int.* **2015**, *46*, 138–146. [CrossRef]
35. Wen, H.; Xiao, Y.; Zhang, L. School district, education quality, and housing price: Evidence from a natural experiment in Hangzhou, China. *Cities* **2017**, *66*, 72–80. [CrossRef]
36. Dubé, J.; Legros, D. Spatial econometrics and the hedonic pricing model: What about the temporal dimension? *J. Prop. Res.* **2014**, *31*, 333–359. [CrossRef]
37. Chen, Y.; Liu, X.; Li, X.; Liu, Y.; Xu, X. Mapping the fine-scale spatial pattern of housing rent in the metropolitan area by using online rental listings and ensemble learning. *Appl. Geogr.* **2016**, *75*, 200–212. [CrossRef]
38. Antipov, E.A.; Pokryshevskaya, E.B. Mass appraisal of residential apartments: An application of Random forest for valuation and a CART-based approach for model diagnostics. *Expert Syst. Appl.* **2012**, *39*, 1772–1778. [CrossRef]
39. Hu, L.; He, S.; Han, Z.; Xiao, H.; Su, S.; Weng, M.; Cai, Z. Monitoring housing rental prices based on social media: An integrated approach of machine-learning algorithms and hedonic modeling to inform equitable housing policies. *Land Use Policy* **2019**, *82*, 657–673. [CrossRef]
40. Stojić, A.; Stanić, N.; Vuković, G.; Stanišić, S.; Perišić, M.; Šoštarić, A.; Lazić, L. Explainable extreme gradient boosting tree-based prediction of toluene, ethylbenzene and xylene wet deposition. *Sci. Total Environ.* **2019**, *653*, 140–147. [CrossRef]
41. Janizek, J.D.; Celik, S.; Lee, S.-I. Explainable machine learning prediction of synergistic drug combinations for precision cancer medicine. *BioRxiv* **2018**, 331769. [CrossRef]

42. Cai, W.; Lu, X. Housing affordability: Beyond the income and price terms, using China as a case study. *Habitat Int.* **2015**, *47*, 169–175. [CrossRef]
43. Shanghai Municipal People's Government. *Shanghai Master Plan (217–2035)*; Shanghai Municipal People's Government: Shanghai, China, 2018.
44. Bray, D. *Social Space and Governance in Urban China: The Danwei System from Origins to Reform*; Stanford University Press: Palo Alto, CA, USA, 2005.
45. Badrinarayanan, V.; Kendall, A.; Cipolla, R. SegNet: A deep convolutional encoder-decoder architecture for image segmentation. *IEEE Trans. Pattern Anal. Mach. Intell.* **2017**, *39*, 2481–2495. [CrossRef] [PubMed]
46. Zhao, H.; Shi, J.; Qi, X.; Wang, X.; Jia, J. Pyramid scene parsing network. In Proceedings of the IEEE Conference on Computer Vision and Pattern Recognition(CVPR), Honolulu, HI, USA, 21–26 July 2017; pp. 2881–2890.
47. Chen, L.-C.; Papandreou, G.; Schroff, F.; Adam, H. Rethinking atrous convolution for semantic image segmentation. *arXiv* **2017**, arXiv:1706.05587.
48. Cordts, M.; Omran, M.; Ramos, S.; Rehfeld, T.; Enzweiler, M.; Benenson, R.; Franke, U.; Roth, S.; Schiele, B. The cityscapes dataset for semantic urban scene understanding. In Proceedings of the IEEE Conference on Computer Vision and Pattern Recognition, Las Vegas, NV, USA, 26 June–1 July 2016; pp. 3213–3223.
49. Sun, B.; Tu, T.; Shi, W.; Guo, Y. Test on the performance of polycentric spatial structure as a measure of congestion reduction in megacities. The case study of Shanghai. *Urban Plan. Forum* **2013**, 69–75.
50. Liaw, A.; Wiener, M. Classification and regression by randomForest. *R News* **2002**, *2*, 18–22.
51. Friedman, J.H. Greedy function approximation: A gradient boosting machine. *Ann. Stat.* **2001**, *29*, 1189–1232. [CrossRef]
52. Chen, T.; Guestrin, C. Xgboost: A scalable tree boosting system. In Proceedings of the 22nd Acm Sigkdd International Conference on Knowledge Discovery and Data Mining, San Francisco, CA, USA, 13–17 August 2016; pp. 785–794.
53. Lundberg, S.; Lee, S.I. A Unified Approach to Interpreting Model Predictions. In Proceedings of the 31st Conference on Neural Information Processing Systems (NIPS 2017), Long Beach, CA, USA, 4–19 December 2017.
54. Ulrich, R.S. Human Responses to Vegetation and Landscapes. *Landsc. Urban Plan.* **1986**, *13*, 29–44. [CrossRef]
55. Wen, H.; Zhang, Y.; Zhang, L. Assessing amenity effects of urban landscapes on housing price in Hangzhou, China. *Urban For. Urban Green.* **2015**, *14*, 1017–1026. [CrossRef]

International Journal of
Geo-Information

MDPI

Article

Quantifying the Spatial Heterogeneity and Driving Factors of Aboveground Forest Biomass in the Urban Area of Xi'an, China

Xuan Zhao [1], Jianjun Liu [1,*], Hongke Hao [2] and Yanzheng Yang [3]

1 College of Landscape Architecture and Art, Northwest A&F University, Xianyang 712100, China; zx666@nwafu.edu.cn
2 College of Forestry, Northwest A&F University, Xianyang 712100, China; haohongke@nwafu.edu.cn
3 State Key Laboratory of Urban and Regional Ecology, Research Center for Eco-Environmental Sciences, Chinese Academy of Sciences, Beijing 100085, China; yangyzh@rcees.ac.cn
* Correspondence: ljj@nwafu.edu.cn

Received: 14 October 2020; Accepted: 10 December 2020; Published: 12 December 2020

Abstract: Investigating the spatial distribution of urban forest biomass and its potential influencing factors would provide useful insights for configuring urban greenspace. Although China is experiencing an unprecedented scale of urbanization, the spatial pattern of the urban forest biomass distribution as a critical component in the urban landscape has not been fully examined. Using the geographic detector method, this research examines the impacts of four geographical factors (GFs)—dominant tree species, forest categories, land types, and age groups—on the aboveground biomass distribution of urban forests in 1480 plots in Xi'an, China. The results indicate that (1) the aboveground biomass and four GFs show obvious heterogeneity regarding their spatial distribution in Xi'an; (2) the dominant tree species and age group which impacts the patterns of aboveground biomass are the primary GFs, with the independent q value (a statistic metric used to quantify the impacts of GFs in this study) reaching 0.595 and 0.202, respectively, while the forest category and land type were weakly linked to the spatial variation of aboveground biomass, with a q value of 0.087 and 0.076, respectively; and (3) the interactions among these four GFs also tend to contribute to the distribution pattern of aboveground biomass. The interactions between GFs achieved a larger impact than the sum of impacts that were independently obtained from the factors. Our results showed that the method of using a geographical detector is a useful tool in the urban area, and can reveal the driver pattern of aboveground biomass and provide a reference for city planning and management.

Keywords: urban forest; forest biomass; biomass distribution; geographic detector

1. Introduction

Due to the rapid urbanization process, the global urban population exceeded the rural population for the first time in 2017 [1], indicating that we had entered a new urban era. There is a universal relationship between development and urbanization—the urbanization pace peaking at the per capita income level of approximately $3000–5000 [2]. The urbanization speed is currently at the highest level in East Asia and has progressed in South Asia and Africa, after the main urbanization growth shifted away from Europe, North America, and Japan [3]. As the largest developing country in the world, China contributes a major portion (837 million) of the global urban population. In the period of 1978–2017, the urbanization level of China increased from 17.92% to 58.52% [1,4], and researchers believe that the urban population proportion of China is projected to increase to over 70% by 2030 and 80% by the middle of this century [5]. Therefore, it is believed that the urbanization of China might play an important role in the world's rapidly urbanizing process [6].

Improving urban ecosystem services, in terms of supply, regulation, habitats, culture, and amenity services, is an important component of measurements that can be used to improve the urbanization quality [7]. Trees in urban areas can provide a carbon sequestration function, as well as a product providing function [8–10]. Close relationships have been reported between the net long-term CO_2 source/sink dynamics and urban forest biomass [11–15]. A higher forest biomass indicates a larger amount of carbon dioxide sequestration in urban forest ecosystems [16–18]. Therefore, a reasonable pattern and community structure of an urban forest offer ecological benefits for urban residents, and could help them to understand that the dynamics and drivers of urban forests are critical for city management.

Spatial heterogeneity refers to uneven distributions of traits, events, or their relationship across a region [19]. This phenomenon can be analyzed and quantified by using the geographical statistical method of employing a geographic detector [20]. The core idea of geographic detectors is based on the hypothesis that the dependent variables should be spatially highly related to the independent variables if the independent variables have major effects on the dependent variables. Therefore, compared to conventional analysis of variance (ANOVA), this method can quantify the impacts of spatial factors on the spatial distribution of a given independent variable [21] and explore spatial (global) stratified heterogeneity within the stratified attribute by the q-statistic. Additionally, this method can detect potential variables that impact the spatial distribution of independent variables, and reveal interactive effects among those variables. It has two significant advantages: Linear assumptions between dependent and independent variables are not required, and it can detect the interactive influence of two independent variables on the dependent variables [22].

In an urban forest, the global spatial heterogeneity of biomass displays an uneven distribution within the whole study area. The driving forces of this phenomenon have been widely studied [9,23–27]. Conventional ANOVA is normally used to explain this relationship [28,29], which only provides a field of view about whether there are significant differences among the subtypes of a certain driving factor (for example, the age group, diameter at breast height (DBH), etc.). The quantitative relations between driving factors and biomass are difficult to directly compare. On the other hand, empirical models, including stepwise regression [14,28], Random Forest regression [30,31], and Artificial Neural Networks [32,33] are normally used to derive quantitative relations between urban forest biomass and driving factors. However, the variation of spatial factors and the impact of interactions between spatial factors on the biomass distribution are generally ignored in such studies, even though these issues are of great interest to urban forest managers.

Overall, the primary objective of this study is to explore the spatial heterogeneity and its driving factors of aboveground forest biomass, in order to estimate and detect potential driving factors based on field inventory data in Xi'an, China. Therefore, this study conducted a statistical analysis with a geographic detector regarding the spatial distribution of urban forests' aboveground biomass to quantitatively evaluate the impacts of factors influencing the distribution. Furthermore, due to Xi'an being a representative Chinese city that has undergone rapid urbanization in recent years and that exhibits significant urban forest changes, it was chosen as the focus in this study. This study addresses two main questions: (1) What are the main driving factors strongly influencing the aboveground forest biomass in Xi'an city? (2) How do the interactions between multiple environmental factors influence the aboveground forest biomass in Xi'an city? These results may help government administrators formulate urban greening strategies in the selection of tree species and spatial configuration of urban forests.

2. Materials and Methods

2.1. Study Area

Xi'an is located between 107°40′–109°49′ E and 33°39′–34°45′ N (Figure 1). The south and southeast sides are bounded by the main ridge of the Qinling Mountains, which serve as a natural boundary between the North and South part of China. The western, northwestern, and eastern sides of Xi'an are bounded by the Taibai Mountains, the Weihe River, and the Weihe Mountain, respectively. Xi'an is located in a river valley far from the sea, which makes the summer heat intense, and the cold air often stagnates on the ground in the winter. Xi'an has a continental climate with four distinct seasons—it is warm in spring, hot and humid in summer, cool in fall, and cold and dry in winter. In the urban green spaces, trees are mainly composed of *Sophora japonica*, *Populus sp.*, *Firmiana platanifolia*, *Cypress sp.*, and *Pinus sp*. The shrubs consist of *Ligustrum quihoui*, *Buxus bodinieri*, *Berberis thunbergii var. atropurpurea*, *Buxus megistophylla*, *Photinia serrulata*, and *Pittosporum tobira*, accounting for more than 80% of the total number of shrubs. The grasses include *Poa annua*, *Festuca elata*, *Trifolium repens*, *Lolium perenne*, and *Ophiopogon japonicus*. The population density of Xi'an city is 1185 per km^2 and the impervious coverage percentage is 31.22% [34].

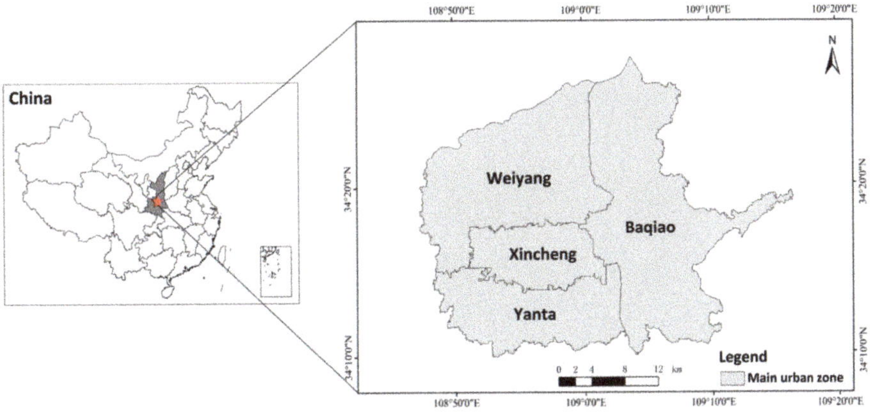

Figure 1. Location of the study area.

2.2. Data Source and Preprocessing

The data used in this study were obtained the Xi'an Urban Forest Resource Survey in 2006, while the field survey was conducted in 2017. In total, there were 1480 plots, covering four administrative districts (Baqiao, Weiyang, Xincheng, and Yanta) in the urban area (Figure 2). Each plot had 20 attributes surveyed in field work, including the forest class, land type, forestland ownership, forest ownership, forest category, authority, protection level, landform, slope, slope position, aspect, origin, dominant tree species, age group, accumulation per hectare, small class accumulation, and area. Of the 20 attributes, five were selected, including the dominant tree species, forest category, land types, age groups, and timber volume, due to these factors being the most relevant to the forest aboveground biomass. The first four attributes were used as potential factors affecting the biomass distribution, and the last one was used in the biomass calculation, which is explained in the following section.

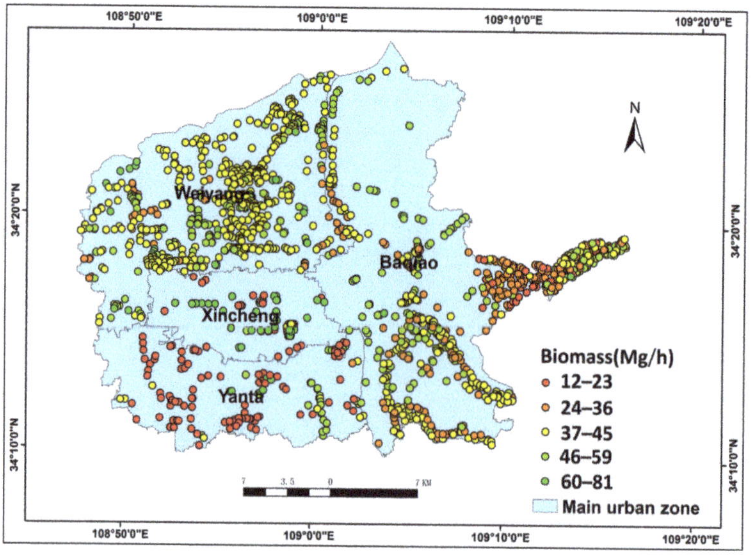

Figure 2. Distribution of biomass grades of 1480 plots.

2.3. Calculation of Aboveground Biomass for Urban Forests

The amount of forest stock comprehensively reflects the site conditions, climatic conditions, forest age, and other forest growth factors. Previous studies have found that the volume can be converted to biomass through a linear regression [35–37] (Equation (1)):

$$B = aV + b, \tag{1}$$

where a and b are model parameter, depending on different tree types, and represent the slope and intercept in the linear regression function, respectively; B is the aboveground biomass, while V is the stock volume. Table 1 summarizes the a and b values for different tree species.

Table 1. Conversion model parameters between the aboveground biomass and stock volume for different tree species [35].

Serial Number	Tree Species	a	b	R^2	Tree Type
1	*Chinese pine*	0.7554	5.0928	0.980	Coniferous tree
2	*Other pine trees*	0.5168	33.2378	0.970	Coniferous tree
3	*Metasequoia glyptostroboides*	0.4158	41.3318	0.980	Coniferous tree
4	*Cypress class*	0.6129	26.1451	0.980	Coniferous tree
5	*Hard broad-leaved*	0.9644	0.8485	0.980	Deciduous tree
6	*Robinia pseudoacacia*	0.7564	8.3103	0.986	Deciduous tree
7	*Poplar class*	0.4754	30.6034	0.930	Deciduous tree
8	*Soft broad-leaved*	0.4754	30.6034	0.930	Deciduous tree
9	*Ginkgo biloba*	0.4158	41.3318	0.980	Deciduous tree

2.4. Spatial Analysis with the Geographical Detector

Geographical detectors (GDs) [38]—selected to study the forest biomass in our research—are widely used to examine geographical phenomena [21,38–43]. This approach can not only evaluate how certain geographical factors impact the spatial variable's distribution, but also reveal the impacts of the interactions between the geographic factors on the spatial variables' distribution.

The basic idea of a GD is to split the study area into subregions according to different categories of geographical factors (GFs). The variances of the dependent variable in each subregion and across the whole study area are compared to derive the impact of geographical factors on the dependent geographical variable's spatial distribution. According to the principle of GD, the forest aboveground biomass, which is calculated by the stock volume, is used as the dependent variable. Moreover, four classes of multiple-level GFs (dominant tree species, forest categories, forestland class, and age groups at the plot level) are used as independent variables, and referred to as geographical factors. Each plot can be categorized into different numbers of subtypes according to different GFs (Table 2). In this study, the analysis focuses on four parts regarding the impacts of GFs on the spatial distribution of aboveground biomass: (a) Investigating whether there is spatial differentiation of biomass in the study area and how much each GF influences biomass; (b) examining the impacts of interactions between GFs; (c) comparing the impacts of different subcategories for each GF; and (d) comparing the impacts between different GFs.

Table 2. Four geographical factors (GFs) and their categories.

GFs	Categories	Number of Categories
Dominant tree species	*Chinese pine, Other pine trees, Metasequoia glyptostroboides, Parker class, Hard broad-leaved, Robinia pseudoacacia, Poplar class, Soft broad-leaved, and Ginkgo biloba*	9
Forest categories	Water conservation forests, Forest for soil and water conservation, Shelter forest for farmland, Protective belt, Shelter belts, Environmental protection forests, Scenic forests, and Historical site forests	8
Forestland types	Coniferous forestland, Broad leaved forestland, Mixed forestland	3
Age groups	Young forest, Half-matured forest, Near-matured forest, Matured forest, Overmatured forest	5

2.4.1. Individual Impacts of GFs on the Spatial Distribution of Aboveground Biomass

To determine the extent of GFs' impacts on the spatial differentiation of aboveground biomass in urban forests, Equation (2) [44] was adopted to calculate q for each GF:

$$q_X = 1 - \frac{\sum_{h=1}^{L_X} N_{h,X}\sigma_{h,X}^2}{N\sigma_{total}^2}, \quad (2)$$

where $h \in (1, 2, 3 \ldots, L_X)$ represents the category index for GF X. The forest categories denote the type of geographical factor. L_X is the number of total categories for GF X (in Table 2), $N_{h,X}$ is the number of plots in category h for geographical factor X, $\sigma_{h,X}^2$ is the variance of biomass for plots in category h of geographical factor X, N is the total number of plots (i.e., 1480 in this study), and σ_{total}^2 is the variance of biomass for all plots.

The range of q_X is [0,1]. A larger q_X value indicates that the variance of the aboveground biomass for plots within a subtype is more diverse between subtypes that are defined by categories of the GF X and vice versa. In extreme cases, a q_X value of 1 indicates that the GF (X) completely controls the spatial distribution of aboveground biomass (Y), and a q_X value of 0 indicates that the GF (X) has no relationship with the aboveground biomass (Y) of the urban trees.

2.4.2. Interaction Impacts of Geographical Factors on the Spatial Distribution of Aboveground Biomass

This study also investigates how the interaction between different GFs influences the spatial distribution of urban trees' aboveground biomass. In other words, we want to reveal whether a given pair of GFs—X_1 and X_2—interact to influence the explanatory power of the aboveground biomass (Y) distribution, or whether the influence of the GFs X_1 and X_2 on aboveground biomass (Y) of the forest are independent.

In this study, the interaction of a given combination of the GFs X_1 and X_2, was written as $X_1 \cap X_2$. Additionally, $q_{X_1 \cap X_2}$ was calculated using Equation (2). The interaction could be classified as one of five groups by comparing $q_{X_1 \cap X_2}$ with the minimum, maximum, and sum of q_{X_1} and q_{X_2} [22].

2.4.3. Comparing the Impacts of Different Categories for Each GF

Given a GF X with two of its subtypes h_1 and h_2, we applied Tukey's Honestly Significant Differences (Tukey's HSD) test to examine whether the average plot's aboveground biomass in subtypes h_1 was significantly different from it in h_2 using Equations (3) and (4):

$$HSD_{0.05}^{(h_1,h_2)} = q_{0.05}(2, n-2)\sqrt{\frac{1}{2}MS_e\left(\frac{1}{r_1}+\frac{1}{r_2}\right)}, \quad (3)$$

$$HSD_{h_1,h_2} = \left|\overline{Y}_{h_1} - \overline{Y}_{h_2}\right|, \quad (4)$$

where n is the total number of plots (i.e., 1480 in this study); $q_{0.05}(2, n-2)$ is the quantile of the Studentized range distribution MS_e stands for the mean sum of squares of deviation within groups in ANOVA; r_1 and r_2 represent the number of plots of subtypes h_1 and h_2, respectively; $\overline{Y}_{h_1}$ and $\overline{Y}_{h_2}$ represent the average aboveground biomass of subtypes h_1 and h_2, respectively. The null hypothesis H_0 for the test is $\overline{Y}_{h_1} = \overline{Y}_{h_2}$. A rejection of H_0 means that there is a significant difference between the average plot aboveground biomass within subregions h_1 and h_2. If $HSD_{h_1,h_2} \le HSD_{0.05}^{(h_1,h_2)}$, H_0 can be accepted, and it is believed that there is no significant difference between the average plot's aboveground biomass within subregions h_1 and h_2.

2.5. Comparing the Impacts for Different GFs

To investigate whether a combination of the two GFs X1 and X2 exhibits significant differences in terms of the spatial distribution of aboveground biomass (Y) in urban forests, a F-statistic was calculated using Equations (5)–(7):

$$F = \frac{N_{X1}(N_{X2}-1)SSW_{X1}}{N_{X2}(N_{X1}-1)SSW_{X2}}, \quad (5)$$

$$SSW_{X1} = \sum_{h=1}^{L1} N_h \sigma_h^2, \quad (6)$$

$$SSW_{X2} = \sum_{h=1}^{L2} N_h \sigma_h^2, \quad (7)$$

where N_{X1} and N_{X2} represent the sample sizes of X1 and X2, respectively; SSW_{X1} and SSW_{X2} represent the sum of the intralayer variances of the layers formed by X1 and X2, respectively; and L1 and L2 represent the number of layers defined by X1 and X2, respectively. The null hypothesis of the F-test is H_0: $SSW_{X1} = SSW_{X2}$. If H_0 is rejected at the level of significance of α, it indicates that X1 and X2 display significant differences in relation to the spatial distribution of aboveground biomass (Y) in urban forests.

3. Results

3.1. The Distributions of Urban Forest Biomass and Its Influencing Factors

The biomass of 1480 plots shows significant spatial differences (Figure 2). The biomass distribution of plots reflects that the urban forests are mainly distributed in the northwestern, southeastern and the eastern part of Xi'an. The biomass in the northwestern part (Weiyang), with 611 plots and 44.77% of the total forest biomass, primarily consists of the urban garden and protected area. The biomass in the southeast and east exhibits a highly positive relationship with rivers. The highest biomass can be observed in the central area of Xi'an (Xincheng) city, with 83 plots and 6.89% of the total forest biomass, and with the average biomass reaching to 59.25 Mg/h. The biomass in the southern part of Xi'an (Yanta), with 143 plots and 6.64% of the total forest biomass, is the lowest (lower than 22.63 Mg/h). This is because Yanta is a newly developing urban area, and the trees there are almost young forest trees. The northwestern part of Xi'an has medium level of biomass. The eastern part of Xi'an (Baqiao), with 643 plots and 41.7% of the total forest biomass, exhibits a relatively lower biomass than southern Xi'an.

Four influencing factors, including the dominant tree species, forest categories, forestland types, and age groups, present spatial heterogeneity (Figure 3). The dominant tree species which are distributed with a patch pattern are mostly located along the road and in the urban garden (Figure 3a). *Pinus* and hardwood forests are mainly distributed in Yanta District. *Populus* is distributed in Weiyang and Baqiao District. *Platycladus orientalis* is distributed in the south of Baqiao District for the most part. Most of the hardwood trees are found in the west of Yanta District. *Robinia pseudoacacia* is commonly found in eastern Baqiao District. The softwood trees display a significant positive relationship with rivers. *Ginkgo biloba* is mainly distributed in the middle of the south of Yanta District, in a small area.

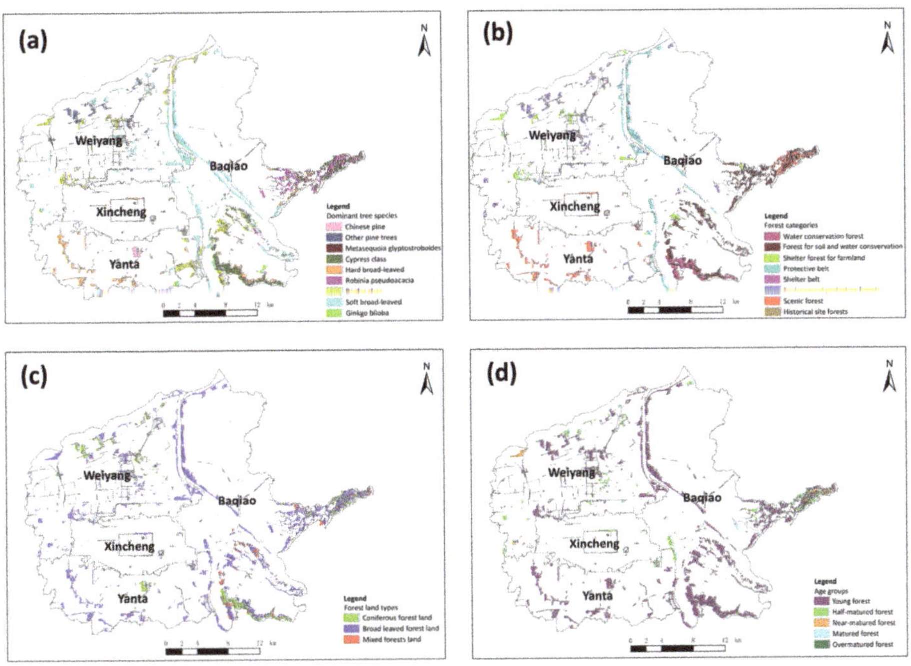

Figure 3. The whole study area was split into different subtypes according to GFs: (**a**) Dominant tree species; (**b**) forest categories; (**c**) forestland types; and (**d**) age groups.

Figure 3b shows the distribution of forest categories. Forests for water conservation are mainly distributed in the south of Baqiao District. Forests for soil conservation are mainly found in the south and east of Baqiao District. Forests for protecting farms are mainly located in the south of Weiyang and Baqiao Districts, with a small area. Forests for shore protection are distributed on both sides of most rivers. Forests for protecting the environment are situated in Weiyang District, while the landscape forests are mainly distributed in Yanta District. Other types of forests exhibit a sporadic distribution, with a small area.

Figure 3c shows the distribution of land types. The needleleaf forestland is mainly distributed in the south of Weiyang and Baqiao Districts. The broadleaf forestland has the largest area and is found everywhere in the study area. The mingled forestland is mainly located in the south and east of Baqiao District. Figure 3d shows the age distribution. Most forests are young in age, while mature and overmatured forests are scarce in the four districts of Xi'an.

3.2. Detecting the Contribution of the Four Influencing Factors

The independent q values of the four influencing factors ranged from 8% to 59% (Table 3). The results of Equation (2) showed that the contribution of each impact factor towards the differentiation of the spatial distribution of aboveground biomass is ordered as follows: Dominant tree species, age group, forest category, and land type. The first two factors (with q value > 0.20) are considered to be the major impact factors.

Table 3. The independent q values of the four GFs.

Dominant Tree Species	Age Group	Forest Category	Land Type
0.595	0.202	0.087	0.076

Ecological detectors can reflect significant differences among the four GFs regarding their impacts on the biomass of forests. As shown in Table 4 (generated by the F-test with Equation (5)), the forest age is significantly different from the other factors. The forestland types only differ from the dominant species, and show no difference from the forest types. The forest type displays a significant difference when compared to the dominant tree species, but shows no significant difference with the forest tree species. The forest tree species is significantly different from the dominant tree species.

Table 4. Significant differences in reflecting forest aboveground biomass among influencing factors.

	Dominant Tree Species	Forest Category	Forestland Type	Age Group
Dominant tree species	-	Y	Y	Y
Forest category	Y	-	N	Y
Forestland type	Y	N	-	Y
Age group	Y	Y	Y	-

Note: Y means the null hypothesis is rejected at a significance level of 0.05, while N means no significant difference between the average plot's biomass.

3.3. Detecting the Contribution of Interactions between the Four Influencing Factors

In the forest environment, the forest aboveground biomass is the result of a combination of multiple factors, and is also influenced by interactions between these factors. The spatial distribution of aboveground biomass in urban forests is always affected by various factors, as well as their interactions with each other, but not by single factors. According to Table 5, our results (Table 6) show that the interaction between GFs mainly involves nonlinear enhancement, indicating that the interaction between GFs' impact is larger than the simple combination of individual factors.

Table 5. Interaction derivation [9].

Comparison Type	Interaction
$q_{X_1 \cap X_2} < \min(q_{X_1}, q_{X_2})$	Weaken, nonlinear
$\min(q_{X_1}, q_{X_2}) < q_{X_1 \cap X_2} < \max(q_{X_1}, q_{X_2})$	Weaken, single factor nonlinear
$q_{X_1 \cap X_2} > \max(q_{X_1}, q_{X_2})$	Enhance, bilinear
$q_{X_1 \cap X_2} = q_{X_1} + q_{X_2}$	Independent
$q_{X_1 \cap X_2} > q_{X_1} + q_{X_2}$	Enhance, nonlinear

Table 6. Comparison of interactions between factors pairs.

Factor Interaction (A)	Factor Combination (B+C)	Comparative Result	Ratio (Interaction/Combination)	Explanation
dominant tree species ∩ forest category = 0.784	dominant tree species (0.595) + forest category (0.087)	A > B+C	1.15	Non-Linear Enhancement
dominant tree species ∩ land types = 0.604	dominant tree species (0.595), land types (0.076)	A > max (B, C)	1.02	Bilinear, Enhancement
dominant tree species ∩ age groups = 0.847	dominant tree species (0.595) + age groups (0.202)	A > B+C	1.06	Non-Linear Enhancement
forest category ∩ land types = 0.269	forest category (0.087) + land types (0.076)	A > B+C	1.65	Non-Linear Enhancement
forest category ∩ age groups = 0.445	forest category (0.087) + age groups (0.202)	A > B+C	1.54	Non-Linear Enhancement
land types ∩ age groups = 0.348	forest category (0.076) + age groups (0.202)	A > B+C	1.25	Non-Linear Enhancement

To quantify the synergistic effects, we combined the ratios of interactions and the combined effect was calculated. A larger ratio value means that stronger synergistic effects exist between GFs. Among all the pairs of GFs, the synergistic effects between the forest category and land types are greater than the rest of the pairs, and show the highest ratio value (1.65). Furthermore, the ratio of the dominant tree species and land type exhibits the weakest synergistic effects.

3.4. Comparing the Difference of the Contribution among Subtypes

The pairwise comparison results, using Tukey's Honestly Significant Differences test for the forestland type (Table 7), show that the average plot's biomass in coniferous forestland was significantly higher than that in broad-leaved forestland and mixed forestland. Furthermore, there was no significant difference between mixed forestland and broad-leaved forestland regarding the average plot's biomass.

Table 7. Tukey's Honestly Significant Differences (Tukey's HSD) test for comparing average plot's biomass for forestland types.

	Coniferous Forestland	Broadleaved Forestland	Mixed Forestland	Average Plot Biomass (Mg/h)
Coniferous forestland	-	Y	Y	46.5
Broad leaved forestland	Y	-	N	38.0
Mixed forestland	Y	N	-	38.2
Average plot biomass	46.5	38.0	38.2	-

Note: Y means the null hypothesis is rejected at a significance level of 0.05, while N means no significant difference between the average plot's biomass.

The Tukey's HSD test, comparing the average plot's biomass for different tree species shows that the plot dominated by *Ginkgo biloba* had a significantly higher average plot's biomass than other species, expect for the Poplar class and Parker class (Appendix A Table A1). These species are the major greening tree species in green spaces in Xi'an city [45]. They exhibit a considerable tolerance for gaseous air pollutants, but are susceptible to damage from acid rain [46–48]. Due to the "Coal to Gas Project" implemented in 1997 [49], the emergence rate of acid rain has obviously decreased [50], providing favorable growth conditions for these species, rather than *Pinus tabuliformis* and *Robinia pseudoacacia*.

A comparison of the average plot's biomass among the eight subtypes defined by forest functionalities showed that the difference between these types is generally not as significant as those between subtypes defined by dominant tree species (Appendix A Table A2). Among the eight subtypes, even though historical site forests retain the largest average plot's biomass, they only displayed significant differences from forest for soil and water conservation and scenic forests. With the lowest mean value of the plot's biomass, scenic forest displayed a significant difference from all other forests, except for the water conservation forest and forest for soil and water conservation.

The investigation of the age group factors shows that all subtypes split by forest age group are significantly different from each other, regarding the average plot's biomass in the subtypes (Appendix A Table A3). If GD is used as a tool to detect the overall picture of impacts for all the GFs, then Tukey's HSD test can be thought of as a magnifier, showing details of how the elements within each GF exerting impacts.

4. Discussion

4.1. The Significance of Studying the Spatial Heterogeneity of Urban Forest Biomass

In this study, we analyzed the spatial heterogeneity of urban forest's aboveground biomass and can conclude that the dominant tree species and age group are the main factors impacting the biomass distribution. These results are consistent with previous studies [51,52]. Detecting the drivers of urban forest biomass is important for the urban forest management. Among the four main drivers, we found that the tree species is the most critical factor affecting the urban forests' aboveground biomass. This result agrees with Shuaifeng Li. et al. [53], who reported that the species richness had a positive impact on aboveground biomass across all forest vegetation layers. This result means that the choice of planted tree species could determine the pattern of urban forest. Therefore, trees with fast growth rates should be considered first. This study also indicates that the interaction effect of two factors is greater than that of a single one, which is also reflected in the nonlinear relation model in urban forest modeling [54]. The results of interactions mean that we should not only focus on the independent role of single driving factors, but also pay more attention to their interaction, which may greatly improve the productivity of urban forests.

Investigating how different GFs drive the distribution of urban forests' aboveground biomass could provide important implications for better urban planning, which responds to urban atmosphere changes and the development of sustainable urbanization. As an important carrier of the urban ecosystem, urban forests offer ecological, economic, and social benefits for human beings. They can not only improve the urban microclimate, alleviate the effects of urban heat islands, increase surface runoff, and play an important role in maintaining the urban carbon and oxygen balance, but also improve the quality of life of residents and provide good places of leisure and entertainment for urban residents [55,56]. Urban forest biomass is an important indicator that can be used to measure the carbon storage, carbon sequestration capacity, and ecological benefits of an urban ecosystem [57]. The accurate and rapid monitoring of urban forest biomass and its spatial pattern are the basis for urban carbon cycle and energy flow research, while they are also the basis for measuring the ecological regulation and environmental protection capacity of urban forests [58]. Analyzing the spatial differentiation of urban forest biomass can provide data for the urban green space planning department, and has great significance for urban ecological space planning and management.

4.2. Challenges and Future Directions

Forest biomass is affected by several variables, including human activities, as well as environmental and biological factors [59]. It shows a certain randomness and distribution with structural differences. The spatial heterogeneity of forest biomass reflects the energy flows and material cycles of forest ecosystems [60,61]. The study area is a plain, and its internal environmental factors (such as its topography and climate) can be considered to be uniform. Based on these conditions, forest resource survey data of the study area were employed, while four qualitative factors (land type, forest category, age group, and dominant tree species) were selected to study the spatial heterogeneity and the influencing factors of urban forest biomass by using the geographic detector method. This approach obtains a quantitative description of qualitative influencing factors and solves the problem of collinearity that has often been ignored in past related research. However, the following aspects should be considered in future related research: (1) In addition to the four factors mentioned above, there are many factors, (i.e., average tree species, average DBH, and human activities) that required further comprehensive analysis in the future; (2) in this study, the spatial heterogeneity of aboveground forest biomass is mainly discussed. However, the biomass of shrubs, herbs, and underground parts of the forest was not considered; and (3) in this study, calculation of the biomass of forestland was obtained from forest resource investigation data. In future related research, using remote sensing technology to retrieve biomass directly is recommended. Therefore, we could quickly analyze the spatial heterogeneity of forest biomass [62].

5. Conclusions

In this study, we conducted spatial statistical analysis by the GD method to systematically study the differentiation of the urban forest biomass distribution of Xi'an. Additionally, we examined how dominant tree species, age groups, forest categories, and forestland types individually and interactively impacted the urban forest biomass distribution. We concluded that: (1) among the four GFs, including dominant tree species, forest species, land types, and age groups, the spatial distribution of aboveground forest biomass in Xi'an is primarily influenced by dominant tree species and forest age. Their combined effects account for 80% of the total impacts; (2) there is no significant difference between forestland and forest categories regarding their impacts on the spatial distribution of aboveground biomass; and (3) all of the pairs of the four GFs have nonlinear enhancement effects, except for the bilinear enhancement effect between dominant tree species and the land type. Among all pairs of GFs, the synergistic effect is most obvious for the interaction between the forest category and land type. Overall, the results of urban forest biomass' spatial heterogeneity among these GFs can help researchers' understanding of urban forest biomass change, which may be applied in future precise forest prediction models on a larger scale and allow for more effective forest management strategies to be developed.

Author Contributions: Conceptualization, Xuan Zhao; data curation, Xuan Zhao; formal analysis, Xuan Zhao; funding acquisition, Jianjun Liu; investigation, Hongke Hao; methodology, Xuan Zhao; project administration, Jianjun Liu; resources, Yanzheng Yang. All authors have read and agreed to the published version of the manuscript.

Funding: This research was primarily funded by Research on Vegetation Restoration Techniques on Steep Loess Slope in Qianyang, Shanxi, grant number K303021613.

Acknowledgments: We thank the funding of "Research on Vegetation Restoration Techniques on Steep Slope of Loess in Qianyang, Shaanxi".

Conflicts of Interest: The authors declare no conflict of interest.

Appendix A

Table A1. Tukey's Honestly Significant Differences (Tukey's HSD) test for the dominant tree species factor (Y means the testing is significant at the 0.05 level, and N represents an insignificant difference).

	Chinese Pine	*Other Pine Trees*	*Metasequoia Glyptostroboides*	*Parker Class*	*Hard Broad-Leaved*	*Robinia Pseudoacacia*	*Poplar Class*	*Soft Broad-Leaved*	*Ginkgo Biloba*	
Chinese pine	-	Y	-	Y	N	Y	Y	Y	Y	18.4
Other pine trees	Y	-	-	Y	Y	Y	Y	Y	N	49.9
Metasequoia glyptostroboides	-	-	-	-	-	-	-	-	-	58.6
Parker class	Y	Y	-	-	Y	Y	Y	Y	Y	44.0
Hard broad-leaved	N	Y	-	Y	-	Y	Y	Y	Y	19.6
Robinia pseudoacacia	Y	Y	-	Y	Y	-	Y	Y	Y	32.9
Poplar class	Y	Y	-	Y	Y	Y	-	Y	N	47.3
Soft broad-leaved	Y	Y	-	Y	Y	Y	Y	-	Y	38.6
Ginkgo biloba	Y	N	-	Y	Y	Y	N	Y	-	56.3
	18.4	49.9	58.6	44.0	19.6	32.9	47.3	38.6	56.3	

Table A2. Tukey's HSD test for the forest category factor (Y means the testing is significant at the 0.05 level, and N represents an insignificant difference).

	Water Conservation Forest	Forest for Soil and Water Conservation	Shelter Forest for Farmland	Protective Belt	Shelter Belt	Environmental Protection Forests	Scenic Forest	Historical Sites Forests	
Water conservation forest	-	N	N	N	N	N	N	N	38.6
Forest for soil and water conservation	N	-	Y	Y	N	Y	N	Y	36.3
Shelter forest for farmland	N	Y	-	N	N	N	Y	N	43.9
Protective belt	N	Y	N	-	N	N	Y	N	41.5
Shelter belt	N	N	N	N	-	N	N	N	41.5
Environmental protection forests	N	Y	N	N	N	-	Y	N	43.3
Scenic forest	N	N	Y	Y	N	Y	-	Y	35.7
Historical sites Forests	N	Y	N	N	N	N	Y	-	64.4
	38.6	36.3	43.9	41.5	41.5	43.3	35.7	64.4	

Table A3. Tukey's HSD test for age group factor (Y means the testing is significant at the 0.05 level, and N represents an insignificant difference).

	Young Forest	Half-Mature Forest	Near-Mature Forest	Mature Forest	Overmature Forest	
Young forest	-	Y	Y	Y	-	37.5
Half-mature forest	Y	-	Y	Y	-	43.1
Near-mature forest	Y	Y	-	Y	-	54.2
Mature forest	Y	Y	Y	-	-	78.6
Overmature forest	-	-	-	-	-	55.7
	37.5	43.1	54.2	78.6	55.7	

References

1. United Nations. *World Urbanization Prospects: The 2018 Revision*; United Nations Publications: New York, NY, USA, 2019.
2. Collier, P.; Venables, A.J. Urbanization in developing economies: The assessment. *Oxf. Rev. Econ. Policy* **2017**, *33*, 355–372. [CrossRef]
3. World Urbanization Prospects. Available online: https://esa.un.org/unpd/wup/publications/fles/wup2014-highlights.Pdf (accessed on 11 November 2014).
4. China Statistical Yearbook. Available online: http://www.stats.gov.cn/tjsj/ndsj/2018/indexch.htm (accessed on 29 November 2018).
5. Zhang, Z.B. *Report on the Healthy Development of China's New Urbanization (2016)*; Social Science Literature Press: Beijing, China, 2017.
6. Orum, A.M.; Iossifova, D. East Asian Urbanization. In *The Wiley Blackwell Encyclopedia of Urban and Regional Studies*; Orum, A.M., Ed.; John Wiley & Sons: Hoboken, NJ, USA, 2019. [CrossRef]
7. Bolund, P.; Hunhammar, S. Ecosystem services in urban areas. *Ecol. Econ.* **1999**, *29*, 293–301. [CrossRef]
8. Nowak, D.J.; Greenfield, E.J.; Hoehn, R.E.; Lapoint, E. Carbon storage and sequestration by trees in urban and community areas of the United States. *Environ. Pollut.* **2013**, *178*, 229–236. [CrossRef] [PubMed]
9. Timilsina, N.; Staudhammer, C.L.; Escobedo, F.J.; Lawrence, A. Tree biomass, wood waste yield, and carbon storage changes in an urban forest. *Landsc. Urban Plan.* **2014**, *127*, 18–27. [CrossRef]
10. Abdi, R.; Endreny, T.; Nowak, D. A model to integrate urban river thermal cooling in river restoration. *J. Environ. Manag.* **2020**, *258*, 110023. [CrossRef] [PubMed]
11. Elmqvist, T.; Fragkias, M.; Goodness, J.; Güneralp, B.; Marcotullio, P.J.; McDonald, R.I.; Parnell, S.; Schewenius, M.; Sendstad, M.; Seto, K.C.; et al. *Urbanization, Biodiversity and Ecosystem Services: Challenges and Opportunities: A Global Assessment*; Springer: Dordrecht, The Netherlands, 2013. [CrossRef]

12. Nowak, D.J.; Daniel, E.C. Carbon storage and sequestration by urban trees in the USA. *Environ. Pollut.* **2002**, *116*, 381–389. [CrossRef]
13. Nowak, D.J. Atmospheric carbon dioxide reduction by Chicago's urban forest. In *Chicago's Urban Forest Ecosystem: Results of the Chicago Urban Forest Climate Project*; Forest Service US: Washington, DC, USA, 1994; pp. 83–94. Available online: https://www.fs.usda.gov/treesearch/pubs/4285 (accessed on 11 December 2020).
14. Li, L.; Zhou, X.; Chen, L.; Chen, L.; Zhang, Y.; Liu, Y. Estimating urban vegetation biomass from Sentinel-2A image data. *Forests* **2020**, *11*, 125. [CrossRef]
15. Hu, S.; Chen, L.; Li, L.; Zhang, T.; Yuan, L.; Cheng, L.; Wang, J.; Wen, M. Simulation of Land Use Change and Ecosystem Service Value Dynamics under Ecological Constraints in Anhui Province, China. *Int. J. Environ. Res. Public Health* **2020**, *17*, 4228. [CrossRef]
16. McPherson, E.G. Atmospheric carbon dioxide reduction by Sacramento's urban forest. *J. Arboric.* **1998**, *24*, 215–223.
17. Aguaron, E.; McPherson, E.G. Comparison of methods for estimating carbon dioxide storage by Sacramento's urban forest. In *Carbon Sequestration in Urban Ecosystems*; Springer: Dordrecht, The Netherlands, 2012; pp. 43–71. [CrossRef]
18. Rowntree, R.A.; Nowak, D.J. Quantifying the role of urban forests in removing atmospheric carbon dioxide. *J. Arboric.* **1991**, *17*, 269–275.
19. Dutilleul, P.R.L. *Spatio-Temporal Heterogeneity: Concepts and Analyses*; Cambridge University Press: Cambridge, UK, 2011.
20. Wang, J.F.; Zhang, T.L.; Fu, B.J. A measure of spatial stratified heterogeneity. *Ecol. Indic.* **2016**, *67*, 250–256. [CrossRef]
21. Wang, J.F.; Li, X.H.; Christakos, G.; Liao, Y.L.; Zhang, T.; Gu, X.; Zheng, X.Y. Geographical detectors-based health risk assessment and its application in the neural tube defects study of the Heshun region, China. *Int. J. Geogr. Inf. Sci.* **2010**, *24*, 107–127. [CrossRef]
22. Wang, J.F.; Xu, C.D. Geodetector: Principle and prospective. *Acta Geogr. Sin.* **2017**, *72*, 116–134.
23. Li, M.; Im, J.; Quackenbush, L.J.; Liu, T. Forest biomass and carbon stock quantification using airborne LiDAR data: A case study over Huntington Wildlife Forest in the Adirondack Park. *IEEE J. Sel. Top. Appl. Earth Obs. Remote Sens.* **2014**, *7*, 3143–3156. [CrossRef]
24. Gleason, C.J.; Im, J. Forest biomass estimation from airborne LiDAR data using machine learning approaches. *Remote Sens. Environ.* **2012**, *125*, 80–91. [CrossRef]
25. Rauf, A. Distribution, above-ground biomass and carbon stock of the vegetation in Taman Beringin Urban Forest, Medan City, North Sumatra, Indonesia. *Malays. For.* **2017**, *80*, 73–84.
26. Pesola, L.; Cheng, X.; Sanesi, G.; Colangelo, G.; Elia, M.; Lafortezza, R. Linking above-ground biomass and biodiversity to stand development in urban forest areas: A case study in Northern Italy. *Landsc. Urban Plan.* **2017**, *157*, 90–97. [CrossRef]
27. Shen, G.; Wang, Z.; Liu, C.; Han, Y. Mapping aboveground biomass and carbon in Shanghai's urban forest using Landsat ETM+ and inventory data. *Urban For. Urban Greening* **2020**, *51*, 126655. [CrossRef]
28. Baker, T.R.; Phillips, O.L.; Malhi, Y.; Almeida, S.; Arroyo, L.; di Fiore, A.; Erwin, T.; Killeen, T.J.; Laurance, S.G.; Laurance, W.F.; et al. Variation in wood density determines spatial patterns in Amazonian forest biomass. *Glob. Chang. Biol.* **2004**, *10*, 545–562. [CrossRef]
29. López-Serrano, P.M.; Corral-Rivas, J.J.; Díaz-Varela, R.A.; Álvarez-González, J.G.; López-Sánchez, C.A. Evaluation of radiometric and atmospheric correction algorithms for aboveground forest biomass estimation using Landsat 5 TM data. *Remote Sens.* **2016**, *8*, 369. [CrossRef]
30. Luo, S.; Wang, C.; Xi, X.; Nie, S.; Fan, X.; Chen, H.; Ma, D.; Liu, J.; Zou, J.; Lin, Y.; et al. Estimating forest aboveground biomass using small-footprint full-waveform airborne LiDAR data. *Int. J. Appl. Earth Obs. Geoinf.* **2019**, *83*, 101922. [CrossRef]
31. Fassnacht, F.E.; Hartig, F.; Latifi, H.; Berger, C.; Hernández, J.; Corvalán, P.; Koch, B. Importance of sample size, data type and prediction method for remote sensing-based estimations of aboveground forest biomass. *Remote Sens. Environ.* **2014**, *154*, 102–114. [CrossRef]
32. Foody, G.M.; Cutler, M.E.; McMorrow, J.; Pelz, D.; Tangki, H.; Boyd, D.S.; Douglas, I.A.N. Mapping the biomass of Bornean tropical rain forest from remotely sensed data. *Glob. Ecol. Biogeogr.* **2001**, *10*, 379–387. [CrossRef]
33. Kimes, D.S.; Nelson, R.F.; Manry, M.T.; Fung, A.K. Attributes of neural networks for extracting continuous vegetation variables from optical and radar measurements. *Int. J. Remote Sens.* **1998**, *19*, 2639–2663. [CrossRef]

34. Xi'an 2015 National Economic and Social Development Statistical Bulletin. Available online: www.xatj.gov.cn (accessed on 12 December 2016).
35. Fang, J.; Liu, G.; Xu, S. Biomass and net production of forest vegetation in China. *Acta Ecol. Sin.* **1996**, *16*, 497–508. (In Chinese)
36. Fang, J.Y.; Wang, G.G.; Liu, G.H.; Xu, S.L. Forest biomass of China: An estimate based on the biomass—Volume relationship. *Ecol. Appl.* **1998**, *8*, 1084–1091.
37. Fang, J.; Chen, A.; Peng, C.; Zhao, S.; Ci, L. Changes in forest biomass carbon storage in China between 1949 and 1998. *Science* **2001**, *292*, 2320–2322. [CrossRef]
38. Luo, W.; Jasiewicz, J.; Stepinski, T.; Wang, J.; Xu, C.; Cang, X. Spatial association between dissection density and environmental factors over the entire conterminous United States. *Geophys. Res. Lett.* **2016**, *43*, 692–700. [CrossRef]
39. Liao, Y.; Wang, J.; Wu, J.; Driskell, L.; Wang, W.; Zhang, T.; Xue, G.; Zheng, X. Spatial analysis of neural tube defects in a rural coal mining area. *Int. J. Environ. Health Res.* **2010**, *20*, 439–450. [CrossRef]
40. Hu, Y.; Wang, J.; Li, X.; Ren, D.; Zhu, J. Geographical detector-based risk assessment of the under-five mortality in the 2008 Wenchuan earthquake, China. *PLoS ONE* **2011**, *6*, e21427. [CrossRef]
41. Zou, B.; Wilson, J.G.; Zhan, F.B.; Zeng, Y.; Wu, K. Spatial-temporal variations in regional ambient sulfur dioxide concentration and source-contribution analysis: A dispersion modeling approach. *Atmos. Environ.* **2011**, *45*, 4977–4985. [CrossRef]
42. Wang, J.F.; Hu, Y. Environmental health risk detection with GeogDetector. *Environ. Model. Softw.* **2012**, *33*, 114–115. [CrossRef]
43. Zhu, Z.; Wang, J.; Hu, M.; Jia, L. Geographical detection of groundwater pollution vulnerability and hazard in karst areas of Guangxi Province, China. *Environ. Pollut.* **2019**, *245*, 627–633. [CrossRef] [PubMed]
44. Wang, F.; Liao, L.; Liu, X. *Spatial Analysis Tutorial*; Science Press: Beijing, China, 2010.
45. Yao, Z.; Liu, J.; Zhao, X.; Long, D.; Wang, L. Spatial dynamics of aboveground carbon stock in urban green space: A case study of Xi'an, China. *J. Arid. Land* **2015**, *7*, 350–360. [CrossRef]
46. Kim, Y.S.; Lee, J.K.; Chung, G.C. Tolerance and susceptibility of Ginkgo to air pollution. In *Ginkgo Biloba A Global Treasure*; Springer: Tokyo, Japan, 1997; pp. 233–242. [CrossRef]
47. Matyssek, R.; Giinthardt-Goerg, M.S.; Schmutz, P.; Saurer, M.; Landolt, W.; Bücher, J.B. Response mechanisms of birch and poplar to air pollutants. *J. Sustain. For.* **1997**, *6*, 3–22. [CrossRef]
48. Barwise, Y.; Prashant, K. Designing vegetation barriers for urban air pollution abatement: A practical review for appropriate plant species selection. *npj Clim. Atmos. Sci.* **2020**, *3*, 1–19. [CrossRef]
49. Miao, J.L. Analysis of climatic characteristics and meteorological conditions of acid rain in Xi'an. *Shaanxi Meteorol.* **2013**, 36–39. (In Chinese) [CrossRef]
50. Xi'an Environmental Status Bulletin. 2017. Available online: http://xaepb.xa.gov.cn/xxgk/hjzkgb/hjzkgb/5d8b5a9cf99d65052290af21.html (accessed on 18 December 2018).
51. Cheng, Y.; Zhang, C.; Zhao, X.; von Gadow, K. Biomass-dominant species shape the productivity-diversity relationship in two temperate forests. *Ann. For. Sci.* **2018**, *75*, 1–9. [CrossRef]
52. Chen, Z.; Yu, G.; Wang, Q. Effects of climate and forest age on the ecosystem carbon exchange of afforestation. *J. For. Res.* **2020**, *31*, 365–374. [CrossRef]
53. Li, S.; Su, J.; Lang, X.; Liu, W.; Ou, G. Positive relationship between species richness and aboveground biomass across forest strata in a primary Pinus kesiya forest. *Sci. Rep.* **2018**, *8*, 1–9. [CrossRef]
54. Zhao, K.; Popescu, S.; Nelson, R. Lidar remote sensing of forest biomass: A scale-invariant estimation approach using airborne lasers. *Remote Sens. Environ.* **2009**, *113*, 182–196. [CrossRef]
55. Singh, K.K.; Gagné, S.A.; Meentemeyer, R.K. Urban Forests and Human Well-Being. *Compr. Remote Sens.* **2018**, 287–305. [CrossRef]
56. Christine, B.; Katrin, R. The role of urban green space for human well-being. *Ecol. Econ.* **2015**, *120*, 139–152.
57. Kang, M.N.; Shawn, L. Aboveground biomass estimation of tropical street trees. *J. Urban Ecol.* **2018**, *4*, 1–6.
58. Wang, Z.; Shen, G.; Zhu, Y.; Liu, C. Spatiotemporai dynamics of urban forest biomass in Shanghai, China. In Proceedings of the 2015 Fourth International Conference on Agro-geoinformatics, Istanbul, Turkey, 20–24 July 2015.
59. Zhang, H.; Song, T.; Wang, K.; Yang, H.; Yue, Y.; Zeng, Z.; Peng, W.; Zeng, F. Influences of stand characteristics and environmental factors on forest biomass and root–shoot allocation in southwest China. *Ecol. Eng.* **2016**, *91*, 7–15. [CrossRef]

60. Edna, R.; Matthias, C.; Jens, H.; Anja, R.; Andreas, H. Spatial heterogeneity of biomass and forest structure of the Amazon rain forest: Linking remote sensing, forest modelling and field inventory. *Glob. Ecol. Biogeogr.* **2017**, *26*, 1–11.
61. Röser, D.; Asikainen, A.; Raulund-Rasmussen, K.; Stupak, I. *Sustainable Use of Forest Biomass for Energy;* Managing Forest Ecosystems: New York, NY, USA, 2008. [CrossRef]
62. Zhang, X.; Ni-meister, W. Remote Sensing of Forest Biomass. *Biophys. Appl. Satell. Remote Sens.* **2013**, 63–98. [CrossRef]

Publisher's Note: MDPI stays neutral with regard to jurisdictional claims in published maps and institutional affiliations.

isprs International Journal of *Geo-Information*

MDPI

Article

The Land-Use Change Dynamics Based on the CORINE Data in the Period 1990–2018 in the European Archipelagos of the Macaronesia Region: Azores, Canary Islands, and Madeira

Rui Alexandre Castanho [1,*], José Manuel Naranjo Gomez [2,3,4], Ana Vulevic [2,5] and Gualter Couto [6]

1 Faculty of Applied Sciences, WSB University, 41-300 Dabrowa Górnicza, Poland
2 VALORIZA—Research Centre for Endogenous Resource Valorization, 7300 Portalegre, Portugal; jnaranjo@unex.es (J.M.N.G.); vulevica@sicip.co.rs (A.V.)
3 Agricultural School, University of Extremadura, 06007 Badajoz, Spain
4 CITUR-Madeira—Centre for Tourism Research, Development and Innovation, 9000-082 Funchal-Madeira, Portugal
5 Department of Urban Planning and Architecture, Institute of Transportation—CIP, 11000 Belgrade, Serbia
6 School of Business and Economics and CEEAplA, University of Azores, 9500-321 Ponta Delgada, Portugal; gualter.mm.couto@uac.pt
* Correspondence: acastanho@wsb.edu.pl; Tel.: +351-912-494-673

Abstract: Islands as peripheral and ultra-peripheral are typically highlighted as ecologically sensitive areas to human activities due to the tremendous biological diversity of beings and the future possibility of habitat loss. In this regard, the comprehension of the land occupation dynamics and trends in the ultra-peripheral territories is crucial to attempt long-lasting regional sustainability, as is the island region's case. Therefore, the present article aims to analyze the trends and dynamics of the land-use changes on the European Archipelagos of the Macaronesia Region over the last three decades, using the CORINE (Coordination of Information on the Environment) data. Some of the obtained results show that about 3.4% of the Azores' surface is characterized mainly by discontinuous urban fabric, representing 67% of the total urban fabric of the Azores over the last thirty years. Additionally, in Madeira Archipelago, the land is mainly occupied by forest and semi-natural areas, representing almost three-thirds of the territory. A similar scenario is verified in the Canary Islands, where forests and semi-natural areas represent approximately three-quarters of the territory. Once more, this study shows the relevance of the island areas' unique character, which should be preserved and protected. Therefore, the priorities must be defined and established management strategies that are significant for the well-being of these highly valued areas. Moreover, the study showed that notable changes had occurred in the period 1990–2018 in this landscape. Hence there is a need for appropriate measures to mitigate these negative impacts on the environment.

Keywords: geographic information systems; land cover; land dynamics; regional studies; sustainable planning; ultra-peripheral territories

Citation: Castanho, R.A.; Naranjo Gomez, J.M.; Vulevic, A.; Couto, G. The Land-Use Change Dynamics Based on the CORINE Data in the Period 1990–2018 in the European Archipelagos of the Macaronesia Region: Azores, Canary Islands, and Madeira. *ISPRS Int. J. Geo-Inf.* **2021**, *10*, 342. https://doi.org/10.3390/ijgi10050342

Academic Editors: Wolfgang Kainz, Weicheng Wu, Yalan Liu and Mingxing Hu

Received: 30 March 2021
Accepted: 14 May 2021
Published: 17 May 2021

1. Introduction

Nowadays, our societies face several challenges. Among these obstacles and requirements, we have regional planning, which is an essential requirement for the so-desired sustainable development [1–9]. In fact, such challenges are even more evident in ultra-peripheral territories [10–13]. Accordingly, the comprehension of the land occupation dynamics and trends in the ultraperipheral territories is crucial to attempt long-lasting regional sustainability, as is the island region's case.

In this regard, the European Archipelagos of the Macaronesia Region (the Azores, Canary Islands, and Madeira) were selected as case studies. Consequently, the land-use changes over the last three decades were analyzed. The Azores and Madeira belong to Portugal, whereas the Canaries belong to Spain.

Changes in land-use can be associated with economic growth and investors' search for profitable locations for the implementation of their projects, supported by local authorities to improve the economic status of local self-government [4]. This is especially harmful to areas of high environmental quality. Exceptionally high ecological values characterize these islands, so it is necessary and vital to explore the fragmentation of the landscape as monitoring the increase of anthropic pressure, especially in the vicinity of highly urbanized areas.

Fragmentation in landscape patterns can compromise its functional integrity by disrupting critical ecological processes to maintain biodiversity and ecosystem health [14]. Many anthropogenic activities (e.g., development, logging) can disrupt the structural integrity of the landscape and can sometimes facilitate ecological flows (e.g., animal movement) across the landscape [15]. Space fragmentation is a phenomenon studied by the scientific community, particularly environmentalists and urban planners (see [16–19]). There are numerous works available that refer to the problem of landscape fragmentation [20–22]. With increased awareness of environmental sustainability issues and intensifying land development, the importance of the CORINE database and assessment of LULC changes (Land-Use Land Change) are examined [23–25].

The Macaronesian region has long been overlooked in comparative LULC research for two reasons. First, compared to other mainland regions more prevalent in academic literature, the small size, and population of the islands denote spatial dynamics of lower magnitude, which may diminish interest in their study. Second, there is a chronic shortage of comparable and uniform geospatial data for this region. Temme and Verburg [26] recognize a "(...) lack of homogeneous modeling, monitoring, and mapping strategies throughout the EU".

Up-to-date advances in geographic information technologies (GIS), where applicable FRAGSTATS as a spatial sample analysis program for categorical maps, can present the landscape through a mosaic model of landscape structure. Such a designated landscape is user-defined and can reproduce any spatial phenomenon. FRAGSTATS quantifies spatial heterogeneity and gives the user the ability to form the basis for defining and scaling landscapes in thematic content and resolution.

Contextually, this investigation is based on research techniques. These techniques and methods allow us to identify the dynamics of land-use changes in these territories. The chosen approach relies on proposing new geographical representations and modeling methods that emphasize the importance of geographical visualization of landscape fragmentation for spatial planning in this environmentally critical area. The research provides a greater understanding of the actors and decision-makers involved in how these ultra-peripheral territories have developed and how new territorial plans should be designed.

This work provides novel techniques in evaluating and managing the landscape of the study areas while considering that they could be applied in other research. Besides, the research results can contribute to the sustainable development of the islands.

Therefore, this paper begins with this initial chapter, followed by a brief overview of the literature related to the protection of ecosystem services: a Geographic Information Systems (GIS) and methodology based on CORINE land cover classes (CLC) and FRAG STATS, a methodological framework regarding the techniques used in the empirical part of the research, followed by the results, as well as their subsequent discussion and conclusions, with the final section on the limitations of the study and future research lines.

2. CORINE Land Cover and Landscape Fragmentation Analysis

The CLC database is a tool for performing complex spatial analyses based on diverse land-use kinds. CORINE land cover classes (CLC) have three levels in their hierarchical organization. The first covers five main types of land-use and land cover: artificial areas, agricultural areas, forest and semi-natural areas, wetlands, and water bodies. The next level has 15 departments. The third level includes 44 departments that note that the methodological scope of the three individual-level three classes is strictly defined [27–29].

The CLC records changes in land cover that are happening gradually and is very useful for research needed at the regional level. The main number of new researches has conducted territorial studies based on GIS Tools and CORINE data methodical approaches to landscape fragmentation. Especially in environmental studies as well as on research about changes in land degradation—i.e., in areas with different types of land cover countries, regions, islands, or cities [30–33], including urban growth monitoring and urban sprawl comparations' land-use forecasts, modeling of road travel speeds or fragmentation of property rights [34].

Islands as peripheral and ultra-peripheral are usually highlighted as ecologically sensitive areas to human activities due to the tremendous biological diversity of beings and the future probability of habitat loss [35]. Many researchers also analyzed the application of land reclamation, as an intensive change of land-use can often cause devastating impacts on the processes of the existing ecosystem and subsequently affect the surrounding environment [36].

CLC is the unique LULC database covering the Macaronesian islands of Portugal and Spain. CLC data sets were used as primary data sources in the study. CLCs are a map of the European landscape based on remote sensing. These public domain data sets provide a list of land cover classes organized hierarchically in three levels as a comparable cartographic product (minimum mapping unit 25 ha). CLC level 1 corresponds to the main categories (i.e., artificial areas, agricultural areas, forests and semi-natural areas, wetlands, water bodies). CLC level 2 covers whole areas with a higher level of detail (15 classes). Finally, level 3 CLC consists of 44 land cover classes (Table 1). Thus, the aggregate level of CLC 1 characterizes land-use, while CLC 2 further characterizes land cover. This research uses CLC level 1 and CLC level 2—which level is used depends on the research question. Table 1 hierarchically organizes the CLC nomenclature according to three levels: CLC level 1 (land-use) and CLC level 2, and CLC level 3 (land cover).

Table 1. Evolution of CORINE Land Cover. (Source: [27]).

	CLC1990	**CLC2000**	**CLC2006**	**CLC2012**	**CLC2018**
Satellite data	Landsat-5 MSS/TM single date	Landsat-7 ETM single date	SPOT-4/5 and IRS P6 LISS III dual date	IRS P6 LISS III and RapidEye dual date	Sentinel-2 and Landsat-8 for gap filling
Time consistency	1986–1998	2000+/−1 year	2006+/−1 year	2011–2012	2017–2018
Geometric accuracy, satellite data	≤50 m	≤25 m	≤25 m	≤25 m	≤10 m (Sentinel-2)
Min. mapping unit/width	25 ha/100 m	25 ha/100 m	25 ha/100 m	25 ha/100 m	25 ha/100 m
Geometric accuracy, CLC	100 m	better than 100 m	better than 100 m	better than 100 m	better than 100 m
Thematic accuracy, CLC	≥85% (probably not achieved)	≥85% (achieved) [13]	≥85%	≥85% (probably achieved)	≥85%
Change mapping (CHA)	not implemented	boundary displacement min. 100 m change area for existing polygons ≥5 ha; for isolated changes ≥25 ha	boundary displacement min. 100 m all changes ≥5 ha are to be mapped	boundary displacement min. 100 m all changes ≥5 ha are to be mapped	boundary displacement min. 100 m all changes ≥5 ha are to be mapped
Thematic accuracy, CHA	–	not checked	≥85% (achieved)	≥85%	≥85%
Production time	10 years	4 years	3 years	2 years	1.5 years
Documentation	incomplete metadata	standard metadata	standard metadata	standard metadata	standard metadata
Access to the data (CLC, CHA)	unclear dissemination policy	dissemination policy agreed from the start	free access for all users	free access for all users	free access for all users
Number of countries involved	26 (27 with late implementation)	30 (35 with late implementation)	38	39	39

Landscape metrics depicting different aspects of the spatial pattern were calculated at class and landscape levels using the software FRAGSTATS [37–40]. Landscape metrics are a unique feature that allows a quantitative assessment of the landscape and its level of fragmentation [19], and an understanding of landscape structure, function, and change. Landscape metrics mainly focus on three characteristics of the landscape [38,39]: (1) Structure: spatial relationships between recognizable ecosystems or elements that are present—more precisely, the distribution of energy, materials, and species concerning the sizes, shapes, numbers, species, and configurations of ecosystems; (2) Function: interactions between spatial elements, i.e., the flow of energy, materials, and species between ecosystem components; and (3) Change: change in the 'Ecological Mosaic' structure and function over time [40].

Landscape metrics quantify landscape patterns and interactions between patch density, several patches, total area, and extensive patch index in a landscape mosaic. In fact, those metrics allow us to see patterns and changes in interaction over time. Landscape composition can be quantified by patch number, patch density, landscape percentage, and highest patch index [38–40].

3. Materials and Methods

The CORINE, Coordination of Information Environment Programme, develops the CLC project, whose main objective is to obtain a European land occupation database for territorial analysis and European policy management. This geographical database, which is the first layer used in the present research, supplies land-uses in the European Union using polygon graphic features.

As for the spatial component, the reference scale is 1:100,000, the Geodesic Reference System is European Reference Terrestrial System 1989 (ETRS89), and the Mapping System is Universal Transverse Mercator (UTM). Additionally, the minimum width recorded for linear phenomena is 100 m. For polygon phenomena, the Minimum Cartographic Unit (MCU) is 25 hectares represented by a square of 5 $\times$ 5 mm or a circle with 2.8 mm. Regarding the thematic component, it offers three hierarchical levels of information.

In this regard, the development of ETRS89 is related to the global International Terrestrial Reference System and Frame (ITRS), which describes procedures for creating reference frames suitable for use with measurements on or near the Earth's surface. Indeed, the continental drift representation in ETRS89 is balanced since continental plates' total apparent angular momentum is about 0.

Besides, the data is in vector format, using polygons that evoke the various land-uses organized into three hierarchical levels using 44 classes, according to the European Environmental Agency (Table 1).

The second layer of information used also consists of polygonal graphical features that evoke administrative divisions at their different levels of the three archipelagoes studied: Autonomous Region of the Azores, Autonomous Region of Madeira, and the Autonomous Region of the Canary Islands.

In the Canary Islands, the information was obtained from the Download Center of the National Center for Geographical Information, belonging to the Ministry of Transport, Mobility, and Urban Agenda of Spain's Government. Specifically, the National Topographic Base was obtained at a 1:100,000 scale [41].

As for the Portuguese archipelagos corresponding to Azores and Madeira, the information was obtained from the National Geographic Information System, obtaining the Official Administrative Charter of Portugal in 2020 [42]. The scale is also 1:100,000.

The geodesic reference systems in the archipelago are different. In the Canary Islands, the projection is UTM and zone 27 and 28, being its EPSG (European Petroleum Survey Group) codes 4082 and 4083. So, for the Azores archipelago that also uses UTM projection, the EPSG code is 5015 in zone 26. In the case of the Madeira Archipelago, the EPSG code corresponds to 5016 in zone 28.

In this regard, EuroGeographics, where the Cartographic and Cadastral Agencies of the various European countries are represented, and the Joint Research Centre of the European Commission (EC), decided in December 2000 to entrust CERCO's No.8 working groups (Commission European des Responsible for Cartographie Officielle (French)) and EUREF (European Reference Frame) transformations with complete development of the technical details of the conventional Coordinate Reference System for Europe, to be adopted by the EC. Accordingly, recommendations to the European Commission [43] turned out to be: (1) Adopt ETRS89 as a geodetic datum and (2) Host, for statistical analysis and presentations, the ETRS89-Azimuth Equiarea coordinate system of Lambert-2001 (ETRS-LAEA) [44]. The ETRS-LAEA is based on the projection of equivalent areas in the territory. In this way, it serves as a reference for homogeneous units for all European countries. As a result, this coordinate system is used for the representation of analytical and statistical data.

In fact, this work intends to compare the area obtained from the uses of CLC in 1990, 2000, 2006, 2012, and 2018 in the archipelagoes of the Canary Islands, Azores, and Madeira. As a result, from Feature Manipulation Engine 2020.2 (FME) software developed by the company SAFE software, all layers of information were transformed to ETRS89-LAEA.

Subsequently, all layers were managed using ArcGIS 10.5 software. Initially, in the administrative divisions' representative layers, the islands corresponding to the archipelagoes to be studied were selected. Three layers of information were obtained, one layer for each archipelago. These layers were then merged into a single layer. This was possible because all layers used the same graphical rendering features, that is, polygons. Thus, a single layer was obtained with the delimitation of the work area. Then, geoprocessing the previous layer corresponding to the islands' administrative delimitations and the CLC layer corresponding to 1990 were related, using the clip tool. In this case, two layers are related to polygonal graphic features. Moreover, a resulting layer with polygons containing the land-use CLC in 1990 included in the various administrative divisions of the islands, including the islands and municipalities was obtained. The associated alpha-numeric information has three fundamental fields: (1) the CLC code of each polygon, (2) the island on which the CLC land-use representative polygon is located, and (3) the municipality where each registered land-use is located. However, there was no field corresponding to the surface occupied by each of the polygons representative of the CLC land-uses within each administrative division. Thus, a geometric measurement on the ETRS89-LAEA projection in hectares of each of the various polygons representing land-uses within each municipality analyzed was carried out. To this aim, a new field of information was generated, and with this, the area was geometrically calculated in hectares.

This procedure linking the islands' administrative divisions and land-uses for 1990 was also repeated for 2000, 2006, 2012, and 2018. In this way, a table of information was obtained for each of the years analyzed. The alphanumeric information obtained for each previous year was then exported and integrated into a Microsoft Access-managed database belonging to the Microsoft Office 365 package.

Moreover, a query was made using a structured query language (SQL) to select land-uses for each of the islands in 1990. Then, the previous query was queried by CLC codes of the registered hectares, also using SQL. So, a table was obtained with the CLC land-uses and the corresponding hectares on each of the islands for 1990, 2000, 2006, 2012, and 2018. Then, to compare the area obtained on each of the islands and archipelagos, it was necessary to normalize the data. Thus, the percentage occupied by each of the land-uses classified according to the CLC code was calculated relative to the total area of the corresponding archipelago for the first and third CLC level.

The calculation of landscape fragmentation analysis was then performed. For this purpose, CLC land-uses were used for the archipelagoes studied in 1990 and 2018. First, for each of the three archipelagoes considered, ArcGIS 10.5 software transformed the polygon vector layer representative of CLC land-uses to a raster file in TIF format with 30 m of cell size, using the CLC level 3 naming value for the output raster file. Subsequently, the TIFF file (.tif) was exported to an ERDAS Imagine grid (.img) file. Then, reviewing the alphanumeric information associated with the latter raster file, a text file was generated with the class descriptors.

Once all this information was prepared, the FRAGSTATS 4.2 software was used to perform patch metrics, class metrics, and landscape metrics using the eight-cell neighborhood rule.

In this regard, the calculations are applied to each fragment individually, to each polygon representing a land-use according to level 3 of the CLC. In this way, indexes obtained with these metrics can be interpreted as fragmentation indexes because they measure the configuration of a particular patch type.

In our case, aggregation measures such as Euclidean Nearest-Neighbor Distance (ENN) equals the distance (*m*) to the nearest-neighboring patch of the same type. ENN $= h_{ij}$, where h_{ij}. is the distance (*m*) from patch to nearest-neighboring patch of the same type (class), based on patch edge-to-edge distance, computed from cell center to cell center.

In addition to the standard patch metrics, FRAGSTATS 4.2 computes several deviation statistics for each patch that measures how much it deviates from the class or landscape norm. For this reason, through the before metric which was obtained by the standard deviations from the landscape mean (LSD): the value of the metric (*x*) for the focal patch (*ij*) minus the mean of the metric across all patches in the landscape divided by the landscape standard deviation:

$$\text{LSD} = \frac{x_{ij} - \overline{x}}{s} \quad (1)$$

where:

x_{ij} = value of a patch metric for patch *ij*.
$\overline{x}$ = mean value of the corresponding patch metric across all patches in the landscape.
s = standard deviation of the corresponding patch metric for all patches in the landscape.

Specifically, the distance between the different patches can be valuable information, based on the basis that greater isolation implies a reduction in the chances of harboring or maintaining a greater degree of biological diversity [44–46]. On the one hand, it provides information about the feasibility for species to survive and travel between the different elements to preserve their ecological value. On the contrary, it can also help eradicate species that have generated a pest. For this reason, corridors that allow the connection between patches play a fundamental role and reduce the distance effect that determines the presence of fewer species in isolated fragments [47].

As for class metrics, the calculations apply to each set of fragments of the same class, that is, those with the same value or that represent the same type of land-use, in our case. It is the appropriate level for calculating which area occupies a specific soil cover, such as forests, or the average extent occupied by forest fragments.

In this case, shape metrics were performed as the arithmetic mean of the shape index equals patch perimeter (given in the number of cell surfaces) divided by the minimum perimeter (given in the number of cell surfaces) possible for a maximally compact patch (in a square raster format) of the corresponding patch area.

$$\text{SHAPE} = \frac{p_{ij}}{min\ p_{ij}} \quad (2)$$

where:

p_{ij} = perimeter of patch *ij* in terms of a number of cell surfaces.
$min\ p_{ij}$ = minimum perimeter of patch *ij* in terms of number of cell surfaces.

If a_{ij} is the area of patch *ij* (in terms of number of cells) and *n* is the side of a largest integer square smaller than a_{ij}, and $m = a_{ij} - n^2$, then the minimum perimeter of patch *ij*, *min*-p_{ii} will take one of the three forms [48,49]:

$min - p_{ii} = 4n$ when $m = 0$, or
$min - p_{ii} = 4n + 2$ when $n^2 < a_{ij} \le n(1+n)$, or
$min - p_{ii} = 4n + 4$ when $a_{ij} \ge n(1+n)$.

In addition, the calculation of the arithmetic mean of the fractal dimension index was executed that calculates the degree of complexity of each fragment from the relationship between area and perimeter:

$$\mathrm{FRAC} = \frac{2\ln(25\,p_{ij})}{\ln a_{ij}} \tag{3}$$

where:

p_{ij} = perimeter (m) of patch ij.

a_{ij} = area (m^2) of patch ij.

The shape of fragments is of paramount importance and is sometimes even considered more relevant than dimension. The form is conditioned by human activity and natural conditions such as topography. Thus, the mastery of natural conditions favors curvilinear and irregular forms. On the contrary, the mastery of human activity promotes the diversification of forms. Intense human activity implies a simplification of variability [50].

Finally, the number of patches was calculated—the number of total fragments and the number of fragments of each class since the number of tiles is the most straightforward metric that can explain the extent to which land-use is divided or fragmented.

Concerning land metrics, calculations apply to the landscape as a whole, that is, to all fragments and classes at once. The result informs us of the degree of heterogeneity or homogeneity of the whole area quantified. In our case, two diversity measures were carried out. Firstly, the Shannon's Diversity Index (SHDI) values landscape diversity, i.e., heterogeneity, based on fragment diversity. Its absolute value is not very significant, but it helps to compare different landscapes or the same landscape at different events of time:

$$\mathrm{SHDI} = -\sum_{i=1}^{m}(P_i \ln P_i)$$

where P_i = proportion of the landscape occupied by patch type (class) i.

SHDI equals, minus the sum, across all patch types, of the proportional abundance of each patch type multiplied by that proportion. Note, P_i is based on total landscape area (A) excluding any internal background present.

Secondly, the Shannon's Everness Index (SHEI) was calculated which is a reverse index of the previous one, both at the calculation and interpretation level, based on landscape homogeneity:

$$\mathrm{SHEI} = \frac{-\sum_{i=1}^{m}(P_i \ln P_i)}{\ln m} \tag{4}$$

where:

P_i = proportion of the landscape occupied by patch type (class) i.

m = number of patch types (classes) present in the landscape, excluding the landscape border if present.

In this case, the SHEI will be applied to assess uniformity in land-uses in each of the archipelagoes.

3.1. The Macaronesia Region

The three archipelagos (Figures 1–3) share regional features: a volcanic origin, a contrasting landscape, and a gentle climate. These features have created an ideal environment for vibrant biodiversity [51]. The name Macaronesia is derived from the Greek words meaning "islands of the fortunate." Ancient Greek geographers first used the name to refer to any islands west of the Strait of Gibraltar. Macaronesia is a collection of four volcanic archipelagos in the North Atlantic Ocean, off Europe and Africa. Each archipelago is made up of several Atlantic oceanic islands formed by seamounts on the ocean floor and have peaks above the ocean's surface.

Figure 1. Macaronesia Region: The Canarias.

Figure 2. Macaronesia Region: The Azores.

Figure 3. Macaronesia Region: Madeira.

Some of the Macaronesian islands belong to Portugal, some belong to Spain, and the rest belong to Cape Verde. Geologically, Macaronesia is part of the African tectonic plate. Some of its islands—the Azores—are situated along the edge of that plate when it abuts the Eurasian and North American plates. According to the European Environment Agency, the three European archipelagos constitute a unique biogeographic realm known as the Macaronesian Region (Figure 4). Entirely volcanic, the Macaronesian islands share a gentle climate and offer a wide variety of landscapes. The large calderas, jagged mountains and cliffs, broad valleys, and sheltered bays are home to various species and habitats. The islands may represent a mere 0.3% of the EU territory, but they host 19% of the habitat types and 28% of all the plants listed in the *Habitats Directive* [51].

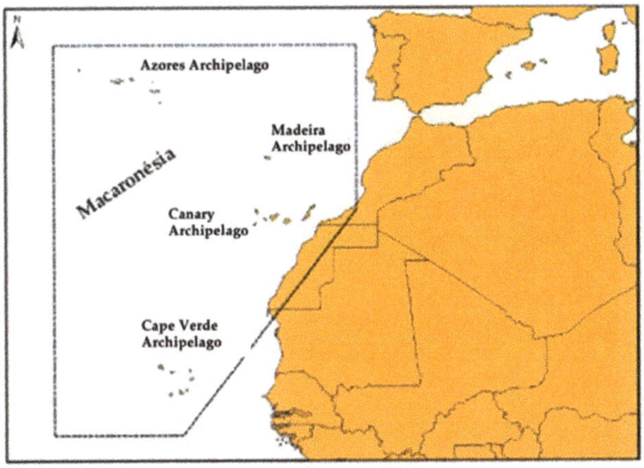

Figure 4. Macaronesia Region location map (adapted from [52]).

Macaronesian islands are volcanic in origin and are thought to be the product of several geologic hotspots. The Macaronesian islands represent a wide range of climates. The annual average maximum temperatures in the Canary Islands range from 24 °C in coastal areas to values below 10 °C in the Pico de Teide. The annual average maximum air temperature in the Azores and Madeira is between 12 °C and 14 °C in higher altitudes and below 8 °C in Ponta do Pico on Pico Island in the Azores. The highest maximum air temperature values in the Azores, above 20 °C, occur in some coastal areas of São Miguel, Santa Maria, Terceira, Graciosa, and Pico. In Madeira, the highest average maximum air temperature values are also above 20 °C in coastal regions of Madeira and in almost the entire island of Porto Santo, where values are even higher than 22 °C in the southern and north-western coastal strip of the island of Madeira [53–58].

There are maritime temperate, the Mediterranean, and subtropical climates in the Azores and Madeira; the Mediterranean and subtropical climates in some of the Canary Islands; arid climates in certain geologically older islands of the Canaries (notably Lanzarote and Fuerteventura) and some of the islands of the Madeira Archipelago (Selvagens and Porto Santo) and Cape Verde (Sal, Boa Vista, and Maio); and a tropical climate in the younger islands of both of the southernmost archipelagos (Santo Antão, Santiago, and Fogo in Cape Verde). In some locations, there are variations in climate due to the rain shadow effect. Macaronesia's laurisilva forests are a type of mountain cloud forest [51] with relict plant species of a vegetation type that originally covered much of the Mediterranean Basin when the climate of that region was more humid. Many of these plant species are endemic and have evolved to adapt to the islands' variable climatic conditions. For example, the Laurisilva of Madeira is the largest surviving relict of a virtually extinct laurel forest type, once widespread in Europe. It is still 90% primary forest and is a center of plant diversity, containing a unique suite of rare and relict plants and animals, especially endemic bryophytes, ferns, vascular plants, and animals, the Madeiran long-toed pigeon, and a vibrant invertebrate fauna [59,60].

Much of the original native vegetation has been displaced because of human activity, including felling forests for timber and firewood, clearing vegetation for grazing and agriculture, and introducing foreign plants and animals into the islands. The laurisilva habitat has been reduced to small disconnected pockets. As a result, many of the endemic biotas of the islands are now seriously endangered or extinct.

Since 2001, the European Union's conservation efforts, mandated by its Natura 2000 regulations [51], have protected large stretches of land and sea in the Azores, Madeira, and the Canary Islands, totaling 5000 km^2.

All archipelagos are outermost regions that are far from the countries or continent to which they belong, which requires special treatment to achieve the connection and development of these territories' economies.

The European Archipelagos of Macaronesia Region

Azores

The Azores are an archipelago formed by nine islands, which constitute an autonomous region of Portugal. Its official language is Portuguese, and it has approximately 250,000 inhabitants in its 2.333 square kilometers of land. The largest island is São Miguel, where more than half of the Azores archipelago population is gathered, whose main city is Ponta Delgada.

The Azorean climate, relief, and rich soils are particularly suitable for agriculture, and the islands are now heavily deforested: only 2% of the original laurel forests remain [40]. The endemic Azores bullfinch (*Pyrrhula murina*) was once a common feature of the native forests and saw its population plummet to 120 pairs. However, it is now on the road to recovery thanks to an EU LIFE project that has now caused the population to treble [51].

Madeira

It is an archipelago that is part of Portugal, formed by only two inhabited islands, Madeira and Porto Santo, with more than 260,000 inhabitants in an extension of 828 square kilometers. Three smaller islands that are not inhabited are also part of the archipelago.

Agriculture is the mainstay of Madeira's economy but has remained mainly small-scale due to the rugged landscape. Tourism is becoming increasingly important, generating 10% of the island's GDP and employing a significant proportion of the 250,000 islanders [51].

Canary Islands

The Canary Islands are formed by seven islands, subdivided into two provinces, being one of Spain's autonomous communities. On the archipelago's whole lives approximately 2,200,000 people in an extension of 7500 square kilometers, the most populated island being that of Gran Canaria, followed by Tenerife.

Tourism is the most important economic activity. Mixed and terraced farming is still practiced inland but has rapidly disappeared, replaced by the tropical and forced crops for the export market, accounting for 75% of the agricultural end production; 18,000 ha of highly fragmented laurel forest remain. Only 6000 ha correspond to mature forest [51].

4. Results

Bearing in mind the land occupation analyzed categories, it was possible to group the results. In the first phase, the number of land-uses has been defined in each of the studied years for each of the European Archipelagos of Macaronesia Region (Sections 4.1–4.3). After, in Section 4.4, it is conceivable to comprehend how the archipelagos' land-use change dynamics could be associated among them.

4.1. The Azores Archipelago

This section presents the analysis of the most relevant and specific land-uses in the Azores archipelago (Tables 2 and 3, and Figures 5–8).

Table 2. Percentage of land-uses according to level 1 of CLC nomenclature in the Autonomous Region of the Azores.

CODE	1990	2000	2006	2012	2018
1. Artificial surfaces	2.90%	3.41%	4.98%	5.13%	**5.22%**
2. Agricultural areas	**57.17%**	56.34%	54.43%	54.22%	54.62%
3. Forests and semi-natural areas	36.04%	36.38%	36.72%	**36.78%**	36.55%
4. Wetlands	**2.36%**	2.33%	2.33%	2.33%	2.08%
5. Water bodies	1.54%	1.54%	1.54%	**1.54%**	1.53%

The highest values found are in bold.

Table 3. Percentage of land-uses according to level 3 of CLC nomenclature in the Autonomous Region of the Azores.

CODE	1990	2000	2006	2012	2018	2018–1990
111	0.05%	0.05%	0.05%	0.05%	**0.05%**	0.00%
112	1.86%	2.16%	3.63%	3.71%	**3.77%**	**1.91%**
121	0.24%	0.35%	0.37%	0.44%	**0.45%**	0.21%
122	0.00%	0.00%	0.00%	0.00%	0.00%	0.00%
123	0.04%	0.04%	0.03%	0.04%	**0.04%**	0.00%
124	0.44%	0.44%	0.45%	0.46%	**0.47%**	0.03%
131	0.07%	0.11%	0.15%	0.16%	**0.18%**	0.11%
132	0.01%	0.07%	0.04%	0.04%	0.03%	0.02%
133	0.01%	0.03%	**0.05%**	0.01%	0.01%	0.00%
141	0.07%	0.07%	0.07%	0.07%	0.07%	0.00%
142	0.10%	0.10%	0.13%	**0.15%**	0.14%	0.04%
211	3.87%	3.83%	4.03%	4.11%	**4.98%**	**1.11%**
212	0.00%	0.00%	0.00%	0.00%	0.00%	0.00%
213	0.00%	0.00%	0.00%	0.00%	0.00%	0.00%
221	0.59%	0.35%	0.35%	0.35%	**0.63%**	0.04%
222	0.01%	0.01%	0.01%	0.01%	0.01%	0.00%
231	24.83%	24.56%	**24.96%**	24.74%	24.73%	−0.10%
241	0.00%	0.00%	0.00%	0.00%	0.00%	0.00%
242	**7.25%**	6.92%	6.12%	6.13%	5.80%	**−1.45%**
243	20.62%	**20.67%**	18.96%	18.89%	18.47%	**−2.15%**
244	0.00%	0.00%	0.00%	0.00%	0.00%	0.00%
311	9.36%	9.57%	9.40%	9.38%	**9.88%**	0.52%
312	3.63%	4.10%	4.35%	**4.45%**	4.11%	0.48%
313	0.89%	0.95%	1.02%	**1.02%**	1.01%	0.12%
321	**8.02%**	7.97%	7.87%	7.76%	7.38%	**−0.64%**
322	8.65%	8.63%	8.42%	8.42%	**9.10%**	0.45%
323	0.00%	0.00%	0.00%	0.00%	0.00%	0.00%
324	4.41%	4.08%	4.62%	**4.72%**	4.10%	−0.31%
331	0.00%	0.00%	0.00%	0.00%	0.00%	0.00%
332	0.25%	0.25%	0.25%	0.25%	**0.25%**	0.00%
333	0.82%	**0.82%**	0.79%	0.78%	0.72%	−0.10%
334	0.00%	0.00%	0.00%	0.00%	0.00%	0.00%
411	0.02%	0.02%	0.02%	0.02%	**0.02%**	0.00%
412	**2.34%**	2.32%	2.32%	2.31%	2.06%	−0.28%
422	0.00%	0.00%	0.00%	0.00%	0.00%	0.00%
512	0.40%	0.40%	0.40%	0.40%	**0.40%**	0.00%
521	0.00%	0.00%	0.00%	0.00%	0.00%	0.00%
523	1.13%	1.13%	1.13%	1.13%	**1.13%**	0.01%

The highest values found are in bold.

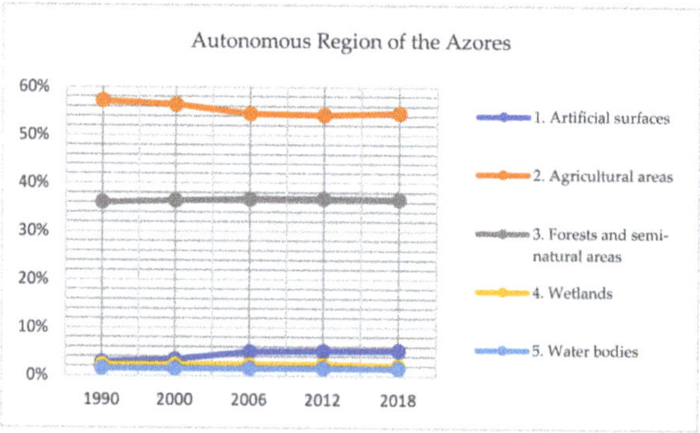

Figure 5. Percentage of land-uses according to level 1 of CLC nomenclature in the Autonomous Region of the Azores.

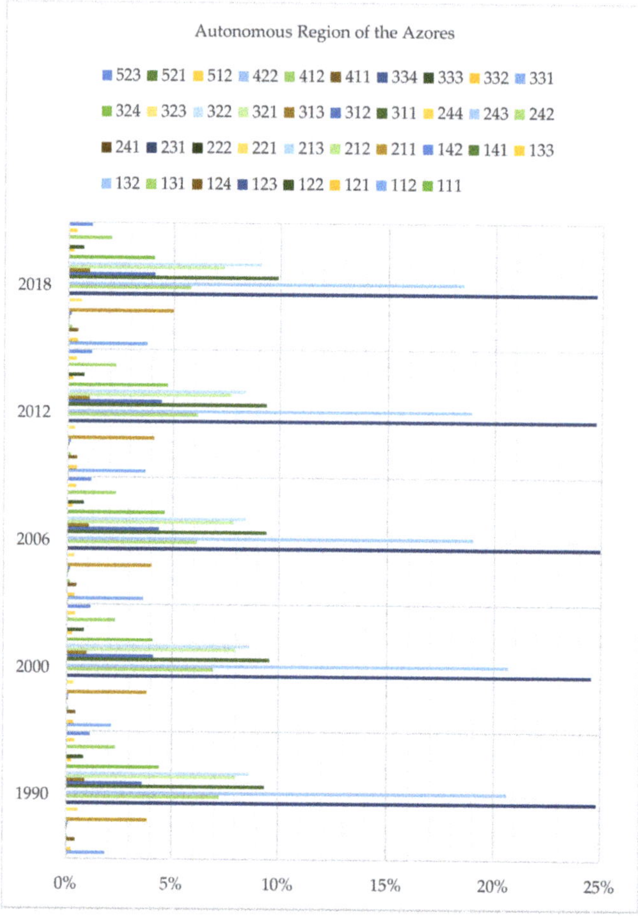

Figure 6. Percentage of land-uses according to CLC nomenclature in the Autonomous Region of the Azores.

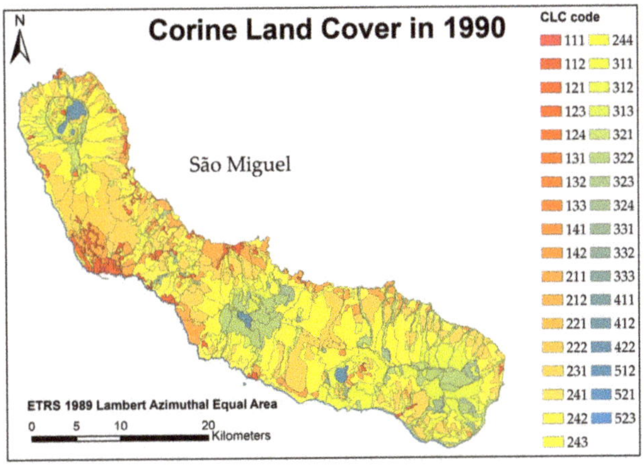

Figure 7. Thematic cartography regarding the land-use changes in the Azores Archipelago Eastern group in the year 1990.

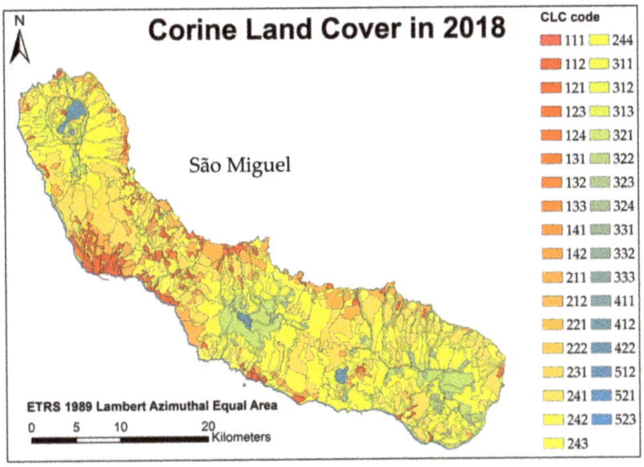

Figure 8. Thematic cartography regarding the land-use changes in the Azores Archipelago Eastern group in the year 2018.

By analyzing Table 2 and Figure 5, it is possible to verify the significant increase in the artificial surfaces (Code 1) in the Azores Region between 1990 and 2018. In fact, the variation of this land occupation around 2.30%. Additionally, the decrease in agricultural areas in the period 1990–2018 is evident—with a variation of 2.55%. A slight increase was noticed in forests and semi-natural areas—a reduction of 0.23% if we consider the highest value in 2012 to the present. A decreased tendency is also found in the land occupation related to wetlands—with a decrease of 0.28%, between 1990 to 2018. Furthermore, an insignificant decrease was identified regarding the water bodies from 1990 to 2018—a variation of 0.01%.

In Table 3 and Figure 6, it is possible to analyze in detail the land-use changes in the Azores Autonomous Region. In this regard, if we consider the period between 1990 and 2018, the most significant difference occurs in CLC-243 (Land principally occupied by agriculture, with significant natural vegetation areas) with a reduction of 2.15%. The second significant difference occurs in CLC-112 (Discontinuous urban fabric), with an increase of 1.91%. The third significant difference corresponds to CLC-242 (Complex cultivation) with a reduction of 1.45%. Finally, the fourth significant difference falls on CLC-211 (Non-irrigated arable land), increasing 1.11%. Besides these, we have CLC-121 (Industrial or commercial units), CLC-124 (Airports), CLC-131 (Mineral extraction sites), CLC-211 (Non-irrigated arable land)—with variations of 1.91%, 0.21%, 0.03%, 0.11%, and 1.11%, respectively. On the other hand, it is also possible to identify other important reductions in the land-use over the years in the Azores Archipelago, as is the case of CLC-133 (Construction sites) and CLC-312 (Coniferous forest)—with decreases of 0.04%, 2.20%, and 0.34%.

For a more accurate analysis of the results, thematic cartography was created for the three regions within the Azorean Archipelago (Western, Central, and Eastern), for the initial period (1990) and the last period (2018) (Appendix A). The following shows the thematic cartography for the Eastern Group of Azores Archipelago, once it is where the Capital Island (São Miguel) is located (Figures 7 and 8).

4.2. The Madeira Archipelago

This section presents the analysis of the most relevant and specific land-uses in the Madeira Archipelago (Tables 4 and 5, Figures 9–12, and Appendix B).

Throughout the analysis of Table 4 and Figure 9, it is possible to find an increase of 4.85% in the artificial surfaces (between 1990 and 2018) in the Madeira Region; however, the highest results were identified in 2012 with more than 0.25% of the surface in comparison with the current period. Additionally, another increase is noticed in forests and semi-natural areas—showing a variation of 1.35%, if we consider the initial period (1990) and the final period (2018). Moreover, the results show a significant reduction of 5.58% in the agricultural surfaces (from 1990 to 2018). Another decrease was found in the land cover classified as water bodies, with a reduction of 0.61% (between 1990 and 2018).

Table 5 and Figure 10 show with greater detail the land-use changes in the Madeira Autonomous Region. Contextually, if we consider the period between 1990 and 2018, the most significant differences in land-use are CLC-322 (Moors and heathland) and CLC-112 (Discontinuous urban fabric) 5.62% and 3.73%, respectively. On the contrary, the most significant decreases are for land-uses of CLC-313 (Mixed forest) and CLC-311 (Broad-leaved forest), with 3.46% and 1.78%, respectively. Besides, the obtained results evidence considerable increases in the surfaces of land occupations in CLC-111 (Continuous urban fabric), CLC-121 (Industrial or commercial units), CLC-131 (Mineral extraction sites), CLC-142 (Sport and leisure facilities), CLC-324 (Transitional woodland shrub), and CLC-332 (Bare rock)—with variations of 0.11%, 0.34%, 0.15%, 0.36%, 5.62%, 1.33%, and 1.55%, respectively. Contrarily, if we focus on the period between 1990 and 2018, it is also possible to find concerning reductions in the land-use over the years in the Madeira Region, as is the case of CLC-222 (Fruit trees and berry plantations), CLC-231 (Pastures), CLC-241 (Annual crops associated with permanent crops), CLC-242 (Complex cultivation), CLC-243 (Land occupied by agriculture), CLC-312 (Coniferous forest), and CLC-523 (Sea and ocean)—with decreases of 0.43%, 0.44%, 0.94%, 1.80%, 1.47%, 1.67%, 3.46%, and 0.61%. Additionally, other reductions should be highlighted as is the case of the CLC-211 and CLC-212 and CLC-333 (from 2000 to 2018); and CLC-334 (from 2012 to 2018).

Additionally, with the results, thematic cartography was created for the Madeira Island for the initial period (1990) and the last period (2018) (Appendix B and Figures 11 and 12).

Table 4. Percentage of land-uses according to level 1 of CLC nomenclature in the Autonomous Region of Madeira.

CODE	1990	2000	2006	2012	2018
1. Artificial surfaces	10.11%	14.30%	15.20%	**15.21%**	14.96%
2. Agricultural areas	**19.41%**	16.25%	14.26%	14.29%	13.83%
3. Forests and semi-natural areas	69.45%	68.45%	70.12%	70.08%	**70.80%**
4. Wetlands	0.00%	0.00%	0.00%	0.00%	0.00%
5. Water bodies	**1.02%**	1.00%	0.42%	0.42%	0.41%

The highest values found are in bold.

Table 5. Percentage of land-uses according to CLC nomenclature in the Autonomous Region of Madeira.

CODE	1990	2000	2006	2012	2018	2018–1990
111	0.21%	0.21%	0.30%	0.30%	**0.32%**	0.11%
112	9.43%	13.05%	13.44%	**13.48%**	13.16%	**3.73%**
121	0.09%	0.22%	0.35%	0.40%	**0.43%**	0.34%
122	0.00%	0.04%	0.04%	0.04%	**0.04%**	0.04%
123	0.02%	0.03%	0.03%	0.03%	**0.04%**	0.02%
124	0.25%	0.30%	0.30%	0.30%	**0.30%**	0.05%
131	0.00%	0.00%	0.15%	0.15%	**0.15%**	0.15%
132	0.00%	0.04%	0.08%	0.06%	**0.06%**	0.06%
133	0.07%	**0.15%**	0.13%	0.06%	0.06%	-0.01%
141	0.04%	0.04%	0.04%	0.04%	**0.04%**	0.00%
142	0.00%	0.21%	0.32%	0.35%	**0.36%**	0.36%
211	0.15%	**0.15%**	0.04%	0.04%	0.04%	−0.11%
212	0.48%	**0.48%**	0.04%	0.04%	0.04%	−0.44%
213	0.00%	0.00%	0.00%	0.00%	0.00%	0.00%
221	0.17%	0.17%	0.18%	**0.21%**	0.19%	0.02%
222	**0.74%**	0.45%	0.31%	0.31%	0.31%	−0.43%
231	**0.95%**	0.92%	0.34%	0.34%	0.51%	−0.44%
241	**0.94%**	0.92%	0.25%	0.25%	0.00%	−0.94%
242	**4.94%**	3.14%	2.83%	2.80%	3.42%	−1.52%
243	**10.80%**	9.56%	10.28%	10.32%	9.33%	−1.47%
244	0.24%	**0.45%**	0.00%	0.00%	0.00%	−0.24%
311	20.42%	20.21%	**20.43%**	18.92%	18.64%	**−1.78%**
312	**5.79%**	5.64%	5.20%	4.65%	4.12%	**−1.67%**
313	**14.06%**	13.81%	13.00%	11.86%	10.60%	**−3.46%**
321	9.03%	8.88%	**9.79%**	9.42%	9.78%	0.75%
322	10.78%	10.70%	11.01%	9.33%	**16.40%**	**5.62%**
323	0.00%	0.00%	0.00%	0.00%	0.00%	0.00%
324	6.52%	6.30%	6.22%	5.56%	**7.85%**	1.33%
331	0.12%	0.12%	0.12%	0.12%	0.06%	−0.06%
332	0.30%	0.30%	1.85%	1.85%	**1.85%**	**1.55%**
333	2.42%	**2.42%**	2.27%	2.27%	1.47%	−0.95%
334	0.00%	0.06%	0.23%	**6.09%**	0.04%	0.04%
411	0.00%	0.00%	0.00%	0.00%	0.00%	0.00%
412	0.00%	0.00%	0.00%	0.00%	0.00%	0.00%
422	0.00%	0.00%	0.00%	0.00%	0.00%	0.00%
512	0.00%	0.00%	0.00%	0.00%	0.00%	0.00%
521	0.00%	0.00%	0.00%	0.00%	0.00%	0.00%
523	**1.02%**	1.00%	0.42%	0.42%	0.41%	−0.61%

The highest values found are in bold.

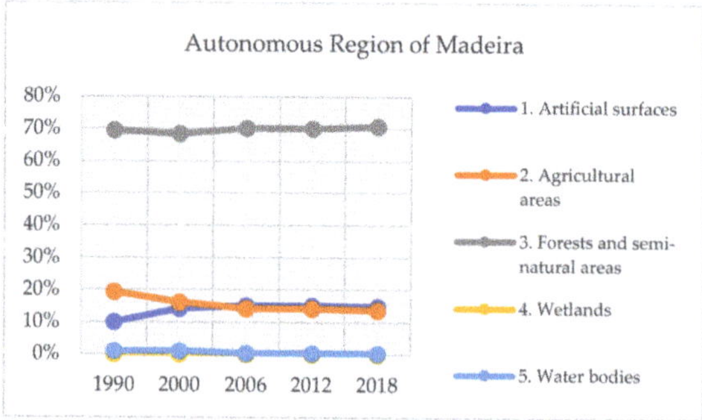

Figure 9. Percentage of land-uses according to level 1 of CLC nomenclature in the Autonomous Region of Madeira.

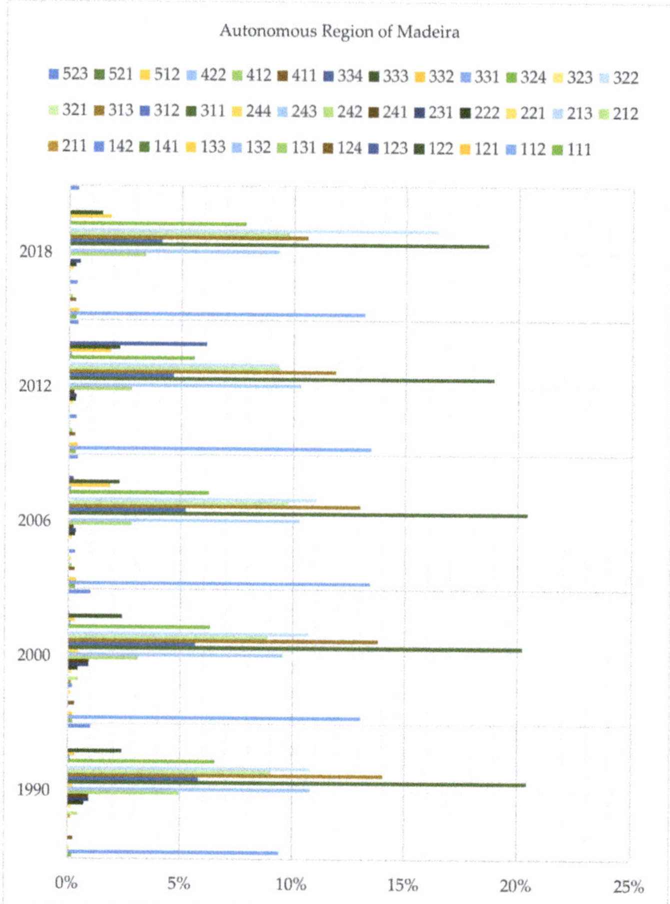

Figure 10. Percentage of land-uses according to CLC nomenclature in the Autonomous Region of Madeira.

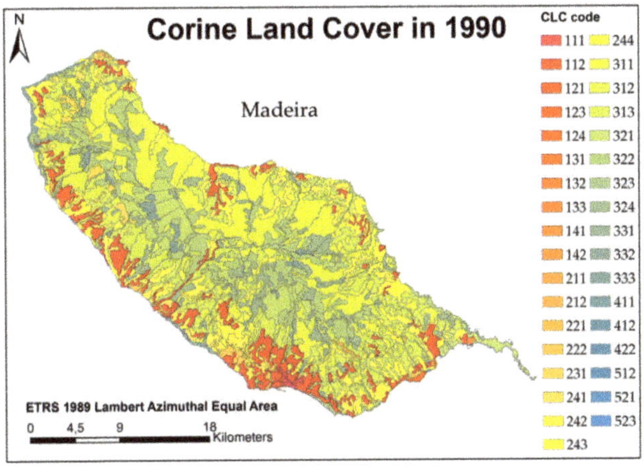

Figure 11. Thematic cartography regarding the land-use changes in the Madeira Archipelago in the year 1990.

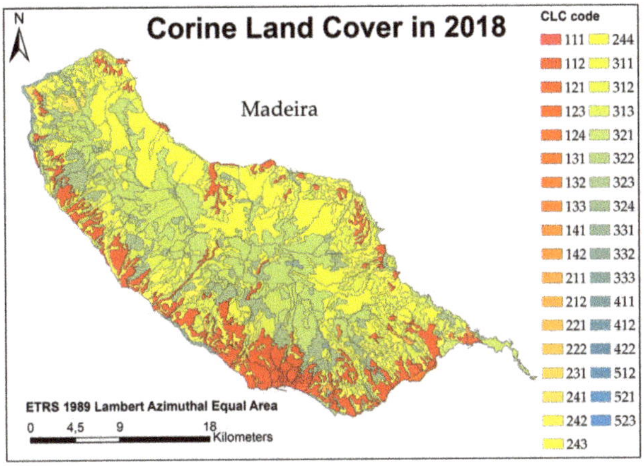

Figure 12. Thematic cartography regarding the land-use changes in the Madeira Archipelago in the year 2018.

4.3. *The Canary Archipelago*

The current section shows the obtained outcomes of the most relevant and specific land-uses in the Canary Archipelago (Tables 6 and 7, Figures 13–16, and Appendix C).

Table 6. Percentage of land-uses according to level 1 of CLC nomenclature in the Autonomous Community of the Canary Islands.

CODE	1990	2000	2006	2012	2018
1. Artificial surfaces	4.14%	4.50%	6.05%	6.11%	**6.23%**
2. Agricultural areas	**22.63%**	22.59%	16.52%	16.51%	19.99%
3. Forests and semi-natural areas	72.12%	71.79%	76.35%	76.29%	**72.69%**
4. Wetlands	0.00%	0.00%	0.00%	0.00%	0.00%
5. Water bodies	1.10%	**1.10%**	1.07%	1.07%	1.08%

The highest values found are in bold.

Table 7. Percentage of land-uses according to CLC nomenclature in the Autonomous Community of the Canary Islands.

CODE	1990	2000	2006	2012	2018	2018–1990
111	2.20%	**2.45%**	1.40%	1.42%	1.39%	−0.81%
112	0.85%	0.89%	2.57%	2.63%	**3.01%**	2.16%
121	0.24%	0.36%	0.57%	0.61%	**0.70%**	0.46%
122	0.00%	0.01%	0.00%	0.00%	**0.05%**	0.05%
123	0.06%	0.07%	0.09%	0.09%	**0.09%**	0.03%
124	0.17%	0.18%	0.23%	**0.23%**	0.22%	0.05%
131	0.15%	0.16%	0.25%	0.25%	**0.29%**	0.14%
132	0.01%	0.01%	0.02%	0.01%	**0.02%**	0.01%
133	0.36%	0.28%	**0.60%**	0.51%	0.11%	−0.25%
141	0.02%	0.02%	0.03%	0.03%	**0.03%**	0.01%
142	0.07%	0.09%	0.29%	0.32%	**0.34%**	0.27%
211	**13.44%**	13.37%	3.19%	3.18%	3.14%	**−10.30%**
212	1.14%	1.20%	2.40%	**2.40%**	1.77%	0.63%
213	0.00%	0.00%	0.00%	0.00%	0.00%	0.00%
221	2.15%	**2.16%**	1.33%	1.33%	1.34%	−0.81%
222	1.75%	1.71%	2.07%	**2.07%**	1.85%	0.10%
231	**3.25%**	3.24%	2.07%	2.05%	1.97%	−1.28%
241	0.00%	0.00%	0.12%	**0.12%**	0.11%	0.11%
242	0.00%	0.00%	1.20%	1.20%	**2.45%**	2.45%
243	0.91%	0.91%	4.15%	4.15%	**7.35%**	**6.44%**
244	0.00%	0.00%	0.00%	0.00%	0.00%	0.00%
311	1.36%	1.36%	2.49%	**2.49%**	2.46%	1.10%
312	10.48%	10.46%	**11.21%**	11.08%	11.03%	0.55%
313	0.00%	0.00%	0.47%	0.47%	**0.59%**	0.59%
321	0.00%	0.00%	2.01%	2.01%	**2.63%**	2.63%
322	2.99%	**2.99%**	0.38%	0.38%	1.07%	−1.92%
323	**44.95%**	44.69%	15.75%	15.72%	7.26%	**−37.69%**
324	0.00%	0.00%	0.81%	0.90%	**1.41%**	1.41%
331	1.68%	1.68%	1.75%	**1.75%**	1.63%	−0.05%
332	4.87%	4.87%	10.57%	**10.57%**	10.51%	5.64%
333	5.78%	5.75%	30.78%	30.76%	**33.92%**	**28.14%**
334	0.00%	0.00%	0.12%	0.18%	**0.18%**	0.18%
411	0.00%	0.00%	0.00%	0.00%	0.00%	0.00%
412	0.00%	0.00%	0.00%	0.00%	0.00%	0.00%
422	0.01%	0.01%	0.01%	0.01%	**0.01%**	0.00%
512	0.01%	0.01%	0.02%	0.02%	**0.02%**	0.01%
521	0.00%	0.00%	0.00%	0.00%	0.00%	0.00%
523	1.09%	**1.09%**	1.05%	1.05%	1.05%	−0.04%

The highest values found are in bold.

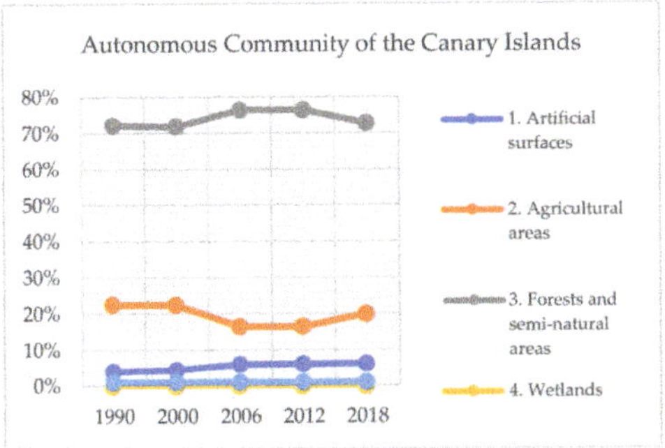

Figure 13. Percentage of land-uses according to level 1 of CLC nomenclature in the Autonomous Community of the Canary Islands.

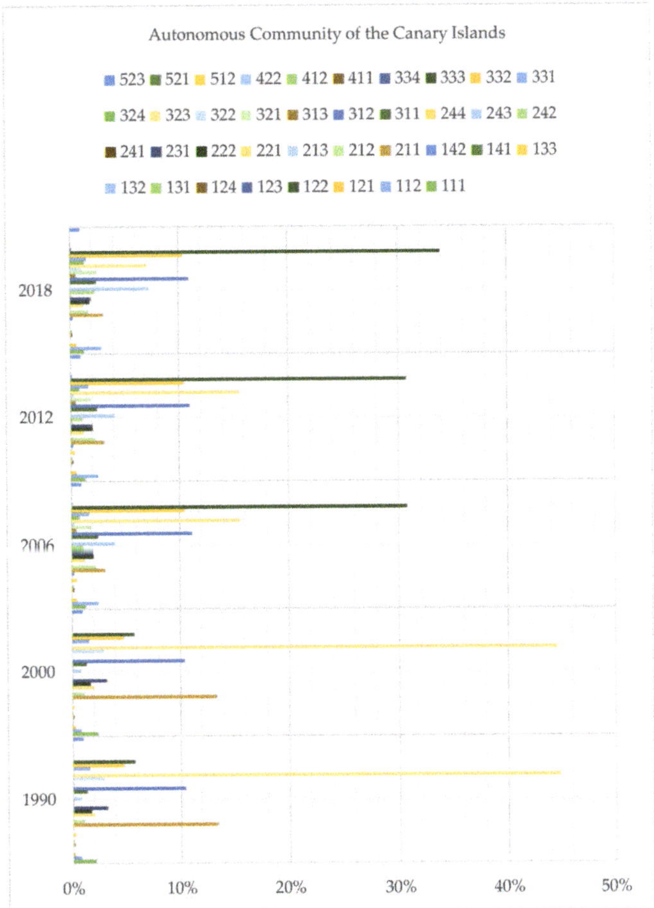

Figure 14. Percentage of land-uses according to CLC nomenclature in the Autonomous Community of the Canary Islands.

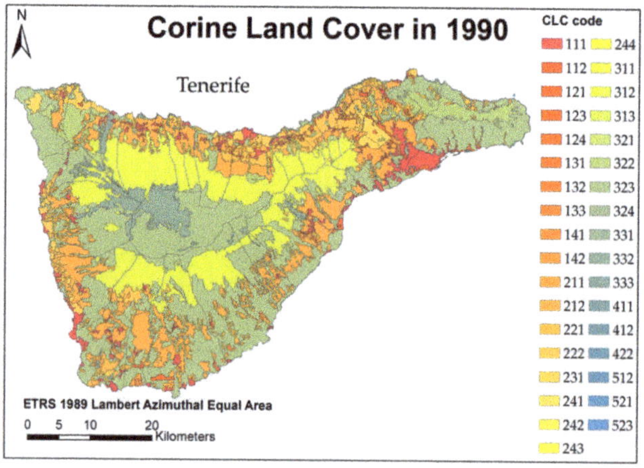

Figure 15. Thematic cartography regarding the land-use changes in the Canary Archipelago in the year 1990.

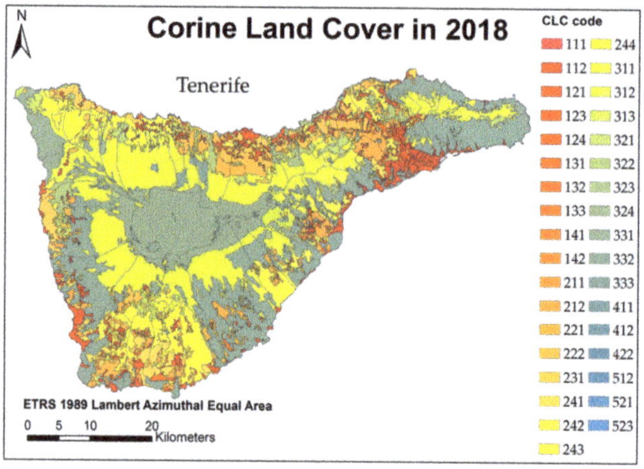

Figure 16. Thematic cartography regarding the land-use changes in the Canary Archipelago in the year 2018.

Table 6 and Figure 13 show the highest increase (2.09%) in the artificial surfaces (between 1990 and 2018) in the Canary Islands. Moreover, in forests and semi-natural areas, another increase was verified—with a variation of 0.57% (in the period 1990–2018). If we focus on reducing land occupation in the same periods, the most significant decrease was identified in agricultural areas (reducing 2.64%). In this sense, there was also a reduction of 0.02% in water bodies from 1990 to 2018.

Contextually, through the analysis of Table 7 and Figure 14, we can perceive the land-use changes in the Canary Archipelago with greater detail. Regarding the period between 1990 and 2018, the most significant differences are produced by a considerable reduction of 37.69% in CLC-323 (Sclerophyllous vegetation) and 10.30% in CLC-211 (Non-irrigated arable land). However, a considerable increase of 28.14% corresponds to CLC-333 (Sparsely vegetated areas) followed by a 6.44% increase equivalent to CLC-243 (Land principally occupied by agriculture). Besides, the collected results show substantial increases in the surfaces of land occupations CLC-112 (Discontinuous urban fabric), in CLC-121 (Industrial or commercial units), CLC-131 (Mineral extraction sites), CLC-142 (Sport and leisure facilities), CLC-242 (Complex cultivation), CLC-313 (Mixed forest), CLC-321 (Natural grassland), CLC-324 (Transitional woodland shrub), and CLC-334 (Burnt areas)—with variations of 2.16%, 0.46%, 0.14%, 0.27%, 2.45%, 6.44%, 0.59%, 2.63%, 1.41%, 28.1%, and 0.18%, respectively. Contrarily, suppose we focus on the land-use cover reductions between 1990 and 2018 on this archipelago. In that case, it is possible to highlight the following CLC-231 (decrease of 1.28%). In this sense, if we consider the period between 2000–2018, some decreases in the land covers should be noticed, as CLC-111 (reduction of 1.06%), CLC-221 (reduction of 0.82%), and CLC-322 (reduction of 1.92%). Moreover, the reduction of 0.49% in CLC-133 (Dump sites) is evident between 2006 and 2018. Finally, between 2012 and 2018 there was also two other reductions in CLC-212 and CLC-222.

Furthermore, based on the results, thematic cartography was created for the Canary Islands for the initial period (1990) and the last period (2018) (Appendix C). Contextually, the thematic cartography for the most populated island of the Canary Archipelago, Tenerife, is shown (Figures 15 and 16).

4.4. Land-Use Change Associations

The present section was created to further comprehend how the archipelagos' land-use change dynamics could be associated.

Therefore, the correlation in the land-uses in each of the archipelagos was performed using the percentage land-use varies according to the CLC code by using R software.

Subsequently, considering a confidence level of 95% and n < 30, the normality of the data was verified using the Shapiro–Wilk test. The values obtained are less than 0.05; as a consequence, normality is discarded. Therefore, a non-parametric test was performed, with ordinal data, considering the covariation hypothesis to determine the Spearman correlation matrix—both for land-uses at level 1.

Based on the previous tables, the strength of negative or positive association between the variables corresponding to the percentage of surface intended for certain level 1 land-uses can be analyzed according to CLC between 1990 and 2018. Initially, it can be observed that the strength of the association is always equal to or greater than moderate (0.40–0.59); there are even strong correlation values (0.60–0.79), and very strong (0.80–1.00), whether we consider strong or very strong partnerships. Firstly, land-use in agricultural areas declined between 1990 and 2012, at 2.95%, equivalent to 6913.45 hectares, as land-uses identified as artificial surfaces and forest and semi-natural areas increased to 2.23% and 0.74%, equivalent to 5231.58 and 1746.21 hectares, respectively. Additionally, artificial surfaces increased slightly between 2012 and 2018, 0.09% equivalent to 214.73 hectares when water bodies were reduced—0.001%, 3.82 hectares. Secondly, artificial surfaces gradually increased throughout the period analyzed—2.32%, 5454.38 hectares, when forest and semi-natural areas decreased between 2012 and 2018—0.51%, 1200.46 hectares, and previously due to the loss of other land-uses. Thirdly, forest and semi-natural areas increased between 1990 and 2012—0.74%, 1746.21 hectares, when the extent of Artificial surfaces land-use increased—2,23%, 5239.65 hectares, over the same period. Fourthly, the surface of the water bodies remains virtually unchanged. Although, it descends slightly between 2012 and 2018—0.002% 4.01 hectares, due to the increase of artificial surfaces. Finally, wetlands have declined— 0.28%, 664.89 hectares, along those analyzed by the increase in artificial surfaces (Table 8 and Appendix D).

Table 8. Spearman's rank correlation coefficient in the Azores Archipelago.

	Agr	Art	For	Wat	Wet	Acronym	CORINE Land Cover Nomenclature
Agr	1.0	−0.7	−1.0	0.7	0.5	Agr	Agricultural areas
Art	−0.7	1.0	0.7	−1.0	−0.9	Art	Artificial surfaces
For	−1.0	0.7	1.0	−0.7	−0.5	For	Forests and semi-natural areas
Wat	0.7	−1.0	−0.7	1.0	0.9	Wat	Water bodies
Wet	0.5	−0.9	−0.5	0.9	1.0	Wet	Wetlands

Regarding the Madeira Archipelago (Table 9 and Appendix E), we have considered strong, and very strong relationships between variables, where the interpretation of Tables 5 and 10 was performed. Firstly, agricultural areas have declined over the years analyzed—5.12%, 4099.94 hectares, when artificial surfaces have increased between 2000 and 2012—0.91%, 726.81 hectares, and forest and semi-natural areas between 2000 and 2006—1.67%, 1338.65 hectares, and between 2012 and 2018—0.72%, 578.45 hectares. Secondly, artificial surfaces would have been dwarfed between 2012 and 2018, and throughout the period analyzed, if the area of agricultural areas and water bodies had increased. However, these last two land-uses have been declining over the years analyzed—5.59% and 0.61%, 4477.59 and 488.12 hectares, respectively. Thirdly, forest and semi-natural areas have increased, as it would have decreased if agricultural areas and water bodies had increased. However, these land-uses have declined—5.59% and 0.61%, 4474.59 and 488.12 hectares, respectively. Finally, water bodies declined over the years and were analyzed mainly due to the increase in artificial surfaces—4.85%, 3882.34 hectares, and forest and semi-natural areas—1.35%, 1080.37 hectares.

Table 9. Spearman's rank correlation coefficient in the Madeira Archipelago.

	Agr	Art	For	Wat	Acronym	CORINE Land Cover Nomenclature
Agr	1.0	−0.9	−0.6	0.9	Agr	Agricultural areas
Art	−0.9	1.0	0.3	−0.7	Art	Artificial surfaces
For	−0.6	0.3	1.0	−0.8	For	Forests and semi-natural areas
Wat	0.9	−0.7	−0.8	1.0	Wat	Water bodies

Table 10. Spearman's rank correlation coefficient in the Canary Archipelago.

	Agr	Art	For	Wat	Wet	Acronym	CORINE Land Cover Nomenclature
Agr	1.0	−0.7	−0.8	1.0	−0.9	Agr	Agricultural areas
Art	−0.7	1.0	0.5	−0.7	0.9	Art	Artificial surfaces
For	−0.8	0.5	1.0	−0.8	0.6	For	Forests and semi-natural areas
Wat	1.0	−0.7	−0.8	1.0	−0.9	Wat	Water bodies
Wet	−0.9	0.9	0.6	−0.9	1.0	Wet	Wetlands

Concerning the Canary Archipelago (Table 10 and Appendix F), firstly, agricultural areas declined in all the years analyzed—2.64%, 19,805.77 hectares, due to the increase in artificial surfaces—2.09%, 15,687.92 hectares, by increasing forest and semi-natural areas—4.55% 34,097.38 hectares, and wetlands—0.001%, 7.49 hectares, between 2000 and 2006. Although, the latter land-use was slightly increased and is not appreciable in Table 7 by using two decimal places. Secondly, artificial surfaces increased in all the years analyzed—2.09%, 15,687.92 hectares. Thirdly, forest and semi-natural areas increased between 2000 and 2006—4.22%, 31,646.37 hectares, and decreased between 2012 and 2018—3.60%, 26,984.02 hectares. Fourthly, water bodies were reduced between 2000 and 2018—0.02%, 182.96 hectares, as artificial surfaces increased—1.73%, 12,929.45 hectares, when forest and semi-natural areas increased between 2000 and 2006—4.55% 34,097 hectares, and between 2012 and 2018—0.12%, 867.34 hectares. Finally, wetlands remain virtually unchanged.

4.5. Results of Landscape Fragmentation Analysis

Regarding the results obtained from landscape fragmentation analysis (FRAGSTATS), it should be noted for the patch analysis level that Euclidean nearest-neighbor distance is perhaps the most straightforward patch context being used extensively to quantify patch isolation. Here, the nearest-neighbor distance is defined using simple Euclidean geometry as the shortest straight-line distance between the focal patch and its nearest-neighbor of the same class. ENN approaches 0 as the distance to the nearest- neighbor decreases. This index provides more information about the structure of the set and some of the vital importance in its dynamics and function. A decrease in its values can result in new fragments in the case of very isolated soil-uses. On the contrary, their increase can mean the aggregation of multiple fragments that were nearby.

The minimum ENN is constrained by the cell size and is equal to twice the cell size when the 8-neighbor patch rule is used. ENN is undefined and reported as "N/A" in the "basename".patch file if the patch has no neighbors.

In the case of the Autonomous Region of Azores (Appendix G and Table 11), in 1990, only four patches had no neighbors; these corresponded to land-uses 222 (permanent crops), 333 (sparsely vegetated areas), and 221 (vineyards). In addition, the patch that had the highest value of ENN and LSD, with values of 338,183.7288 and 243,076, respectively, corresponded to a patch with land-use CLC 512 (water bodies). Therefore, it is this land-use that has the greatest insulation. In contrast, there are several patches whose ENN value is equal to 60 and their standard deviation is −0.1388, corresponding to 322 (Moors and heathland), 324 (Transitional woodland/shrub), 321 (Natural grassland), 112 (Discontinuous urban fabric), 211 (Non-irrigated arable land), 412 (Peat bogs), and 133 (construction sites)—these soils are the least isolated.

Table 11. Class metrics for the Autonomous Region of Azores.

1990				2018			
TYPE	NP	SHAPE_MN	FRAC_MN	TYPE	NP	SHAPE_MN	FRAC_MN
111	2	1.9487	1.1032	111	2	1.9487	1.1032
112	155	2.0913	1.0951	112	112	2.324	1.1202
121	3	1.2476	1.0356	121	21	1.6568	1.0774
123	11	1.7607	1.0875	123	14	1.8397	1.0893
124	7	1.9395	1.0986	124	8	1.9887	1.1026
131	67	1.7617	1.0796	131	10	1.4863	1.0624
132	6	1.4548	1.0594	132	2	1.2541	1.0373
133	17	1.87	1.0911	133	1	1.5556	1.0716
141	98	1.7825	1.0829	141	3	1.2415	1.0352
142	7	1.5219	1.0607	142	7	1.4744	1.0591
211	226	1.9385	1.0906	211	121	1.6826	1.0753
221	1	1.8824	1.1049	221	23	1.8637	1.0919
222	1	1.2059	1.0321	222	1	1.8824	1.1049
231	21	1.9689	1.0934	231	179	2.0588	1.0945
242	110	1.621	1.0704	242	183	1.8611	1.0894
243	108	2.0105	1.094	243	244	1.925	1.0903
311	202	1.9902	1.0989	311	128	2.059	1.0987
312	10	1.9605	1.088	312	83	1.7925	1.0816
313	4	1.4511	1.0569	313	27	2.013	1.0968
321	45	2.1118	1.1019	321	42	2.1045	1.0987
322	145	2.0762	1.1007	322	150	2.0874	1.1016
324	801	1.6223	1.085	324	105	1.7656	1.0818
332	11	2.4216	1.1279	332	5	2.3809	1.1143
333	1	1.6944	1.0857	333	6	2.5457	1.131
411	1	1.2857	1.0422	411	1	1.2857	1.0422
412	75	2.1509	1.1113	412	16	1.7019	1.0722
512	5	2.3809	1.1143	512	7	1.5219	1.0607
523	14	1.8039	1.0877	523	799	1.6235	1.0852

TYPE = land-use according to CLC nomenclature at level 3; NP = number of patches present in the landscape; SHAPE_MN = arithmetic mean of the Shape Index; FRAC_MN = arithmetic mean of the Fractal Dimension Index.

In 2018, there were only two patches without neighbors corresponding to land-uses 133 (Construction sites) and 222 (Fruit trees and berry plantations). In addition, the patch with the highest ENN and LSD corresponded to the use of Soil 148, whose values were 85,939.4863 and 18.1421, respectively. On the contrary, and as in 1990, there were several land-uses whose ENN values are the lowest and equal. The value is as before—60 for ENN and LSD equal to −0.142, and land-uses are 322 (Moros and heathland), 523 (Sea and ocean), 321 (Natural grassland), 231 (Pastures), 243 (Land principally occupied by agriculture, with significant areas of natural vegetation), and 112 (Discontinuous urban fabric).

About the Autonomous Region of Madeira (Appendix H and Table 12), in 1990, only three patches had no neighbors; these corresponded to land-uses 133 (Construction sites), 141 (Green urban areas), and 111 (Continuous urban fabric). In addition, the patch that had the highest value of ENN and LSD, with values of 66,992.1227 and 15.6451, respectively, corresponded to a patch with land-use CLC 241 (Annual crops associated with permanent crops), presenting the greatest insulation. In contrast, there are several patches whose ENN value is equal to 60, and their standard deviation is −0.1691, corresponding to 321 (Natural grassland), 322 (Moors and heathland), and 523 (Sea and ocean); these are the less isolated soils. In 2018, there were seven patches without neighbors corresponding to land-uses 122 (Road and rail networks and associated land), 132 (Dump sites), 133 (Construction sites), 141 (Green urban areas), 211 (Non-irrigated arable land), 212 (Permanently irrigated land), and 334 (Burnt areas). In addition, the patch with the highest ENN and LSD corresponded to the use of Soil 231, whose values were 85,939.4863 and 18.1421, respectively. On the contrary, and as in 1990, there are several land-uses whose ENN values are the lowest and equal. The value is as before 60 for ENN and LSD equal to −0.1515 and land-uses are 112 (discontinuous urban fabric), 123 (Port areas), 243 (Land principally occupied by agriculture, with significant areas of natural vegetation), 321 (Natural grassland), 322 (Moors and heathland), and 523 (sea and ocean).

Finally, in the case of the Autonomous Community of the Canary Islands (Appendix I and Table 13), in 1990, only four patches had no neighbors; these corresponded to land-uses 132 (Dump sites), 243 (Land principally occupied by agriculture, with significant areas of natural vegetation), 422 (Salines) and 521 (Coastal lagoons). In addition, the patch that had the highest value of ENN and LSD, with values of 242,710.3533 and 26.9532, respectively, corresponded to a patch with land-use CLC 512 (Water bodies), presenting the greatest insulation. In contrast, there are several patches whose ENN value is equal to 60 and their standard deviation is −0.1205, corresponding to 111 (Continuous urban fabric), 112 (Discontinuous urban fabric), 123 (Port areas), and 523 (Sea and ocean)—these floors being the least isolated. In 2018, there was only one patch without neighbors corresponding to land-use 521 (Coastal lagoons). In addition, the patch with the highest ENN and LSD corresponded to the use of soil 512 (Water bodies), whose values were 238,421.3833 and 28.576, respectively. On the contrary, and as in 1990, there were several land-uses whose ENN values are the lowest and equal. The value is as before—60 for ENN and LSD equal to −0.1466, and land-uses 111 (Continuous urban fabric), 112 (Discontinuous urban fabric), 123 (Port areas), 211 (Non-irrigated arable land), 212 (Permanently irrigated land), 221 (Vineyards), 222 (Fruit trees and berry plantations), 242 (Complex cultivation), 243 (Land principally occupied by agriculture with significant areas of natural vegetation), 312 (Coniferous forest), 322 (Moors and heathland), 323 (Sclerophyllous vegetation), 324 (Transitional woodland/shrub), 331 (Beaches, dunes, and sand plains), 332 (Bare rock), 333 (Sparsely vegetated areas), and 523 (Sea and ocean).

Each of the soil-uses previously classified according to the CLC nomenclature for level 3 are quantified in class metrics. In the case of the shape index, the arithmetic mean was determined. The range of values in this index can be greater than 1 and no limit. Additionally, if the value is equal to 1, then the patch is maximally compact and increases without limit as the patch shape becomes more unpredictable.

Table 12. Class metrics for the Autonomous Region of Madeira.

1990				2018			
TYPE	NP	SHAPE_MN	FRAC_MN	TYPE	NP	SHAPE_MN	FRAC_MN
111	1	2.8721	1.1489	111	3	2.0427	1.096
112	65	2.4168	1.1224	112	64	2.5642	1.1259
121	2	1.9167	1.1027	121	7	2.0104	1.1064
123	7	1.6142	1.0748	122	1	2.3684	1.1397
-	-	-	1.1059	123	22	1.3203	1.0537
133	1	1.5833	1.0701	124	2	2.3378	1.1216
141	1	1.3846	1.0516	131	4	1.8503	1.0992
211	1	2.3649	1.1246	132	1	1.766	1.088
212	7	1.9693	1.0972	133	1	1.8261	1.0955
221	4	2.1828	1.1239	141	1	2.3243	1.1346
222	7	2.1932	1.1139	142	4	2.1027	1.111
231	10	2.002	1.1013	211	1	1.5135	1.07
241	16	2.0334	1.1069	212	1	1.8108	1.0963
242	53	2.0648	1.1047	221	3	1.8528	1.0909
243	89	2.3001	1.1179	222	5	2.5269	1.1397
244	1	2.9783	1.1513	231	5	1.9274	1.094
311	36	2.4974	1.1204	242	61	1.9635	1.1012
312	38	2.0665	1.1014	243	111	2.2117	1.1165
313	66	2.4705	1.1227	311	49	2.1474	1.104
321	36	2.1283	1.1019	312	31	2.071	1.1014
322	54	2.2327	1.1103	313	64	2.4338	1.1234
324	66	1.9424	1.0962	321	34	2.216	1.1077
331	2	4.0851	1.1935	322	45	2.432	1.1195
332	5	2.4709	1.1368	324	51	2.0125	1.0958
333	20	2.436	1.1252	331	2	3.5745	1.1396
523	588	1.2611	1.0499	332	6	2.6791	1.1389
				333	15	2.4625	1.1217
				334	1	1.3333	1.0481
				523	695	1.258	1.05

TYPE = land-use according to CLC nomenclature at level 3; NP = number of patches present in the landscape; SHAPE_MN = arithmetic mean of the Shape Index; FRAC_MN = arithmetic mean of the Fractal Dimension Index.

Moreover, the fractal index dimension reflects the complexity of the shape; in this case, the arithmetic mean was also determined. This index can yield values between 1 and 2. In this way, when the obtained value is close to 1, it indicates that the patches have very simple perimeters, such as squares. However, if the value is close to 2, then the shapes are highly convoluted.

In addition, the number of patches for each of the CLC land-uses was determined for both indexes.

In the case of Azores (Table 11), firstly, we can see that the number of classes remains unchanged in the two years analyzed. Therefore, there has been no increase or decrease in the diversity of land-uses. Likewise, SHAPE_MN values in 1990 are between 1.0321 as the most compact form, corresponding to land-use 222 (Fruit trees and berry plantations) and 2.4216 corresponding to 332 (Sparsely vegetated areas) as the least compact form.

Additionally, the values in 2018 for FRAC_MN range from 1.0321 to 1.1279 for land-uses 222 and 332, again. It can then be said that these are relatively compact forms.

Table 13. Class metrics for the Autonomous Community of the Canary Islands.

1990				2018			
TYPE	NP	SHAPE_MN	FRAC_MN	TYPE	NP	SHAPE_MN	FRAC_MN
111	183	2.0936	1.1051	111	145	2.5214	1.1253
112	86	2.6815	1.1411	112	389	2.7045	1.1389
121	17	1.8542	1.0894	121	82	2.3258	1.1202
123	44	1.6309	1.0658	122	8	3.2236	1.1733
124	8	1.9131	1.0928	123	21	2.4724	1.1349
131	24	1.7422	1.0829	124	10	2.0101	1.0953
132	1	2	1.1022	131	44	1.8618	1.0916
133	18	1.7222	1.078	132	3	1.6195	1.0749
141	2	2.2718	1.1263	133	21	1.8954	1.0936
142	7	1.6707	1.0772	141	7	2.108	1.1153
211	446	2.5147	1.1222	142	39	1.9945	1.1033
212	84	2.1102	1.106	211	213	2.6764	1.1368
221	55	2.6381	1.1255	212	121	2.5278	1.1298
222	97	2.2752	1.1131	221	57	2.6044	1.1278
231	61	2.4163	1.1143	222	107	2.5701	1.1309
243	1	9.098	1.245	231	109	2.3129	1.1161
311	28	2.3986	1.1153	241	5	2.8412	1.1395
312	61	2.4341	1.11	242	148	2.8061	1.1418
322	39	2.9538	1.1399	243	383	2.7984	1.1413
323	175	3.0574	1.1303	311	70	3.0191	1.1537
331	40	2.8667	1.147	312	100	2.6721	1.1209
332	64	2.7405	1.1272	313	7	3.1368	1.1485
333	141	2.1837	1.1022	321	106	2.5026	1.1249
422	1	1.8936	1.0995	322	31	2.8597	1.1429
512	2	1.5015	1.0626	323	216	2.8621	1.1349
521	1	1.3784	1.0546	324	59	2.4474	1.1208
523	2276	1.7549	1.0962	331	54	2.3242	1.1253
				332	141	2.6453	1.1266
				333	209	2.71	1.1218
				334	11	2.1419	1.1046
				422	2	1.8473	1.0939
				512	5	2.2716	1.1262
				521	1	1.5	1.0657
				523	2354	1.7242	1.0946

TYPE = land-use according to CLC nomenclature at level 3; NP = number of patches present in the landscape; SHAPE_MN = arithmetic mean of the Shape Index; FRAC_MN = arithmetic mean of the Fractal Dimension Index.

As for the number of patches in 1990, the values range from 1 for land-uses 222 (Fruit trees and berry plantations), 411 (Inland marshes), 333 (Sparsely vegetated areas), and 221 (Vineyards), and 801 patches for land-use 324 (Transitional woodland shrub), and in 2018, values range from 1 for land-uses 222 to 411 again and 133 (Construction sites) with 1 patch to 799 patches for land-use 523 (be they and ocean-zone seaward of the lowest tide limit). While in 2018, 1 patch is registered for land-uses 133 (construction sites) and again for land-uses 222 and 411. In contrast, the maximum number of patches with 799 is collected for soil-uses 523 also again; in 2018 with 1 patch, again for land-uses 133, 222, and 411, and even for the maximum number of 799 patches corresponding to land-use 523.

In addition, if we look at Table 14, we can see a high variation in the number of patches for certain land-uses, highlighting the most significant decrease for 324 (Transitional woodland/shrub) and the most significant increase for 523. Additionally, considering the total patches in 1990 were 2154, in 2018, it increased to 2300. Therefore, there has been an increase in land-uses.

Table 14. Shannon's Diversity Index and Shannon's Everness Index values for the archipelagos.

Archipelago	Year	SHDI	SHEI
Azores	1990	0.0972	0.0289
Azores	2018	0.0979	0.0291
Madeira	1990	1.7607	0.5342
Madeira	2018	1.6776	0.4932
Canarias	1990	0.571	0.1714
Canarias	2018	0.5998	0.1687

In the case of Madeira (Table 12), if there has been a slight variation in the number of classes, in 1990 there were 26 and in 2018, 29 classes. Therefore, there has been a slight increase in the diversification of land-uses. While it is true that soil-uses 241 (Annual crops associated with permanent crops) and 244 (Agroforestry areas) disappear in 2018, soil-uses 122 (road and rail networks and associated land), 131 (Mineral extraction sites), 132 (Dump sites) and 334 (Burnt areas). In addition, SHAPE_MN values in 1990 are between 1.2611 as the most compact form, corresponding to the use of soil 523 (sea and ocean) and 4.0851 corresponding to 331 (Beaches, dunes, and plains) as the least compact form. Similarly, values in 2018 for FRAC_MN range from 1.0499 to 1.1935 for misused land-uses. So, it can be said that there are relatively compact shapes in general, but also un-compact shapes if we look at Table 14. Additionally, if we compare these results with those obtained for Azores (Table 11), we can say that there are fewer compact soils in Madeira than in the Azores. As for the number of patches in 1990, the values range from 1 for land-use 141 (Green urban areas), 133 (construction sites), 211 (Non-irrigated arable land), 111 (Continuous urban fabric), and 244 A areas), and with the maximum number of patches, that is, 588 for soil-use 523 (Sea and ocean) which would be the most fragmented. While in 2018, with 1 patch are land-uses 122 (Road and rail networks and associated land), 132 (Dump sites), 133 (Construction sites), 141 (Green urban areas), 211 (Non-irrigated arable land), 212 (Permanently irrigated land), and 334 (Burnt areas), and the maximum number of patches with 695 correspondings to land-use 523 (Sea and ocean).

In addition, if we look at Table 14, we can see a high variation in the number of patches for certain land-uses; additionally, considering the total patches in 1990 is 1178, while in 2018, it increased to 1290. Therefore, there has been an increase in soil fragmentation.

In the case of the Canary Islands (Table 13), there has also been an increase in the number of land-uses. As a result, there has also been greater diversification in land-uses. Land-uses did not disappear in 2018, and soil-uses 122 (Road and rail networks associated land), 241 (Annual crops associated with permanent crops), 242 (Complex cultivation), 313 (Mixed forest), 321 (Natural grassland), and 334 (Burnt areas) appear in 2018. Likewise, SHAPE_MN values in 1990 are between 1.3784 as the most compact form, corresponding to land-use 521 (Coastal lagoons) and 9.098 corresponding to 243 (Land principally occupied by agriculture, with significant areas of natural vegetation) as the least compact form. Additionally, the values in 2018 for FRAC_MN range from 1.0546 to 1.245 for the same land-uses. So, it can be said that there are relatively compact shapes in general. Moreover, if we compare these results with those obtained for Azores and Madeira (Tables 11 and 12), we can say that there are fewer compact soils in the Canary Islands than in Madeira and Azores. As for the number of patches in 1990, the values range from 1 for use 521 (Coastal lagoons), 422 (Salines), 132 (Dump sites), 243 (Land principally occupied by agriculture, with significant areas of natural vegetation) being the least fragmented land-uses, and with 2276 patches for 523 (Sea and ocean); and in 2018 with 1 patch for 521 (Coastal lagoons) and a maximum of 2354 patches for 523 again. In addition, if we look at Table 14, we can see that there is the most significant variation in the number of patches of all the archipelagoes analyzed, in 1990 from 3962 to 5278 in 2018. Therefore, in the Autonomous Community of the Canary Islands, it is where there has been the greatest fragmentation in land-uses. Finally, each of the archipelagoes' whole of the geographic space was analyzed using metrics landscape. Precisely, the SHDI was calculated, which is a popular measure of diversity in community ecology, which applies to landscapes. In addition, this index is somewhat more sensitive to rare patch types than Simpson's diversity index. Thus it is possible to display values greater than or equal to zero to infinity. However, when the value is 0, then the geographic space studied contains only a single patch, and there is no diversity. However, as the number of different types of patches increases, the diversity is greater, and/or the proportional distribution of the area between patch types becomes more equitable. The last calculated index was the SHEI, whose values range from 0 to 1. So, when it is equal to 0, the landscape also contains a single patch, and just as before, there is also no diversity. Additionally, as the area's distribution between the different types of polygons representative of land-uses approaches 0, it becomes increasingly unequal. On the contrary, when the index value is 1, the distribution of the area between the types of polygons representative of the land-uses is perfectly uniform. Thus, a uniform distribution of the area between patch types results in maximum uniformity. The whole of the geographic space determined by each of the archipelagoes was analyzed using metrics landscape. Specifically, the SHDI was calculated which is a popular measure of diversity in community ecology, which applies to landscapes. In addition, this index is somewhat more sensitive to rare patch types than Simpson's diversity index. You can display values greater than or equal to zero to infinity. However, when the value is 0 then the geographic space studied contains only a single patch and there is no diversity. However, as the number of different types of patches increases, the diversity is greater, and/or the proportional distribution of the area between patch types becomes more equitable. The last calculated index was the SHEI whose values range from 0 to 1. So when it is equal to 0, the landscape also contains a single patch and just as before, there is also no diversity. Moreover, as the area's distribution between the different types of polygons representative of land-uses approaches 0, it becomes increasingly unequal. On the contrary, when the index value is 1, the distribution of the area between the types of polygons representative of the land-uses is perfectly uniform. Thus, a uniform distribution of the area between patch types results in maximum uniformity.

The values obtained for each of the archipelagoes in the years analyzed (Table 14) allows us to observe that the greatest diversity occurs in Madeira, and on the contrary, the lowest in the Canary Islands. Furthermore, if we look at the temporal evolution, it is possible to verify how in the Azores and Canary Islands, there is a tendency to greater diversity. On the contrary, in Madeira, there is a reduction. In addition, the distribution of the area between the polygons representative of land-uses is more irregular in the Azores, followed by the Canary Islands and finally, Madeira. This inequality also decreases in the years analyzed in the Azores but increases in Madeira and the Canary Islands.

5. Discussion and Conclusions

Considering the study's complexity and, consequently, the large amount of data and results obtained regarding the three archipelagos under analysis, this section was divided into Subsections (Sections 5.1–5.5).

5.1. Azores Archipelago

Land-use shows similar patterns in all the Azores' islands, emphasizing the installation of urban areas next to coastal regions. In fact, some of the obtained results could be explained by the specific geomorphology of each island.

Nevertheless, the predominance of areas related to agricultural and pasture activities, and forests and natural environments between these areas and the islands' interior is evident.

The territorial areas in which the forest and natural vegetation have greater representativeness are those where there is a protection status, attributed under the Regional Network of Protected Areas or the Natura 2000 Network, contributing to the conservation of biodiversity, strengthening the claim of Azores as a nature destination. Here, the Western Group islands assume a considerable weight, and São Jorge and Pico, with the surface's occupations by agricultural areas as natural pastures and landscapes, and the forest and natural vegetation of around 60%. The lagoons, usually constituting points of relevant interest for tourist activity, are only represented on three islands: Corvo, São Miguel, and Flores. However, except Graciosa alone, all other islands have inland water bodies of appreciable beauty.

There is a significant increase in urban areas in evolutionary terms, reflecting the urban growth that has been witnessed in recent years. The agricultural and pasture areas have decreased in recent years, considering that in the 1990s, they represented more than 50% of the Azores area. On the other hand, there was an increase in forest areas and natural environments, when in the middle of the 1990s, they represented nearly 30% of the Azores' territory.

Regarding the artificial occupation of the territory—the urban occupation—about 3.4% of the surface of the territory of the Azores is characterized mainly by discontinuous urban fabric, representing 67% of the total urban fabric of the Azores. Only in the largest island—São Miguel—the continuous urban fabric is predominant, with around 59%. Industry, commerce, general equipment, and infrastructure only represent 0.44% of the Azores' surface total occupation. The islands with the greatest relative implantation of this economic activity are São Miguel and Terceira—the Azorean economy engines.

5.2. Madeira Archipelago

Identifying the different land-uses in the years analyzed in the Madeira archipelago allows for differentiating these land-uses according to their extension, determining which are the most extensive land-uses, and therefore, more hegemonic. Under this criterion, the forest and semi-natural areas constitute the predominant land-use, since it occupies, in all the years analyzed, approximately 70% of the surface of the archipelago, increasing progressively even in all the last years analyzed.

The second most predominant land-use alternates throughout the analyzed years. Between 1990 and 2000 corresponds to agricultural areas. However, between 2006 and 2018, artificial surfaces were the second most predominant land-use, always with percentages of approximately 15%.

In this regard, both agricultural and artificial surfaces between 2012 and 2018 show a downward trend. Likewise, these land-uses are directly related to human activity. Therefore, it could be said that the incidence of humans in the archipelago is becoming less and less impactful in terms of land-uses.

The presence in the archipelago of the remaining land-uses is practically non-existent. Regarding the water bodies, they occupy an area consistently below 1.5%. Besides, they tend to shrink as the years go by. As for the latest use of wetlands, soil practically does not exist. However, both water bodies and wetlands, due to their importance, predominantly environmental, must be monitored in order not to reduce their presence, and thus disappear in years to come.

Regarding the location of the different land-uses on the islands that make up the archipelago, Desert Islands always have the same land-uses; as the name suggests, they are deserted, and the lack of human activity makes the variation in land-use virtually non-existent. The same is not valid on the other islands where other patterns are observed. Thus, on the island of Porto Santo, the artificial surfaces are concentrated in the central region. Besides, these land-uses are usually surrounded by agricultural areas, and forest and semi-natural areas. In this regard, the accumulation of artificial surfaces in the same region could indicate that they are under tourist pressure. As for Madeira's island, the pattern of location of land-uses is reversed, as artificial surfaces are grouped into outlying coastal areas, mainly in the south and also in the north but to a lesser extent. Additionally, the central part is dominated by land-uses unrelated to human activity. Therefore, this area registers less anthropic pressure.

5.3. Canary Archipelago

The predominant land-use corresponds to forests and semi-natural areas, as it occupies approximately three-quarters of the territory. Therefore, this land-use should be used to conserve existing ecosystems in the archipelago. However, over the last three years analyzed (2006, 2012, and 2016), a slight downward trend could continue over the years. However, there is little chance that it will lose its hegemony in the territory analyzed. The second predominant land-use corresponds to agricultural areas, occupying approximately one-fifth of the territory. The third majority use refers to artificial surfaces, occupying approximately 6% of the area analyzed. These last two land-uses are directly related to human activity on the territory, taking into account that their surface area increases slightly over the years analyzed; so, it can be said that the most impactful activity of the human being on the territory analyzed is also increasing slightly. As a result, all activities related to these two land-uses should be monitored. Moreover, water bodies' existence is residual, and curiously, there are no significant records for wetlands.

As for the location of the various land-uses, more significant variation is observed in areas located on some islands' peripheries, where there is an increase of artificial surfaces. Among all the islands stand out Tenerife and Gran Canaria, as they are the most populated areas and there is a more intense tourist activity. There is also less anthropic pressure in the internal regions. Therefore, these regions are less exploited by both population pressure and the tourism industry. In fact, in these areas, the use of forests and semi-natural areas predominate, and are spaces that should serve to preserve the eco-systems of the archipelago.

If we analyze this territory more closely and from different perspectives, it is possible to understand the considerable impact of tourism over all the regional economy. In 2017, economic activity represented 85.7% in services (related to tourism), 7.6% in industry and energy, 5.4% in construction, 3.7% in the manufacturing industry, and 1.3% in agriculture and animal breeding and fishing. In the Canary Archipelago, tourism activity obtains a meaningful relevance in numerous contexts due to its economic potential.

In the Canary Archipelago, tourism is a cultural event in which people move freely for a certain period for various reasons, i.e., recreation, rest, culture, or health [53]. According to Santana [54]: "(...) tourism is a complex and dynamic activity in close relationship with society and its nuances—behavior, motivations, values, history, traditions, and beliefs." In fact, the Canary territory is a tourist destination in the reorientation phase [55]. Contextually, the Canary Islands' Government is committed to "the renewal, innovation and regeneration" of the tourist area, always having northern tourism of higher quality [56]. Additionally, tourism planning inevitably includes planning both in terms of tourism and spatial planning [37]. Therefore, it is common to have land-use policies, namely legislation on the territory's division into fields varying from environmental protection categories to control tourism development spaces [57]. Nonetheless, spatial planning usually results in the "post" when several destinations are previously openly overloaded or threatened [58].

5.4. General Conclusions

The remote islands face many challenges today as land-use changed from 1990 to 2018, along with the increase in the number of visitors, when the demand for living space increased and the resource capacities were limited. As we have seen, it has not been possible to maintain the islands' situation as it was before, and change will continue to be made. What is essential is that the living environment and the island areas' unique character should be preserved and protected; the priorities must be defined, and management strategies established which are significant for the well-being of these highly valued areas. Forestry, as the traditional land-use and predominant land-use in island areas, especially in Madeira Archipelago and Canary Islands (over the last three years analyzed (2006, 2012, and 2016), a slight downward trend) should support sustainable forestry as an appropriate form of land-use. With responsible management, the natural resources of island areas can be used for a long time [61–65]. Degradation of water quantity and quality cannot be overlooked, particularly on some small islands where groundwater conditions have completely changed. It is possible to identify a notable increase in the artificial surfaces (Code 1) in the Azores and Canary islands during the study period. In addition to recognizing that agricultural land is traditional, and agriculture is a valuable activity, it is necessary to encourage agricultural management practices that are compatible with the sustainable development of islands area [65–68]. As examples of these practices, it is possible to name some as: use of renewable energy sources; integrated pest management; hydroponics and aquaponics; crop rotation; polyculture farming; permaculture; avoiding soil erosion; crop diversity; among several others. According to ESPON Program for the Develpment of the Islands [68]: "Due to the relatively small land masses and isolation, islands are typically land-resource constrained. This limits living space, space for infrastructure, waste disposal, agricultural production, industrial development, water resource availability, among several others. Additionally, it results in very vulnerable ecosystems with high endemism." Thus, pressures resulting from anthropogenic factors can have more critical consequences on insular environments, invaliding their capability to supply goods and services, and sustain life.

Detailed analysis metrics have been presented throughout this work as a helpful instrument in the analysis, monitoring, and monitoring of changes in land-uses, through different levels of analysis until their application as a comparison tool in different temporal situations.

In this sense, patch analysis shows that there is greater fragmentation in each of the regions analyzed. In addition, the class analysis makes it possible to establish that there has been no increase in land-uses in the Azores Autonomous Region. On the contrary, in the Autonomous Region of Madeira and the Autonomous Community of the Canary Islands, diversity in land-uses was increased. It could therefore be said that in the latter two regions, anthropic activity has been greater. More compact shapes are generally observed in the latter two regions. In addition, landscape metrics make it possible to state that the Madeira Autonomous Region, despite being the smallest in the extension of the areas analyzed, is the one with the greatest diversity, although in decline in the years analyzed. On the contrary, there is less diversity in the Autonomous Community of the Canary Islands despite being the most extensively analyzed region.

5.5. Study Limitations and Prospective Research Lines

The limitations of this study are directly related to the technical limitations of CLC. In this regard, if we consider the three analysis components that we can deal with a GIS, it is possible to describe each of them.

As for the spatial component, the geometric accuracy has been going down over the years, analyzed from 50 m to be below 10 m. While it has improved considerably, this could have been slightly better in the years leading up to 2018, and it would be optimal for it to be improved over the years to come. Moreover, while it is true that the minimum mapping unit and minimum mapping width have always been the same over the years, 25 hectares and 100 m respectively, it would have been optimal to have lower minimum mapping units. However, the minimum mapping unit is considered appropriate for the extent of the analyzed terrain and the scale of data processing geometric accuracy. Regarding the thematic component, the accuracy in identifying the various land-uses has always been equal to or greater than 85%. Therefore, although there is a slight range of improvement, this accuracy has always been high.

Concerning the landscape metrics recorded, a more specific analysis focused on more specific objectives would be desirable to determine the evolution of some of the land-uses analyzed.

Finally, concerning the temporary component, the last update to land-uses occurred in 2018. It would therefore be optimal if there were more up-to-date land-use data. Besides, note that the products offered by CLC are of high quality and rigorous. Furthermore, these products are open and cover a large area of territory.

Author Contributions: Conceptualization, Rui Alexandre Castanho; methodology, José Manuel Naranjo Gomez and Ana Vulevic; software, José Manuel Naranjo Gomez; validation, Rui Alexandre Castanho, José Manuel Naranjo Gomez and Gualter Couto; formal analysis, Ana Vulevic; investigation, José Manuel Naranjo Gomez; resources, Gualter Couto; data curation, Rui Alexandre Castanho and Ana Vulevic; writing—original draft preparation, Rui Alexandre Castanho and Ana Vulevic; writing—review and editing, Rui Alexandre Castanho; visualization, Gualter Couto; supervision, Rui Alexandre Castanho; project administration, José Manuel Naranjo Gomez; funding acquisition, Rui Alexandre Castanho, José Manuel Naranjo Gomez and Gualter Couto. All authors have read and agreed to the published version of the manuscript.

Funding: This publication has been made possible thanks to funding granted by the Consejería de Economía, ciencia y Agenda Digital from Junta de Extremadura and by the European Regional Development Fund of the European Union through the reference grant GR18052. This paper is financed by Portuguese National Funds through FCT—Fundação para a Ciência e a Tecnologia, I.P., project number UIDB/00685/2020; also, by the program of the Minister of Science and Higher Education titled "Regional Initiative of Excellence" in 2019–2022, project number 018/RID/2018/19, the amount of funding PLN 10 788 423,16.

Data Availability Statement: Data sharing not applicable.

Acknowledgments: This paper is financed by Portuguese National Funds through FCT—Fundação para a Ciência e a Tecnologia, I.P., project number UIDB/00685/2020 and also by the project GREAT—Genuine Rural Experiences in the Azores Tourism with the code: ACORES-01-0145-FEDER-000089. Likewise, this work has been possible thanks to the support of the mobility grant for teaching and research staff of the University of Extremadura and the Technological Centers of the Autonomous Community of Extremadura, in Foreign Centers of Higher Education and/or research for the year 2019—File B23.

Conflicts of Interest: The authors declare no conflict of interest.

Appendix A

Thematic cartography regarding the land-use changes in the Azores Archipelago in the years 1990 and 2018.

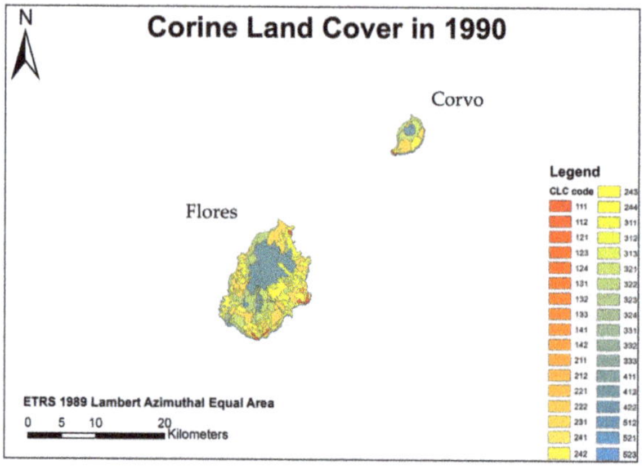

Figure A1. Thematic cartography regarding the land-use changes in the Western Group of Azores Archipelago in the year 1990.

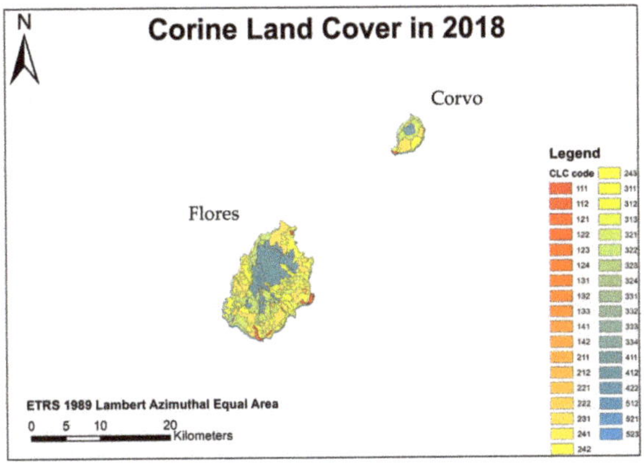

Figure A2. Thematic cartography regarding the land-use changes in the Western Group of Azores Archipelago in the year 2018.

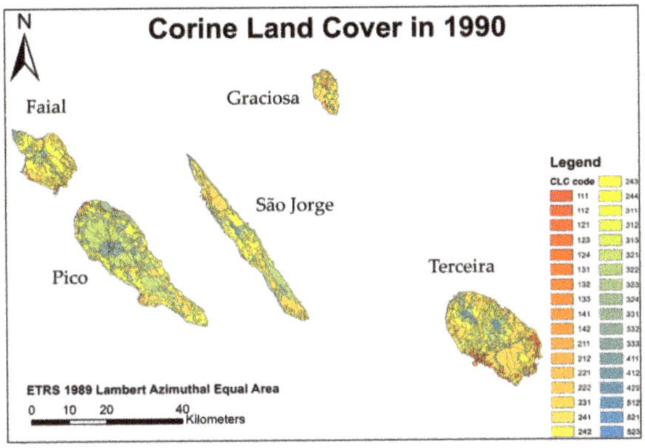

Figure A3. Thematic cartography regarding the land-use changes in the Central Group of Azores Archipelago in the year 1990.

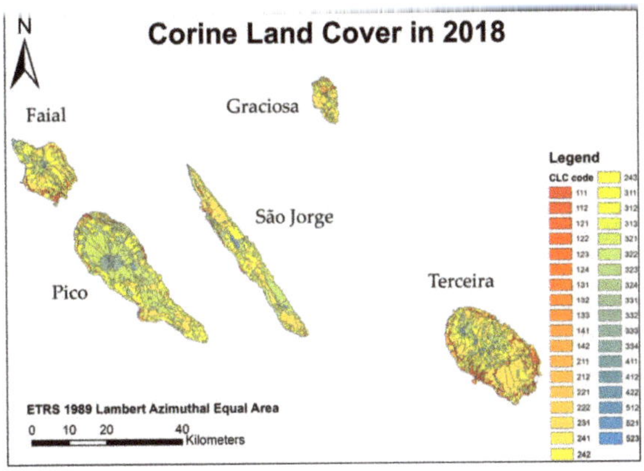

Figure A4. Thematic cartography regarding the land-use changes in the Central Group of Azores Archipelago in the year 2018.

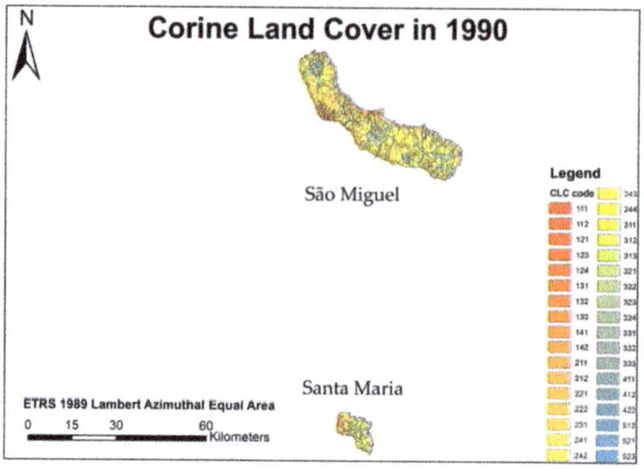

Figure A5. Thematic cartography regarding the land-use changes in the Eastern Group of Azores Archipelago in the year 1990.

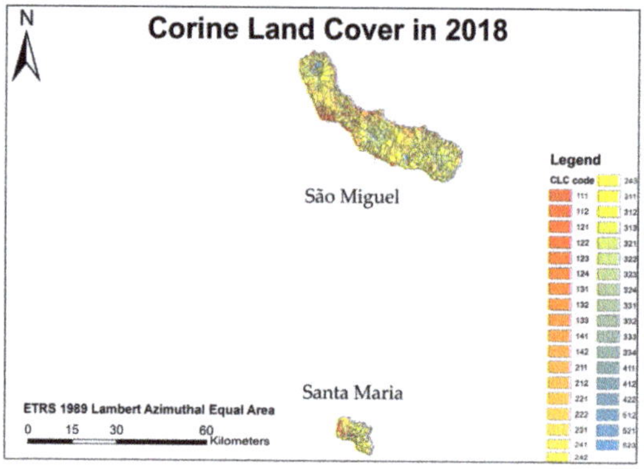

Figure A6. Thematic cartography regarding the land-use changes in the Eastern Group of Azores Archipelago in the year 2018.

Appendix B

Thematic cartography regarding the land-use changes in the Madeira Archipelago in the years 1990 and 2018.

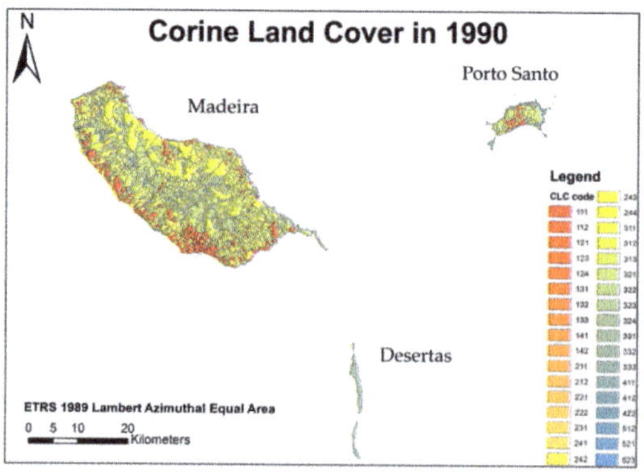

Figure A7. Thematic cartography regarding the land-use changes in the Madeira Archipelago in the year 1990.

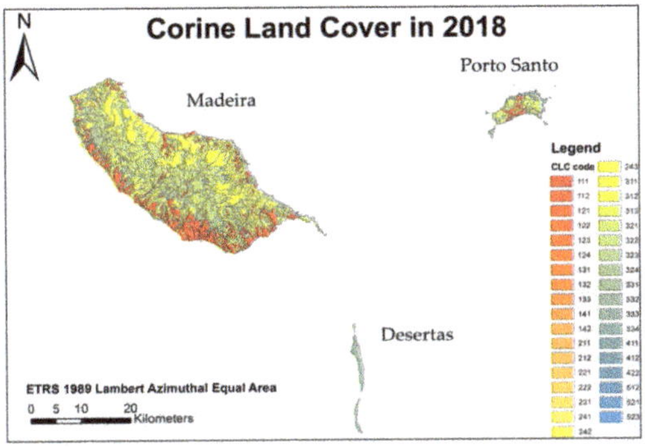

Figure A8. Thematic cartography regarding the land-use changes in the Madeira Archipelago in the year 2018.

Appendix C

Thematic cartography regarding the land-use changes in the Canary Archipelago in the years 1990 and 2018.

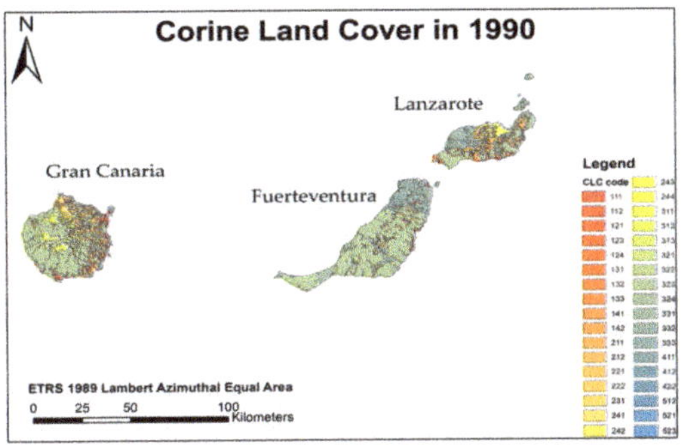

Figure A9. Thematic cartography regarding the land-use changes in the North of Canary Archipelago in the year 1990.

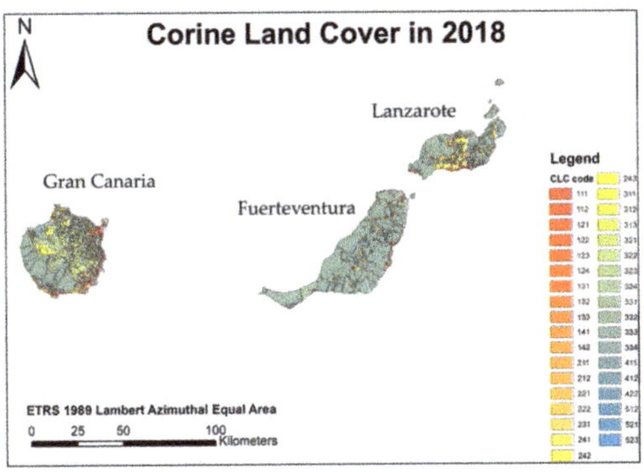

Figure A10. Thematic cartography regarding the land-use changes in the North of Canary Archipelago in the year 2018.

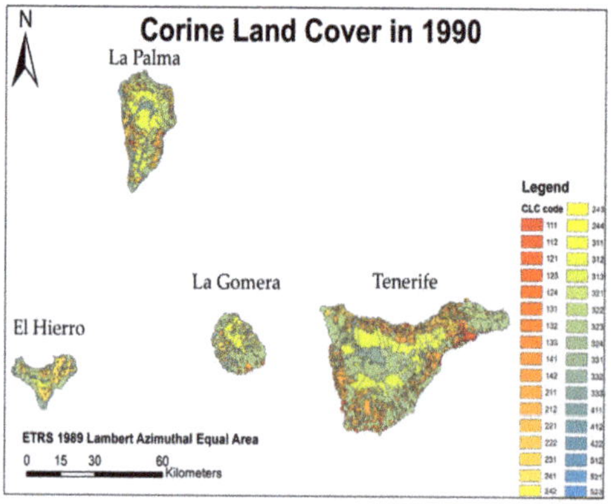

Figure A11. Thematic cartography regarding the land-use changes in the South of Canary Archipelago in the year 1990.

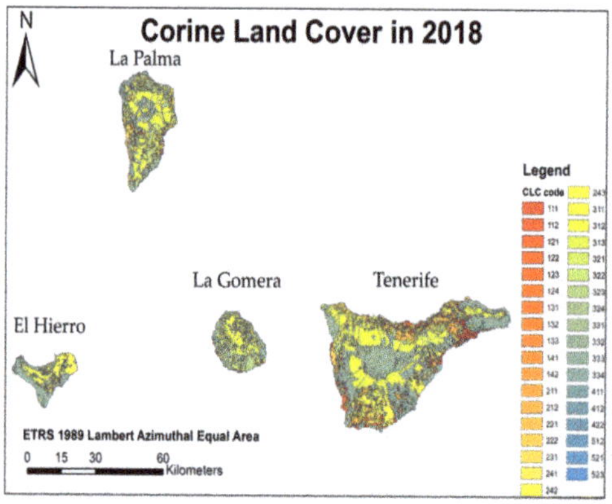

Figure A12. Thematic cartography regarding the land-use changes in the South of Canary Archipelago in the year 2018.

Appendix D

Table A1. Spearman's rank correlation coefficient values in the Azores archipelago for land-uses at level 3 according to the CLC nomenclature between 1990 and 2018.

	c111	c112	c121	c123	c124	c131	c132	c133	c141	c142	c211	c221	c222	c231	c242	c243	c311	c312	c313	c321	c322	c324	c332	c333	c411	c412	c512	c523
c111	1.0	−0.1	−0.1	0.3	0.3	−0.1	−0.7	−0.8	0.3	0.0	0.3	0.7	0.3	0.4	0.3	−0.3	−0.7	−0.1	−0.1	0.1	0.1	0.5	−0.6	−0.3	0.7	0.2	0.1	0.1
c112	−0.1	1.0	1.0	−0.6	0.9	1.0	0.1	0.1	−0.1	0.9	0.9	0.4	−0.6	−0.2	−0.9	−0.9	0.6	0.7	0.7	−1.0	0.0	0.2	0.3	−0.9	−0.5	−0.9	−1.0	−1.0
c121	−0.1	1.0	1.0	−0.6	0.9	1.0	0.1	0.1	−0.1	0.9	0.9	0.4	−0.6	−0.2	−0.9	−0.9	0.6	0.7	0.7	−1.0	0.0	0.2	0.3	−0.9	−0.5	−0.9	−1.0	−1.0
c123	0.3	−0.6	−0.6	1.0	−0.5	−0.6	−0.3	−0.7	−0.5	−0.7	−0.5	0.2	1.0	−0.4	0.7	0.5	−0.2	−0.9	−0.9	0.6	0.6	−0.6	−0.1	0.5	0.7	0.3	0.6	0.6
c124	0.3	0.9	0.9	−0.5	1.0	0.9	−0.3	−0.2	0.0	0.8	1.0	0.7	−0.5	0.1	−0.8	−1.0	0.3	0.6	0.6	−0.9	0.1	0.4	−0.1	−1.0	−0.3	−0.7	−0.9	−0.9
c131	−0.1	1.0	1.0	−0.6	0.9	1.0	0.1	0.1	−0.1	0.9	0.9	0.4	−0.6	−0.2	−0.9	−0.9	0.6	0.7	0.7	−1.0	0.0	0.2	0.3	−0.9	−0.5	−0.9	−1.0	−1.0
c132	−0.7	0.1	0.1	−0.3	−0.3	0.1	1.0	0.7	0.2	0.3	−0.3	−0.8	−0.3	−0.5	0.0	0.3	0.3	0.4	0.4	−0.1	−0.6	−0.1	0.9	0.3	−0.2	−0.3	−0.1	−0.1
c133	−0.8	0.1	0.1	−0.7	−0.2	0.1	0.7	1.0	0.3	0.2	−0.2	−0.8	−0.7	0.1	−0.3	0.2	0.3	0.5	0.5	−0.1	−0.6	0.1	0.4	0.2	−0.7	0.0	−0.1	−0.1
c141	0.3	−0.1	−0.1	−0.5	0.0	−0.1	0.2	0.3	1.0	0.3	0.0	−0.3	−0.5	0.6	0.2	0.0	−0.7	0.6	0.6	0.1	−0.9	0.9	−0.1	0.0	0.2	0.3	0.1	0.1
c142	0.0	0.9	0.9	−0.7	0.8	0.9	0.3	0.2	0.3	1.0	0.8	0.2	−0.7	−0.1	−0.7	−0.8	0.3	0.9	0.9	−0.9	−0.4	0.5	0.4	−0.8	−0.3	−0.8	−0.9	−0.9
c211	0.3	0.9	0.9	−0.5	1.0	0.9	−0.3	−0.2	0.0	0.8	1.0	0.7	−0.5	0.1	−0.8	−1.0	0.3	0.6	0.6	−0.9	0.1	0.4	−0.1	−1.0	−0.3	−0.7	−0.9	−0.9
c221	0.7	0.4	0.4	0.2	0.7	0.4	−0.8	−0.8	−0.3	0.2	0.7	1.0	0.2	0.1	−0.3	−0.7	0.0	−0.1	−0.1	−0.4	0.6	0.1	−0.5	−0.7	0.2	−0.3	−0.4	−0.4
c222	0.3	−0.6	−0.6	1.0	−0.5	−0.6	−0.3	−0.7	−0.5	−0.7	−0.5	0.2	1.0	−0.4	0.7	0.5	−0.2	−0.9	−0.9	0.6	0.6	−0.6	−0.1	0.5	0.7	0.3	0.6	0.6
c231	0.4	−0.2	−0.2	−0.4	0.1	−0.2	−0.5	0.1	0.6	−0.1	0.1	0.1	−0.4	1.0	0.0	−0.1	−0.6	0.2	0.2	0.2	−0.3	0.7	−0.8	−0.1	−0.1	0.6	0.2	0.2
c242	0.3	−0.9	−0.9	0.7	−0.8	−0.9	0.0	−0.3	0.2	−0.7	−0.8	−0.3	0.7	0.0	1.0	0.8	−0.7	−0.6	−0.6	0.9	−0.1	−0.1	−0.1	0.8	0.8	0.7	0.9	0.9
c243	−0.3	−0.9	−0.9	0.5	−1.0	−0.9	0.3	0.2	0.0	−0.8	−1.0	−0.7	0.5	−0.1	0.8	1.0	−0.3	−0.6	−0.6	0.9	−0.1	−0.4	0.1	1.0	0.3	0.7	0.9	0.9
c311	−0.7	0.6	0.6	−0.2	0.3	0.6	0.3	0.3	−0.7	0.3	0.3	0.0	−0.2	−0.6	−0.7	−0.3	1.0	0.1	0.1	−0.6	0.4	−0.6	0.5	−0.3	−0.7	−0.7	−0.6	−0.6
c312	−0.1	0.7	0.7	−0.9	0.6	0.7	0.4	0.5	0.6	0.9	0.6	−0.1	−0.9	0.2	−0.6	−0.6	0.1	1.0	1.0	−0.7	−0.7	0.7	0.3	−0.6	−0.4	−0.5	−0.7	−0.7
c313	−0.1	0.7	0.7	−0.9	0.6	0.7	0.4	0.5	0.6	0.9	0.6	−0.1	−0.9	0.2	−0.6	−0.6	0.1	1.0	1.0	−0.7	−0.7	0.7	0.3	−0.6	−0.4	−0.5	−0.7	−0.7
c321	0.1	−1.0	−1.0	0.6	−0.9	−1.0	−0.1	−0.1	0.1	−0.9	−0.9	−0.4	0.6	0.2	0.9	0.9	−0.6	−0.7	−0.7	1.0	0.0	−0.2	−0.3	0.9	0.5	0.9	1.0	1.0
c322	0.1	0.0	0.0	0.6	0.1	0.0	−0.6	−0.6	−0.9	−0.4	0.1	0.6	0.6	−0.3	−0.1	−0.1	0.4	−0.7	−0.7	0.0	1.0	−0.7	−0.3	−0.1	0.0	−0.1	0.0	0.0
c324	0.5	0.2	0.2	−0.6	0.4	0.2	−0.1	0.1	0.9	0.5	0.4	0.1	−0.6	0.7	−0.1	−0.4	−0.6	0.7	0.7	−0.2	−0.7	1.0	−0.3	−0.4	0.1	0.1	−0.2	−0.2
c332	−0.6	0.3	0.3	−0.1	−0.1	0.3	0.9	0.4	−0.1	0.4	−0.1	−0.5	−0.1	−0.8	−0.1	0.1	0.5	0.3	0.3	−0.3	−0.3	−0.3	1.0	0.1	−0.1	−0.6	−0.3	−0.3
c333	−0.3	−0.9	−0.9	0.5	−1.0	−0.9	0.3	0.2	0.0	−0.8	−1.0	−0.7	0.5	−0.1	0.8	1.0	−0.3	−0.6	−0.6	0.9	−0.1	−0.4	0.1	1.0	0.3	0.7	0.9	0.9
c411	0.7	−0.5	−0.5	0.7	−0.3	−0.5	−0.2	−0.7	0.2	−0.3	−0.3	0.2	0.7	−0.1	0.8	0.3	−0.7	−0.4	−0.4	0.5	0.0	0.1	−0.1	0.3	1.0	0.3	0.5	0.5
c412	0.2	−0.9	−0.9	0.3	−0.7	−0.9	−0.3	0.0	0.3	−0.8	−0.7	−0.3	0.3	0.6	0.7	0.7	−0.7	−0.5	−0.5	0.9	−0.1	0.1	−0.6	0.7	0.3	1.0	0.9	0.9
c512	0.1	−1.0	−1.0	0.6	−0.9	−1.0	−0.1	−0.1	0.1	−0.9	−0.9	−0.4	0.6	0.2	0.9	0.9	−0.6	−0.7	−0.7	1.0	0.0	−0.2	−0.3	0.9	0.5	0.9	1.0	1.0
c523	0.1	−1.0	−1.0	0.6	−0.9	−1.0	−0.1	−0.1	0.1	−0.9	−0.9	−0.4	0.6	0.2	0.9	0.9	−0.6	−0.7	−0.7	1.0	0.0	−0.2	−0.3	0.9	0.5	0.9	1.0	1.0

The Spearman's rank correlation coefficient value 1.0 is highlighted when the values correspond to the same variable.

Appendix E

Table A2. Spearman's rank correlation coefficient values in the Madeira archipelago for land-uses at level 3 according to the CLC nomenclature between 1990 and 2018.

	c111	c112	c121	c122	c123	c124	c131	c132	c133	c141	c142	c211	c212	c221	c222	c231	c241	c242	c243	c244	c311	c312	c313	c321	c322	c324	c331	c332	c333	c334	c523
c111	1.0	0.7	1.0	0.6	0.7	0.7	1.0	0.8	−0.3	−1.0	0.9	−1.0	−1.0	0.7	−0.8	−0.9	−1.0	−0.6	−0.3	−0.9	−0.7	−0.9	−0.9	0.9	0.4	0.2	−0.6	1.0	−0.7	0.4	−1.0
c112	0.7	1.0	0.8	1.0	1.0	1.0	0.7	0.9	0.3	−0.7	0.9	−0.7	−0.7	0.5	−1.0	−0.7	−0.6	−1.0	−0.6	−0.4	−0.4	−0.6	−0.5	0.5	0.1	−0.1	−0.2	0.7	−0.3	0.3	−0.7
c121	1.0	0.8	1.0	0.8	0.8	0.8	0.9	0.9	−0.2	−0.9	1.0	−0.9	−0.9	0.7	−0.9	−0.9	−0.9	−0.8	−0.5	−0.7	−0.7	−0.9	−0.9	0.8	0.4	0.2	−0.5	0.9	−0.7	0.4	−0.9
c122	0.6	1.0	0.8	1.0	1.0	1.0	0.6	0.9	0.3	−0.6	0.9	−0.6	−0.6	0.4	−0.9	−0.6	−0.6	−1.0	−0.7	−0.3	−0.4	−0.6	−0.5	0.5	0.2	0.0	−0.3	0.6	−0.3	0.3	−0.6
c123	0.7	1.0	0.8	1.0	1.0	1.0	0.6	0.9	0.3	−0.6	0.9	−0.6	−0.6	0.4	−0.9	−0.6	−0.6	−0.9	−0.8	−0.3	−0.5	−0.6	−0.6	0.5	0.3	0.1	−0.4	0.6	−0.5	0.2	−0.6
c124	0.7	1.0	0.8	1.0	1.0	1.0	0.6	0.9	0.3	−0.6	0.9	−0.6	−0.6	0.4	−0.9	−0.6	−0.6	−0.9	−0.7	−0.3	−0.5	−0.6	−0.6	0.5	0.2	0.0	−0.3	0.6	−0.4	0.2	−0.6
c131	1.0	0.7	0.9	0.6	0.6	0.6	1.0	0.8	−0.3	−1.0	0.9	−1.0	−1.0	0.7	−0.8	−1.0	−1.0	−0.6	−0.2	−0.9	−0.6	−0.8	−0.8	0.9	0.3	0.1	−0.4	1.0	−0.6	0.4	−1.0
c132	0.8	0.9	0.9	0.9	0.9	0.9	0.8	1.0	0.2	−0.8	0.9	−0.8	−0.8	0.5	−1.0	−0.9	−0.8	−0.9	−0.4	−0.6	−0.4	−0.6	−0.6	0.8	0.2	0.0	−0.3	0.8	−0.4	0.3	−0.9
c133	−0.3	0.3	−0.2	0.3	0.3	0.3	−0.3	0.2	1.0	0.3	0.0	0.3	0.3	−0.6	−0.1	0.2	0.3	−0.4	−0.2	0.5	0.7	0.5	0.5	−0.2	−0.3	−0.3	0.4	−0.3	0.5	−0.4	0.3
c141	−1.0	−0.7	−0.9	−0.6	−0.6	−0.6	−1.0	−0.8	0.3	1.0	−0.9	1.0	1.0	−0.7	0.8	1.0	1.0	0.6	0.2	0.9	0.6	0.8	0.8	−0.9	−0.3	−0.1	0.4	−1.0	0.6	−0.4	1.0
c142	0.9	0.9	1.0	0.9	0.9	0.9	0.9	0.9	0.0	−0.9	1.0	−0.9	−0.9	0.7	−1.0	−0.8	−0.9	−0.9	−0.6	−0.6	−0.7	−0.8	−0.8	0.7	0.3	0.1	−0.4	0.9	−0.5	0.4	−0.9
c211	−1.0	−0.7	−0.9	−0.6	−0.6	−0.6	−1.0	−0.8	0.3	1.0	−0.9	1.0	1.0	−0.7	0.8	1.0	1.0	0.6	0.2	0.9	0.6	0.8	0.8	−0.9	−0.3	−0.1	0.4	−1.0	0.6	−0.4	1.0
c212	−1.0	−0.7	−0.9	−0.6	−0.6	−0.6	−1.0	−0.8	0.3	1.0	−0.9	1.0	1.0	−0.7	0.8	1.0	1.0	0.6	0.2	0.9	0.6	0.8	0.8	−0.9	−0.3	−0.1	0.4	−1.0	0.6	−0.4	1.0
c221	0.7	0.5	0.7	0.4	0.4	0.4	0.7	0.5	−0.6	−0.7	0.7	−0.7	−0.7	1.0	−0.6	−0.7	−0.6	−0.5	0.0	−0.7	−0.8	−0.7	−0.7	0.4	−0.1	−0.3	−0.1	0.7	−0.3	0.9	−0.7
c222	−0.8	−1.0	−0.9	−0.9	−0.9	−0.9	−0.8	−1.0	−0.1	0.8	−1.0	0.8	0.8	−0.6	1.0	0.8	0.8	0.9	0.6	0.6	0.6	0.7	0.7	−0.7	−0.2	0.0	0.4	−0.8	0.5	−0.4	0.8
c231	−0.9	−0.7	−0.9	−0.6	−0.6	−0.6	−1.0	−0.9	0.2	1.0	−0.8	1.0	1.0	−0.7	0.8	1.0	0.9	0.7	0.1	0.9	0.5	0.7	0.7	−0.9	−0.1	0.1	0.2	−1.0	0.4	−0.5	1.0
c241	−1.0	−0.6	−0.9	−0.6	−0.6	−0.6	−1.0	−0.8	0.3	1.0	−0.9	1.0	1.0	−0.6	0.8	0.9	1.0	0.6	0.3	0.9	0.7	0.9	0.9	−0.9	−0.5	−0.3	0.6	−1.0	0.7	−0.3	1.0
c242	−0.6	−1.0	−0.8	−1.0	−0.9	−0.9	−0.6	−0.9	−0.4	0.6	−0.9	0.6	0.6	−0.5	0.9	0.7	0.6	1.0	0.5	0.3	0.3	0.5	0.4	−0.4	0.1	0.3	0.0	−0.6	0.1	−0.4	0.6
c243	−0.3	−0.6	−0.5	−0.7	−0.8	−0.7	−0.2	−0.4	−0.2	0.2	−0.6	0.2	0.2	0.0	0.6	0.1	0.3	0.5	1.0	−0.1	0.5	0.5	0.5	−0.2	−0.7	−0.6	0.7	−0.2	0.6	0.2	0.2
c244	−0.9	−0.4	−0.7	−0.3	−0.3	−0.3	−0.9	−0.6	0.5	0.9	−0.6	0.9	0.9	−0.7	0.6	0.9	0.9	0.3	−0.1	1.0	0.5	0.7	0.7	−0.9	−0.3	−0.1	0.4	−0.9	0.5	−0.4	0.9
c311	−0.7	−0.4	−0.7	−0.4	−0.5	−0.5	−0.6	−0.4	0.7	0.6	−0.7	0.6	0.6	−0.8	0.6	0.5	0.7	0.3	0.5	0.5	1.0	0.9	0.9	−0.5	−0.5	−0.3	0.7	−0.6	0.8	−0.5	0.6
c312	−0.9	−0.6	−0.9	−0.6	−0.6	−0.6	−0.8	−0.6	0.5	0.8	−0.8	0.8	0.8	−0.7	0.7	0.7	0.9	0.5	0.5	0.7	0.9	1.0	1.0	−0.8	−0.6	−0.4	0.8	−0.8	0.9	−0.3	0.8
c313	−0.9	−0.5	−0.9	−0.5	−0.6	−0.6	−0.8	−0.6	0.5	0.8	−0.8	0.8	0.8	−0.7	0.7	0.7	0.9	0.4	0.5	0.7	0.9	1.0	1.0	−0.7	−0.7	−0.5	0.8	−0.8	0.9	−0.3	0.8
c321	0.9	0.5	0.8	0.5	0.5	0.5	0.9	0.8	−0.2	−0.9	0.7	−0.9	−0.9	0.4	−0.7	−0.9	−0.9	−0.4	−0.2	−0.9	−0.5	−0.8	−0.7	1.0	0.5	0.3	−0.6	0.9	−0.7	0.1	−0.9
c322	0.4	0.1	0.4	0.2	0.3	0.2	0.3	0.2	−0.3	−0.3	0.3	−0.3	−0.3	−0.1	−0.2	−0.1	−0.5	0.1	−0.7	−0.3	−0.5	−0.6	−0.7	0.5	1.0	1.0	−1.0	0.3	−0.9	−0.5	−0.3
c324	0.2	−0.1	0.2	0.0	0.1	0.0	0.1	0.0	−0.3	−0.1	0.1	−0.1	−0.1	−0.3	0.0	0.1	−0.3	0.3	−0.6	−0.1	−0.3	−0.4	−0.5	0.3	1.0	1.0	−0.9	0.1	−0.8	−0.6	−0.1
c331	−0.6	−0.2	−0.5	−0.3	−0.4	−0.3	−0.4	−0.3	0.4	0.4	−0.4	0.4	0.4	−0.1	0.4	0.2	0.6	0.0	0.7	0.4	0.7	0.8	0.8	−0.6	−1.0	−0.9	1.0	−0.4	1.0	0.2	0.4
c332	1.0	0.7	0.9	0.6	0.6	0.6	1.0	0.8	−0.3	−1.0	0.9	−1.0	−1.0	0.7	−0.8	−1.0	−1.0	−0.6	−0.2	−0.9	−0.6	−0.8	−0.8	0.9	0.3	0.1	−0.4	1.0	−0.6	0.4	−1.0
c333	−0.7	−0.3	−0.7	−0.3	−0.5	−0.4	−0.6	−0.4	0.5	0.6	−0.5	0.6	0.6	−0.3	0.5	0.4	0.7	0.1	0.6	0.5	0.8	0.9	0.9	−0.7	−0.9	−0.8	1.0	−0.6	1.0	0.1	0.6
c334	0.4	0.3	0.4	0.3	0.2	0.2	0.4	0.3	−0.4	−0.4	0.4	−0.4	−0.4	0.9	−0.4	−0.5	−0.3	−0.4	0.2	−0.4	−0.5	−0.3	−0.3	0.1	−0.5	−0.6	0.2	0.4	0.1	1.0	−0.4
c523	−1.0	−0.7	−0.9	−0.6	−0.6	−0.6	−1.0	−0.9	0.3	1.0	−0.9	1.0	1.0	−0.7	0.8	1.0	1.0	0.6	0.2	0.9	0.6	0.8	0.8	−0.9	−0.3	−0.1	0.4	−1.0	0.6	−0.4	1.0

The Spearman's rank correlation coefficient value 1.0 is highlighted when the values correspond to the same variable.

Appendix F

Table A3. Spearman's rank correlation coefficient values in the Canary Islands archipelago for land-uses at level 3 according to the CLC nomenclature between 1990 and 2018.

	c111	c112	c121	c122	c123	c124	c131	c132	c133	c141	c142	c211	c212	c213	c221	c222	c231	c241	c242	c243	c311	c312	c313	c321	c322	c323	c324	c331	c332	c333	c334	c422	c512	cC523
c111	1.0	−0.8	−0.8	−0.2	−0.8	−0.5	−0.8	−1.0	0.0	−0.7	−0.8	0.8	−0.5	−0.2	0.7	−0.7	0.8	−0.7	−1.0	−0.8	−0.6	−0.7	−0.9	−1.0	0.6	0.8	−0.9	0.1	−0.6	−1.0	−0.9	−0.6	−0.8	0.8
c112	−0.8	1.0	1.0	0.6	1.0	0.7	1.0	0.8	−0.2	0.5	1.0	−1.0	0.7	0.3	−0.5	0.5	−1.0	0.6	0.9	1.0	0.6	0.5	1.0	0.9	−0.6	−1.0	1.0	−0.1	0.6	0.8	1.0	0.9	1.0	−1.0
c121	−0.8	1.0	1.0	0.6	1.0	0.7	1.0	0.8	−0.2	0.5	1.0	−1.0	0.7	0.3	−0.5	0.5	−1.0	0.6	0.9	1.0	0.6	0.5	1.0	0.9	−0.6	−1.0	1.0	−0.1	0.6	0.8	1.0	0.9	1.0	−1.0
c122	−0.2	0.6	0.6	1.0	0.6	0.1	0.6	0.2	−0.8	−0.4	0.6	−0.6	0.1	−0.3	0.4	−0.4	−0.6	−0.2	0.3	0.6	−0.1	−0.4	0.5	0.3	0.1	−0.6	0.5	−0.7	−0.1	0.2	0.5	0.4	0.6	−0.6
c123	−0.8	1.0	1.0	0.6	1.0	0.7	1.0	0.8	−0.2	0.5	1.0	−1.0	0.7	0.3	−0.5	0.5	−1.0	0.6	0.9	1.0	0.6	0.5	1.0	0.9	−0.6	−1.0	1.0	−0.1	0.6	0.8	1.0	0.9	1.0	−1.0
c124	−0.5	0.7	0.7	0.1	0.7	1.0	0.7	0.5	0.5	0.8	0.7	−0.7	1.0	0.9	−0.8	0.8	−0.7	0.9	0.6	0.7	0.9	0.8	0.7	0.6	−0.9	−0.7	0.7	0.6	0.9	0.5	0.7	0.9	0.7	−0.7
c131	−0.8	1.0	1.0	0.6	1.0	0.7	1.0	0.8	−0.2	0.5	1.0	−1.0	0.7	0.3	−0.5	0.5	−1.0	0.6	0.9	1.0	0.6	0.5	1.0	0.9	−0.6	−1.0	1.0	−0.1	0.6	0.8	1.0	0.9	1.0	−1.0
c132	−1.0	0.8	0.8	0.2	0.8	0.5	0.8	1.0	0.0	0.7	0.8	−0.8	0.5	0.2	−0.7	0.7	−0.8	0.7	1.0	0.8	0.6	0.7	0.9	1.0	−0.6	−0.8	0.9	−0.1	0.6	1.0	0.9	0.6	0.8	−0.8
c133	0.0	−0.2	−0.2	−0.8	−0.2	0.5	−0.2	0.0	1.0	0.7	−0.2	0.2	0.5	0.8	−0.7	0.7	0.2	0.7	−0.1	−0.2	0.6	0.7	−0.2	−0.1	−0.6	0.2	−0.2	0.9	0.6	0.0	−0.2	0.1	−0.2	0.2
c141	−0.7	0.5	0.5	−0.4	0.5	0.8	0.5	0.7	0.7	1.0	0.5	−0.5	0.8	0.8	−1.0	1.0	−0.5	1.0	0.7	0.5	0.9	1.0	0.6	0.7	−0.9	−0.5	0.6	0.6	0.9	0.7	0.6	0.6	0.5	−0.5
c142	−0.8	1.0	1.0	0.6	1.0	0.7	1.0	0.8	−0.2	0.5	1.0	−1.0	0.7	0.3	−0.5	0.5	−1.0	0.6	0.9	1.0	0.6	0.5	1.0	0.9	−0.6	−1.0	1.0	−0.1	0.6	0.8	1.0	0.9	1.0	−1.0
c211	0.8	−1.0	−1.0	−0.6	−1.0	−0.7	−1.0	−0.8	0.2	−0.5	−1.0	1.0	−0.7	−0.3	0.5	−0.5	1.0	−0.6	−0.9	−1.0	−0.6	−0.5	−1.0	−0.9	0.6	1.0	−1.0	0.1	−0.6	−0.8	−1.0	−0.9	−1.0	1.0
c212	−0.5	0.7	0.7	0.1	0.7	1.0	0.7	0.5	0.5	0.8	0.7	−0.7	1.0	0.9	−0.8	0.8	−0.7	0.9	0.6	0.7	0.9	0.8	0.7	0.6	−0.9	−0.7	0.7	0.6	0.9	0.5	0.7	0.9	0.7	−0.7
c213	−0.2	0.3	0.3	−0.3	0.3	0.9	0.3	0.2	0.8	0.8	0.3	−0.3	0.9	1.0	−0.8	0.8	−0.3	0.8	0.2	0.3	0.8	0.8	0.3	0.2	−0.8	−0.3	0.3	0.9	0.8	0.2	0.3	0.7	0.3	−0.3
c221	0.7	−0.5	−0.5	0.4	−0.5	−0.8	−0.5	−0.7	−0.7	−1.0	−0.5	0.5	−0.8	−0.8	1.0	−1.0	0.5	−1.0	−0.7	−0.5	−0.9	−1.0	−0.6	−0.7	0.9	0.5	−0.6	−0.6	−0.9	−0.7	−0.6	−0.6	−0.5	0.5
c222	−0.7	0.5	0.5	−0.4	0.5	0.8	0.5	0.7	0.7	1.0	0.5	−0.5	0.8	0.8	−1.0	1.0	−0.5	1.0	0.7	0.5	0.9	1.0	0.6	0.7	−0.9	−0.5	0.6	0.6	0.9	0.7	0.6	0.6	0.5	−0.5
c231	0.8	−1.0	−1.0	−0.6	−1.0	−0.7	−1.0	−0.8	0.2	−0.5	−1.0	1.0	−0.7	−0.3	0.5	−0.5	1.0	−0.6	−0.9	−1.0	−0.6	−0.5	−1.0	−0.9	0.6	1.0	−1.0	0.1	−0.6	−0.8	−1.0	−0.9	−1.0	1.0
c241	−0.7	0.6	0.6	−0.2	0.6	0.9	0.6	0.7	0.7	1.0	0.6	−0.6	0.9	0.8	−1.0	1.0	−0.6	1.0	0.7	0.6	1.0	1.0	0.6	0.7	−1.0	−0.6	0.6	0.6	1.0	0.7	0.6	0.7	0.6	−0.6
c242	−1.0	0.9	0.9	0.3	0.9	0.6	0.9	1.0	−0.1	0.7	0.9	−0.9	0.6	0.2	−0.7	0.7	−0.9	0.7	1.0	0.9	0.7	0.7	0.9	1.0	−0.7	−0.9	0.9	−0.2	0.7	1.0	0.9	0.7	0.9	−0.9
c243	−0.8	1.0	1.0	0.6	1.0	0.7	1.0	0.8	−0.2	0.5	1.0	−1.0	0.7	0.3	−0.5	0.5	−1.0	0.6	0.9	1.0	0.6	0.5	1.0	0.9	−0.6	−1.0	1.0	−0.1	0.6	0.8	1.0	0.9	1.0	−1.0
c311	−0.6	0.6	0.6	−0.1	0.6	0.9	0.6	0.6	0.6	0.9	0.6	−0.6	0.9	0.8	−0.9	0.9	−0.6	1.0	0.7	0.6	1.0	0.9	0.6	0.7	−1.0	−0.6	0.6	0.5	1.0	0.6	0.6	0.7	0.6	−0.6
c312	−0.7	0.5	0.5	−0.4	0.5	0.8	0.5	0.7	0.7	1.0	0.5	−0.5	0.8	0.8	−1.0	1.0	−0.5	1.0	0.7	0.5	0.9	1.0	0.6	0.7	−0.9	−0.5	0.6	0.6	0.9	0.7	0.6	0.6	0.5	−0.5
c313	−0.9	1.0	1.0	0.5	1.0	0.7	1.0	0.9	−0.2	0.6	1.0	−1.0	0.7	0.3	−0.6	0.6	−1.0	0.6	0.9	1.0	0.6	0.6	1.0	0.9	−0.6	−1.0	1.0	−0.1	0.6	0.9	1.0	0.9	1.0	−1.0
c321	−1.0	0.9	0.9	0.3	0.9	0.6	0.9	1.0	−0.1	0.7	0.9	−0.9	0.6	0.2	−0.7	0.7	−0.9	0.7	1.0	0.9	0.7	0.7	0.9	1.0	−0.7	−0.9	0.9	−0.2	0.7	1.0	0.9	0.7	0.9	−0.9
c322	0.6	−0.6	−0.6	0.1	−0.6	−0.9	−0.6	−0.6	−0.6	−0.9	−0.6	0.6	−0.9	−0.8	0.9	−0.9	0.6	−1.0	−0.7	−0.6	−1.0	−0.9	−0.6	−0.7	1.0	0.6	−0.6	−0.5	−1.0	−0.6	−0.6	−0.7	−0.6	0.6
c323	0.8	−1.0	−1.0	−0.6	−1.0	−0.7	−1.0	−0.8	0.2	−0.5	−1.0	1.0	−0.7	−0.3	0.5	−0.5	1.0	−0.6	−0.9	−1.0	−0.6	−0.5	−1.0	−0.9	0.6	1.0	−1.0	0.1	−0.6	−0.8	−1.0	−0.9	−1.0	1.0
c324	−0.9	1.0	1.0	0.5	1.0	0.7	1.0	0.9	−0.2	0.6	1.0	−1.0	0.7	0.3	−0.6	0.6	−1.0	0.6	0.9	1.0	0.6	0.6	1.0	0.9	−0.6	−1.0	1.0	−0.1	0.6	0.9	1.0	0.9	1.0	−1.0
c331	0.1	−0.1	−0.1	−0.7	−0.1	0.6	−0.1	−0.1	0.9	0.6	−0.1	0.1	0.6	0.9	−0.6	0.6	0.1	0.6	−0.2	−0.1	0.5	0.6	−0.1	−0.2	−0.5	0.1	−0.1	1.0	0.5	−0.1	−0.1	0.3	−0.1	0.1
c332	−0.6	0.6	0.6	−0.1	0.6	0.9	0.6	0.6	0.6	0.9	0.6	−0.6	0.9	0.8	−0.9	0.9	−0.6	1.0	0.7	0.6	1.0	0.9	0.6	0.7	−1.0	−0.6	0.6	0.5	1.0	0.6	0.6	0.7	0.6	−0.6
c333	−1.0	0.8	0.8	0.2	0.8	0.5	0.8	1.0	0.0	0.7	0.8	−0.8	0.5	0.2	−0.7	0.7	−0.8	0.7	1.0	0.8	0.6	0.7	0.9	1.0	−0.6	−0.8	0.9	−0.1	0.6	1.0	0.9	0.6	0.8	−0.8
c334	−0.9	1.0	1.0	0.5	1.0	0.7	1.0	0.9	−0.2	0.6	1.0	−1.0	0.7	0.3	−0.6	0.6	−1.0	0.6	0.9	1.0	0.6	0.6	1.0	0.9	−0.6	−1.0	1.0	−0.1	0.6	0.9	1.0	0.9	1.0	−1.0
c422	−0.6	0.9	0.9	0.4	0.9	0.9	0.9	0.6	0.1	0.6	0.9	−0.9	0.9	0.7	−0.6	0.6	−0.9	0.7	0.7	0.9	0.7	0.6	0.9	0.7	−0.7	−0.9	0.9	0.3	0.7	0.6	0.9	1.0	0.9	−0.9
c512	−0.8	1.0	1.0	0.6	1.0	0.7	1.0	0.8	−0.2	0.5	1.0	−1.0	0.7	0.3	−0.5	0.5	−1.0	0.6	0.9	1.0	0.6	0.5	1.0	0.9	−0.6	−1.0	1.0	−0.1	0.6	0.8	1.0	0.9	1.0	−1.0
c523	0.8	−1.0	−1.0	−0.6	−1.0	−0.7	−1.0	−0.8	0.2	−0.5	−1.0	1.0	−0.7	−0.3	0.5	−0.5	1.0	−0.6	−0.9	−1.0	−0.6	−0.5	−1.0	−0.9	0.6	1.0	−1.0	0.1	−0.6	−0.8	−1.0	−0.9	−1.0	1.0

The Spearman's rank correlation coefficient value 1.0 is highlighted when the values correspond to the same variable.

Appendix G

Patch metrics for the Autonomous Region of Azores. Due to the size of this Appendix, the link below is provided for download: https://www.dropbox.com/s/lbwd1q688ngk2ug/APPENDIX%20G%2C%20H%2C%20and%20I.docx.zip?dl=0, accessed on 3 January 2021.

Appendix H

Patch metrics for the Autonomous Region of Madeira. Due to the size of this Appendix, the link below is provided for download: https://www.dropbox.com/s/lbwd1q688ngk2ug/APPENDIX%20G%2C%20H%2C%20and%20I.docx.zip?dl=0, accessed on 3 January 2021.

Appendix I

Patch metrics for the Autonomous Region of Canary Islands. Due to the size of this Appendix, the link below is provided for download: https://www.dropbox.com/s/lbwd1q688ngk2ug/APPENDIX%20G%2C%20H%2C%20and%20I.docx.zip?dl=0, accessed on 3 January 2021.

References

1. Fadigas, L. *Urbanismo e Território: As Políticas Públicas*; Edições Sílabo: Lisbon, Portugal, 2015.
2. Loures, L.; Panagopoulos, T.; Burley, J.B. Assessing user preferences on post-industrial redevelopment. *Environ. Plan. B Plan. Des.* **2016**, *43*, 871–892. [CrossRef]
3. Baptista, T.; Cabezas, J.; Fernandez, L.; Pinto-Gomes, C.; IDE-OTALEX, C. The first crossborder SDI between Portugal and Spain: Background and development. *J. Earth Sci. Eng.* **2013**, *3*, 393.
4. Gómez, J.M.N.; Castanho, R.A.; Loures, L. Evolutionary Dynamics in Mediterranean Landscapes: The Changes in Forests and Semi-Natural Areas in the Iberian Peninsula: A Study From 1990–2018. In *Management and Conservation of Mediterranean Environments*; IGIGLOBA: Hershey, PA, USA, 2021; pp. 14–21. ISBN 139781799873914.
5. Vulevic, A.; Castanho, R.A.; Naranjo Gómez, J.M.; Loures, L.; Cabezas, J.; Fernández-Pozo, L.; Martín Gallardo, J. Accessibility Dynamics and Regional Cross-Border Cooperation (CBC) Perspectives in the Portuguese-Spanish Borderland. *Sustainability* **2020**, *12*, 1978. [CrossRef]
6. Nunes, J.R.; Ramos-Miras, J.; Lopez-Piñeiro, A.; Loures, L.; Gil, C.; Coelho, J.; Loures, A. Concentrations of Available Heavy Metals in Mediterranean Agricultural Soils and their Relation with Some Soil Selected Properties: A Case Study in Typical Mediterranean Soils. *Sustainability* **2014**, *6*, 9124–9138. [CrossRef]
7. Loures, L.; Crawford, P. Democracy in progress: Using public participation in post-industrial landscape (re)-development. *WSEAS Trans. Environ. Dev.* **2008**, *4*, 794–803.
8. Ferreira, V.; Barreira, A.; Loures, L.; Antunes, D.; Panagopoulos, T. Stakeholders' Engagement on Nature-Based Solutions: A Systematic Literature Review. *Sustainability* **2020**, *12*, 640. [CrossRef]
9. Loures, L.; Panagopoulos, T. Reclamation of Derelict Industrial Land in Portugal: Greening is not Enough. *Int. J. Sustain. Dev. Plan.* **2010**, *5*, 343–350. [CrossRef]
10. Couto, G.; Castanho, R.A.; Pimentel, P.; Carvalho, C.B.; Sousa, Á. The Potential of Adventure Tourism in the Azores: Focusing on the Regional Strategic Planning. In *Advances in Tourism, Technology and Sys-tems. ICOTTS 2020. Smart Innovation, Systems and Technologies*; Abreu, A., Liberato, D., González, E.A., Garcia Ojeda, J.C., Eds.; Springer: Singapore, 2021; Volume 209. [CrossRef]
11. Gómez, J.M.N.; Lousada, S.; Velarde, J.G.G.; Castanho, R.A.; Loures, L. Land-Use Changes in the Canary Archipelago Using the CORINE Data: A Retrospective Analysis. *Land* **2020**, *9*, 232. [CrossRef]
12. Pimentel, P.; Oliveira, A.; Couto, G.; Ponte, J.C.; Castanho, R.A. The Azores Archipelago as a Region with Vast Potential for the Development of Adventure and Slow Tourism. In *Peripheral Territories, Tourism, and Regional Development*; IntechOpen: London, UK, 2020. [CrossRef]
13. Castanho, R.A.; Couto, G.; Lousada, P.; Carvalho, C.; Sousa, A. Princípios de Planeamento Estratégico e Gestão de Turismo Rural em Territórios Ultraperiféricos: O Caso de Estudo do Arquipélago dos Açores. *Rev. Ibér. Sist. Tecnol. Inf.* **2020**, *E36*, 30–41.
14. With, K.A. Is landscape connectivity necessary and sufficient for wildlife management? In *Forest Fragmentation: Wildlife and Management Implications*; Rochelle, J.A., Lehmann, L.A., Wisniewski, J., Eds.; Brill: Leiden, The Netherlands, 1999; pp. 97–115.
15. Gardner, R.H.; O'Neill, R.V.; Turner, M.G. Ecological implications of landscape fragmentation. In *Humans as Components of Ecosystems: Subtle Human Effects and Ecology of Population Areas*; Pickett, S.T.A., McDonnell, M.G., Eds.; Springer: New York, NY, USA, 1993; pp. 208–226.
16. Turner, M.G. Spatial and temporal analysis of landscape patterns. *Landsc. Ecol.* **1990**, *4*, 21–30. [CrossRef]
17. Turner, M.G.; Gardner, R.H. *Quantitative Methods in Landscape Ecology*; Springer Verlag: New York, NY, USA, 1991.
18. Baker, W.L.; Cai, Y. The r.le programs for multiscale analysis of landscape structure using the GRASS geographical information system. *Landsc. Ecol.* **1992**, *7*, 291–302. [CrossRef]
19. McGarigal, K.; Marks, B.J. FRAGSTATS: Spatial pattern analysis program for quantifying landscape structure. In *General Technical Report*; U.S. Department of Agriculture, Forest Service, Pacific Northwest Research Station: Portland, OR, USA, 1995.

20. Fischer, J.; Lindenmayer, D.B. Landscape modification and habitat fragmentation: A synthesis. *Glob. Ecol. Biogeogr.* **2007**, *16*, 265–280. [CrossRef]
21. Didham, R.K. Ecological Consequences of Habitat Fragmentation. *eLS* **2010**. [CrossRef]
22. EEA-FOEN. *Landscape Fragmentation in Europe*; EEA Report, No 2/2011; EEA: Copenhagen, Denmark, 2011.
23. Wickham, J.D.; O'Neill, R.V.; Jones, K.B. Forest fragmentation as an economic indicator. *Landsc. Ecol.* **2000**, *15*, 171–179. [CrossRef]
24. Jaimes, N.B.P.; Sendra, J.B.; Delgado, M.G.; Plata, R.F. Exploring the driving forces behind deforestation in the state of Mexico (Mexico) using geographically weighted regression. *Appl. Geogr.* **2010**, *30*, 576–591. [CrossRef]
25. Peneva-Reed, E. Understanding land-cover change dynamics of a mangrove ecosystem at the village level in Krabi Province, Thailand, using Landsat data. *GIScience Remote Sens.* **2014**, *51*, 403–426. [CrossRef]
26. Temme, A.J.A.M.; Verburg, P.H. Mapping and modelling of changes in agricultural intensity in Europe. *Agric. Ecosyst. Environ.* **2011**, *140*, 46–56. [CrossRef]
27. CORINE Land Cover—CLC (2019). Available online: http://clc.gios.gov.pl/index.php/o-clc/program-clc (accessed on 15 November 2019).
28. Martínez-Fernández, J.; Ruiz-Benito, P.; Bonet, A.; Gómez, C. Methodological variations in the production of of CORINE land cover and consequences for long-term land cover change studies. The case of Spain. *Int. J. Remote Sens.* **2019**, *40*, 1–19. [CrossRef]
29. Rysz, K. Zakres pojeciowy kategorii pokrycia i u˙zytkowania ziemi stosowany w programie CORINE. In *Analiza Zmian I Prognoza Przyrostu Zabudowy Mieszkaniowej Na Obszarze Polski Do 2020 Roku*; Gibas, P., Ed.; BoguckiWydawnictwo Naukowe: Poznań, Poland, 2017; pp. 31–35.
30. Pasca, A.; Nasui, D. The use of Corine Land Cover 2012 and Urban Atlas 2012 databases in agricultural spatial analysis. Case study: Cluj County, Romania. *Res. J. Agric. Sci.* **2016**, *48*, 314–322.
31. Weng, Q. Remote Sensing for Sustainability. In *The Efects of Land Use and Land Cover Geoinformation Raster23*; Meneses, B., Reis, E., Reis, R., Vale, M.J., Eds.; Routledge: London, UK, 2016; p. 357.
32. Meneses, B.; Reis, E.; Reis, R.; Vale, M.J. The Efects of Land Use and Land Cover Geoinformation Raster Generalization in the Analysis of LUCC in Portugal. *ISPRS Int. J. Geo-Inf.* **2018**, *7*, 390. [CrossRef]
33. Hartvigsen, M. Land reform and land fragmentation in Central and Eastern Europe. *Land Use Policy* **2014**, *36*, 330–341. [CrossRef]
34. Sleszynski, P. Expected trac speed in Poland using Corine land cover, SRTM-3 and detailed population places data. *J. Maps* **2015**, *11*, 245–254. [CrossRef]
35. Allen, J.M.; Leininger, T.J.; Hurd, J.D.; Civco, D.L.; Gelfand, A.E.; Silander, J.A., Jr. Socioeconomics drive woody invasive plant richness in New England, USA through forest fragmentation. *Landsc. Ecol.* **2013**, *28*, 1671–1686. [CrossRef]
36. Castilla, G.; Larkin, K.; Linke, J.; Hay, G.J. The impact of thematic resolution on the patch-mosaic model of natural landscapes. *Landsc. Ecol.* **2008**, *24*, 15–23. [CrossRef]
37. Maes, J.; Barbosa, A.P.; Baranzelli, C.; Zulian, G.; Silva, F.B.E.; Vandecasteele, I.; Hiederer, R.; Liquete, C.; Paracchini, M.L.; Mubareka, S.; et al. More green infrastructure is required to maintain ecosystem services under current trends in land-use change in Europe. *Landsc. Ecol.* **2015**, *30*, 517–534. [CrossRef]
38. Forman, R.T.T.; Godron, M. *Landscape Ecology*; Wiley: New York, NY, USA, 1986.
39. Turner, M.G. Landscape ecology: The effect of pattern on process. *Annu. Rev. Ecol. Evol. Syst.* **1989**, *20*, 171–197. [CrossRef]
40. Singh, S.K.; Pandey, A.C.; Singh, D. Land Use Fragmentation Analysis Using Remote Sensing and Fragstats. In *Remote Sensing Applications in Environmental Research, Society of Earth Scientists*; Srivastava, P.K., Mukherjee, S., Gupta, M., Islam, T., Eds.; Springer International Publishing: Cham, Switzerland, 2014.
41. National Geographic Information Center. Download Center of the National Geographic Information Center (2021). Available online: http://centrodedescargas.cnig.es/CentroDescargas/locale?request_locale=en (accessed on 3 January 2021).
42. Directorate-General for Territory of Portugal. National Geographic Information System. Available online: https://snig.dgterritorio.gov.pt/ (accessed on 3 January 2021).
43. Joint Research Centre (European Commission). *Map Projections for Europe*; Institute for Environment and Sustainability: Ispra, Italy, 2001.
44. National Geographic Institute of Spain. *Necesidad de un Nuevo "Datum". ETRS89, Working Group for the Transition to ETRS89*; Higher Geographic Council of Spain: Madrid, Spain, 2007.
45. Forman, R.T.T. *Land Mosaics: The Ecology of Landscapes and Regions*; Cambridge University Press: Cambridge, UK, 1995; ISBN 978-0-521-47462-7.
46. Hilty, J.A.; Lidicker, W.Z.; Merenlender, A.M. *Corridor Ecology: The Science and Practice of Linking Landscapes for Biodiversity Conservation*; Island Press: Washington, DC, USA, 2006.
47. Wilson, E.O. *The Diversity of Life*; Harvard Univesity Press: Cambridge, MA, USA, 1992.
48. Milne, B.T. Lessons from applying fractal models to landscape patterns. In *Quantitative Methods in Landscape Ecology–Then Analysis and Interpretation of Landscape Heterogeneity*; Turner, M.G., Gardner, R.H., Eds.; Springer Verlag: New York, NY, USA, 1991; pp. 199–235.
49. Bogaert, J.; Rousseau, R.; van Hecke, P.; Impens, I. Alternative area-perimeter ratios for measurement of 2D shape compactness of habitats. *Appl. Math. Comput.* **2000**, *111*, 71–85. [CrossRef]
50. Subirós, V.J.; Linde, V.D.; Pascual, A.L.; Palom, R.A. Conceptos y métodos fundamentales en ecología del paisaje (landscape ecology). Una interpretación desde la geografía. *Doc. Anàl. Geogr.* **2006**, *48*, 151–166.

51. Sundseth, K. *Natura 2000 in the Macaronesian Region*; Publications Office of the European Union: Luxembourg, 2009.
52. ESRI. Esri Data & Maps. 2020. Available online: https://www.esri.com/arcgis-blog/products/product/mapping/esri-data-maps/ (accessed on 30 November 2020).
53. Ignarra, R. *Fundamentos do Turismo*; Pioneira Thomson Learning: São Paulo, Brazil, 1998.
54. Santana, A. *La Antropología y el Turismo*; Filho, S., Ed.; Ariel: Barcelona, Spain, 1997.
55. Gomes, C.; Pereira, J. O Produto Turístico All Inclusive na Ilha de Tenerife, Espanha: Características, problematizações e desafios. *Rev. Tur. Anál.* **2016**, *27*, 108–130. [CrossRef]
56. Tavares, C.A. O Ordenamento e a Gestão do Território em Cabo Verde: Constrangimentos e Desafios. In *A Juventude e a Promoção da Cultura de Investigação*; AJIC: Lisboa, Portugal, 2007; pp. 97–115.
57. Bardolet, E.; Pauline, S. Tourism in archipelagos: Hawaii and the Balearics. *Ann. Tour. Res.* **2008**, *35*, 900–923. [CrossRef]
58. Arévalo, J.R.; Delgado, J.D.; Otto, R.; Naranjo, A.; Salas, M.; Fernández-Palacios, J.M. Distribution of alien vs. native plant species in roadside communities along an altitudinal gradient in Tenerife and Gran Canaria (Canary Islands). *Perspect. Plant Ecol. Evol. Syst.* **2005**, *7*, 185–202. [CrossRef]
59. Fernández-Palacios, J.M.; de Nascimento, L.; Otto, R.; Delgado, J.D.; García-Del-Rey, E.; Arévalo, J.R.; Whittaker, R.J. A reconstruction of Palaeo-Macaronesia, with particular reference to the long-term biogeography of the Atlantic island laurel forests. *J. Biogeogr.* **2010**, *38*, 226–246. [CrossRef]
60. Rodrigues, M. Representing coastal land use in the island of Gran Canaria. *J. Maps* **2015**, *12*, 311–315. [CrossRef]
61. Foley, J.A.; DeFries, R.; Asner, G.P.; Barford, C.; Bonan, G.; Carpenter, S.R.; Chapin, F.S.; Coe, M.T.; Daily, G.C.; Gibbs, H.K.; et al. Global Consequences of Land Use. *Science* **2005**, *309*, 570–574. [CrossRef]
62. Rosina, K.; Silva, F.B.E.; Vizcaino, P.; Herrera, M.M.; Freire, S.; Schiavina, M. Increasing the detail of European land use/cover data by combining heterogeneous data sets. *Int. J. Digit. Earth* **2018**, *13*, 602–626. [CrossRef]
63. AlQurashi, A.F.; Kumar, L. Investigating the Use of Remote Sensing and GIS Techniques to Detect Land Use and Land Cover Change: A Review. *Adv. Remote Sens.* **2013**, *2*, 193–204. [CrossRef]
64. Climate Atlas of the Archipelagos of Canary Islands, Madeira and Azores. 2011. Available online: https://www.ipma.pt/export/sites/ipma/bin/docs/publicacoes/atlas.clima.ilhas.iberico.2011.pdf (accessed on 3 January 2021).
65. World Heritage Datasheets. 1999. Available online: http://world-heritage-datasheets.unep-wcmc.org/datasheet/output/site/laurisilva-of-madeira/. (accessed on 3 January 2021).
66. Melchiorri, M.; Florczyk, A.J.; Freire, S.; Schiavina, M.; Pesaresi, M.; Kemper, T. Unveiling 25 Years of Planetary Urbanization with Remote Sensing: Perspectives from the Global Human Settlement Layer. *Remote Sens.* **2018**, *10*, 768. [CrossRef]
67. Benz, U.C.; Hofmann, P.; Willhauck, G.; Lingenfelder, I.; Heynen, M. Multi-resolution, object-oriented fuzzy analysis of remote sensing data for GIS-ready information. *ISPRS J. Photogramm. Remote Sens.* **2004**, *58*, 239–258. [CrossRef]
68. ESPON. The ESPON Programme. The Development of the Islands—European Islands and Cohesion Policy. EUROISLANDS Targeted Analysis 2013/2/2. 2013. Available online: https://europeansmallislands.files.wordpress.com/2017/03/espon-euroislands-report-2013.pdf (accessed on 3 January 2021).

International Journal of
Geo-Information

MDPI

Article

Spatiotemporal Variation of NDVI in the Vegetation Growing Season in the Source Region of the Yellow River, China

Mingyue Wang [1,2], Jun'e Fu [2,3,*], Zhitao Wu [1] and Zhiguo Pang [2,3]

1 Institute of Loess Plateau, Shanxi University, Taiyuan 030006, China; sxwmyue@163.com (M.W.); wuzhitao@sxu.edu.cn (Z.W.)
2 China Institute of Water Resources and Hydropower Research, Beijing 100038, China; pangzg@iwhr.com
3 Research Center on Flood & Drought Disaster Reduction of the Ministry of Water Resources, Beijing 100038, China
* Correspondence: fuje@iwhr.com

Received: 27 February 2020; Accepted: 22 April 2020; Published: 24 April 2020

Abstract: Research on vegetation variation is an important aspect of global warming studies. The quantification of the relationship between vegetation change and climate change has become a central topic and challenge in current global change studies. The source region of the Yellow River (SRYR) is an appropriate area to study global change because of its unique natural conditions and vulnerable terrestrial ecosystem. Therefore, we chose the SRYR for a case study to determine the driving forces behind vegetation variation under global warming. Using the Normalized Difference Vegetation Index (NDVI) and climate data, we investigated the NDVI variation in the growing season in the region from 1998 to 2016 and its response to climate change based on trend analysis, the Mann–Kendall trend test and partial correlation analysis. Finally, an NDVI–climate mathematical model was built to predict the NDVI trends from 2020 to 2038. The results indicated the following: (1) over the past 19 years, the NDVI showed an increasing trend, with a growth rate of 0.00204/a. There was an upward trend in NDVI over 71.40% of the region. (2) Both the precipitation and temperature in the growing season showed upward trends over the last 19 years. NDVI was positively correlated with precipitation and temperature. The areas with significant relationships with precipitation covered 31.01% of the region, while those with significant relationships with temperature covered 56.40%. The sensitivity of the NDVI to temperature was higher than that to precipitation. Over half (56.58%) of the areas were found to exhibit negative impacts of human activities on the NDVI. (3) According to the simulation, the NDVI will increase slightly over the next 19 years, with a linear tendency of 0.00096/a. From the perspective of spatiotemporal changes, we combined the past and future variations in vegetation, which could adequately reflect the long-term vegetation trends. The results provide a theoretical basis and reference for the sustainable development of the natural environment and a response to vegetation change under the background of climate change in the study area.

Keywords: vegetation; partial correlation analysis; trend prediction; the source region of the Yellow River

---

1. Introduction

Global environmental change, which is marked by "global warming", has possible serious impacts on ecosystems and has attracted great attention from scientists around the world [1,2]. Vegetation cover is an important component of the environment, and is also the best indicator of the regional ecological environment [3]. The variation in vegetation cover is the direct result of environmental change [4]. As the main component of terrestrial ecosystems, vegetation is a sensitive indicator of climate change. Therefore, in the context of global climate change, it is of great significance to identify the spatiotemporal characteristics of vegetation cover to regulate ecological processes and ensure ecological security.

The Normalized Difference Vegetation Index (NDVI) can be used to measure the improvement and degradation of vegetation cover. NDVI is a good satellite-based indicator of vegetation at the landscape scale [5,6]. The NDVI time series intuitively reflects the vegetation growth and coverage status. The NDVI is widely used in global and regional vegetation change research. Kawabata et al. found that vegetation activities increased remarkably in the northern middle-high latitudes [7]. Relevant research has shown that the vegetation in China has exhibited the same trend. Liu et al. analysed the vegetation changes in China from 1982 to 2012. The results showed that the NDVI exhibited a slowly increasing trend with obvious regional characteristics. The increasing trend slowed after 1997 [8]. Piao and Fang used global inventory modeling and mapping studies (GIMMS) NDVI data to analyse the vegetation cover in China from 1982 to 1999, and showed that 86.2% of China's area exhibited an increasing trend in vegetation. The changes in NDVI were significantly affected by climate fluctuations and had obvious regional differences [9]. Xu et al. analysed the vegetation coverage from 2000 to 2015 and showed that the area with increased vegetation coverage accounted for 83.34% of the area in China [10]. Other scholars have performed extensive research on vegetation cover changes in the Huang-Huai-Hai River basin [11], in the Yangtze River basin [12], on the Qinghai–Tibet Plateau [13], and in the southwestern karst region [14]. The above research showed that the vegetation increased in regional areas or throughout whole countries. The Qinghai–Tibet Plateau was shown to be more sensitive to the effects of climate warming than other regions.

The source region of the Yellow River (SRYR) is located on the sensitive margin of the northeastern Qinghai–Tibet Plateau [15]. Most of this region are located between 4200 and 5000 m above sea level, and the percentage of the area above 5000 m is less than 1% [16]. The main vegetation is grassland, including alpine grassland and alpine meadow, which cover 74.55% of this region. The region is an important part of the terrestrial ecosystem of the Qinghai–Tibet Plateau. The SRYR is also a water conservation area and a key protected area in the Yellow River basin [17]. With the rise of ecological protection in the Yellow River basin as a major national strategy [18], it is of great significance to dynamically monitor the spatiotemporal evolution of surface vegetation cover. Over the past decades, the region has experienced severe climate change. Many studies have indicated that the temperature and precipitation in the study area have increased [19,20], and the vegetation coverage has exhibited a tendency of restoration because the climate has become gradually warm and wet [21]. Some researchers have studied the relationships between vegetation coverage and environmental variation [22]. Guo et al. found that vegetation cover changes in the SRYR showed very impressive correlations with climatic factors [23]. Liang et al. reported that local hydrological conditions directly influenced vegetation variations, and overgrazing can be a leading cause of localized vegetation degradation [24]. Most of the studies on the vegetation coverage in the study area have focused on current interannual changes, while few studies have focused on different regions and the future.

Land-use types represent the ongoing challenges of environmental variation and the impacts of human activities [25]. Therefore, based on the NDVI data and climate data from 1998 to 2016, this study analysed the spatiotemporal changes in the NDVI and the response mechanisms of different land-use types in the growing season and predicted them for the next 20 years. This study provides a scientific basis for the development of countermeasures to protect vegetation in the SRYR under the background of global warming.

2. Data and Methods

2.1. Study Area

The SRYR (32°09–36°34 N, 95°54–103°24 E) is located on the northeastern Qinghai–Tibet Plateau and covers an area of 131,400 km^2, accounting for 16.2% of the Yellow River basin area. The study areas in this paper are shown in Figure 1. The administrative area mainly includes 17 counties in the three provinces of Qinghai, Gansu and Sichuan. The elevation in the SRYR decreases from west to east with a maximum altitude of 6253 m and a minimum of 2410 m. The lowest elevation of the study area exists in Longyangxia Reservoir, and the highest elevation is in the Anyemaqen Mountains. The region has a continental plateau climate, which is obviously affected by the southwest monsoon. The temperature and precipitation decrease from the southeast to the northwest. The annual average rainfall is approximately 530 mm yr^{-1} [26]. From the southeast to the northwest, the annual average daily temperature varies between 2 °C and −4 °C [27]. There are many glaciers and extensive permafrost as well as a large number of lakes and rivers, which feed a large number of marsh wetlands; these areas provide more than 40% of the runoff in the Yellow River basin [28]. The SRYR is an important water conservation area and is also known as a "plateau water tower".

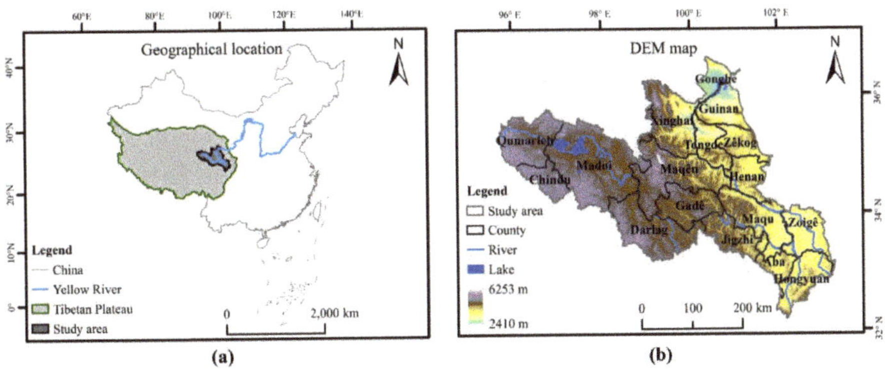

Figure 1. The source region of the Yellow River (SRYR): (**a**) geolocation location; (**b**) digital elevation model (DEM) map.

2.2. Data

The NDVI data used in this paper were provided by the Resources and Environment Science Data Center of the Chinese Academy of Sciences (http://www.resdc.cn). The data were based on NDVI time series data obtained from satellite remote-sensing images by SPOT/VEGETATION and moderate-resolution imaging spectroradiometer (MODIS). The data can reflect the distribution and change of vegetation cover in various regions of China on the spatiotemporal scales effectively, which is of great reference significance to the monitoring of vegetation variation, the rational utilization of vegetation resources and other researches on ecological environment [29]. The data were processed with atmospheric, radiation, and geometric corrections. The data were synthesized by using maximum value composites (MVC) with a spatial resolution of 1 km and a temporal resolution of one month. After verification, the accuracy was found to meet the requirements. The data have been used in research on monitoring vegetation dynamic changes. The average monthly NDVI values from May to September were used to obtain the annual NDVI values during the growing season from 1998 to 2016. The daily meteorological grid data were provided by the National Climate Center [30], including the CN05.1 data from 1998–2016 and the regional climate model version 4 (RegCM4) data from 2020–2038, which had spatial resolutions of 0.25° and 0.0625°, respectively. The climate elements included precipitation (Pre) and temperature (Tm). The daily values of the two climate elements were statistically estimated from May to September, and these data were resampled to a spatial resolution of 1 km to be consistent with the spatial resolution of the NDVI dataset. The data have been heavily cited in scientific papers [31]. The land-use types in 2015 were obtained from the Resources and Environment Science Data Center of the Chinese Academy of Sciences (http://www.resdc.cn), with a spatial resolution of 100 m. The land-use types were integrated into six categories: cropland, woodland, grassland, water bodies, built-up land and unused land.

2.3. Methods

The linear trend method [32] was used to analyse the NDVI trends. A positive slope indicated that the vegetation tended to improve with increasing NDVI. A negative slope indicated that the vegetation tended to deteriorate with decreasing NDVI. The statistic e_{slope} was calculated as in Equation (1) as follows:

$$e_{slope} = \frac{n \times \sum_{i=1}^{n} i \times NDVI_i - \sum_{i=1}^{n} i \times \sum_{i=1}^{n} NDVI_i}{n \times \sum_{i=1}^{n} i^2 - (\sum_{i=1}^{n} i)^2} \tag{1}$$

where e_{slope} represents the slope of the NDVI trend, i represents the year serial number, and n represents the time series length.

The Mann–Kendall abrupt test [33] was used to determine the year of the NDVI change. The statistics were defined under the assumption that the time series were random and independent. The $\mathrm{UF_k}$ statistic was calculated with the following equations:

$$UF_k = \frac{d_k - E(d_k)}{\sqrt{\mathrm{Var}(d_k)}} \tag{2}$$

where

$$d_k = \sum_{i=1}^{k} m_i, \ (2 \le k \le n) \tag{3}$$

$$m_i = \begin{cases} 1, \ NDVI_i > NDVI_j \\ 0, \ else \end{cases}, \ (1 \le j \le i) \tag{4}$$

$$E(d_k) = \frac{k(k-1)}{4} \tag{5}$$

$$\mathrm{Var}(d_k) = \frac{k(k-1)(2k+5)}{72} \tag{6}$$

where UF_1 is equal to 0. The d_k statistic is reduced to that given in Equations (3) and (4), which indicates that the value at time i was greater than the value at time j. Equation (5) calculates the mean of the UF_k statistic, and Equation (6) calculates the variance. Then, the order of column d_k is calculate as the reverse time series. The UB_k value was calculated according to the above equation. Given the significance level $\alpha = 0.05$, the critical value of the UB_k statistic was $|1.96|$. A sequence was constructed for 19 samples and the UF_k and UB_k curves and significant horizontal lines were drawn. If UF_k was greater than 0, it meant that the sequence showed an increasing trend, and a value of less than 0 indicated a decline. When the threshold exceeded $|1.96|$, it indicated that the trend was significant. If the UF_k and UB_k curves had intersection points within the confidence interval, the time corresponding to the intersection point was the possible change point.

In addition, this study used partial correlation analysis [20] to analyse the relationship between the climatic factors and the NDVI. The partial correlation coefficient is an index that measures the degree of linear correlation between the two variables by controlling the effects of multiple other variables. Moreover, this study used the residual analysis method [34,35] to analyse the impacts of human activities on the NDVI. Based on the NDVI, precipitation and temperature values from 1998 to 2016, the residual method was used to simulate the relationship between the NDVI and the climate elements for each pixel. The changes in the residuals in the NDVI predictions and observations reflected the contributions of human impacts to the actual changes in NDVI. Positive residuals indicated that the human impacts on vegetation were positive, and negative residuals indicated that the human impacts on vegetation were negative.

3. Results and Analysis

3.1. Spatiotemporal Variation in Normalized Difference Vegetation Index (NDVI)

Table 1 shows the basic situation of the main counties, which shows that water bodies and built-up land accounted for a small proportion, so vegetation variations in these land-use types will not be subsequently analysed. Among the different land-use types, the proportion of grassland in the region was the highest. Among the different counties, NDVI in Zoigê had the highest values.

Table 1. Normalized Difference Vegetation Index (NDVI) value and proportion of land-use types in different counties.

Countie	Mean NDVI Value	Land-Use Types/%					
		Crop-Land	Wood-Land	Grass-Land	Water Body	Built-Up Land	Unused Land
The source region of the Yellow River (SRYR)	0.486	0.94	6.77	74.45	2.27	0.13	15.44
Zoigê	0.674	0.03	0.75	70.44	1.07	0.27	27.44
Hongyuan	0.674	0.03	7.26	79.62	0.03	0.23	12.83
Aba	0.666	/	8.37	85.57	0.09	/	5.98
Henan	0.655	/	15.75	78.45	0.80	0.12	4.87
Maqu	0.640	/	8.10	73.75	1.72	0.09	16.34
Jigzhi	0.602	/	10.18	86.04	0.71	0.03	3.03
Zêkog	0.593	1.49	2.90	78.88	0.44	0.05	16.24
Gadê	0.558	/	12.72	82.55	0.62	0.03	4.08
Tongde	0.540	7.91	22.89	58.34	1.08	0.17	9.61
Maqên	0.485	0.02	16.29	68.29	1.88	0.12	13.40
Darlag	0.476	/	2.23	88.94	0.95	0.01	7.88
Xinghai	0.424	0.50	10.98	67.47	0.87	0.08	20.11
Guinan	0.410	11.14	2.43	65.73	3.00	0.21	17.49
Chindu	0.396	/	/	84.65	1.44	/	13.91
Madoi	0.324	/	0.20	74.21	7.22	0.01	18.36
Qumarleb	0.290	/	/	62.51	1.50	0.01	35.98
Gonghe	0.257	3.70	1.20	63.47	7.57	2.50	21.56

The spatial distribution of the NDVI in the growing season in the SRYR from 1998 to 2016 exhibited obvious regional differences. The spatial variability analysis showed an increasing gradient of NDVI from northwest to southeast in Figure 2a. The maximum NDVI value was 0.76, which was located in the Zoigê wetland. By referring to related studies [36], the NDVI values were classified into 5 levels. The NDVI distribution was analysed in combination with the land-use types (Figure 2c). The multiyear average NDVI in the growing season was 0.486, of which the area where NDVI was <0.1 covered 1.17% of the total area, mainly including water bodies represented by Eling Lake, Zaling Lake, and Longyangxia Reservoir and permanent glacial snow on the Anyemaqen Mountains. The area with NDVI values between 0.1 and 0.3 covered 15.12% of the area, mainly including unused land that was dominated by sand; the Gobi Desert; the marshlands in northern Qumarlêb, northern Madoi, and western Xinghai around Longyangxia Reservoir; and the sandy area in Huangshatou. The area where the NDVI was between 0.3 and 0.6 covered 50.77% of the area and was mainly distributed in Qumarleb, Madoi, Chindu, Maqên, Xinghai and Guinan, with medium- and low-coverage grassland. In addition, this area also included cultivated land in parts of Guinan. The NDVI values between 0.6 and 0.7 covered 27.90% of the total area. These areas were mainly located in the central counties of the region, which are dominated by medium- and low-coverage grasslands. The areas where NDVI was >0.7 covered 5.04% of the total area, mainly including Aba, Maqu, Zoigê, and Hongyuan, which have high-coverage grasslands and some medium-coverage grasslands.

The spatial distribution of the mean NDVI values can represent the overall trend of the vegetation, but there were opposite changes in different regions, and they can offset each other. Therefore, based on the unitary regression model, the trend of NDVI over the 19 years was analysed at the pixel scale. According to Figure 2b, the NDVI in the SRYR increased in most areas and decreased in some local areas.

According to the statistics, from 1998 to 2016, the area where the NDVI increased covered 71.40% of the total area. Among the areas with NDVI increases, the rapidly increasing area covered 33.12% of the total area and was mainly distributed in the southeast. The NDVI values did not change significantly in 19.41% of the areas. These areas were mainly distributed in Madoi, Gadê and Huangshatou in Guinan. As a typical aeolian sand control area, the trend of NDVI remained basically unchanged, which reflected the long-term and arduous nature of sandy land management. The reduced NDVI area covered 9.19% and was mainly distributed in Qumarleb (grasslands with medium and low coverage, unused land with bare rock), Maqên, and an urban area of Gonghe. The above studies indicated that while the state of vegetation in the SRYR had improved, some areas experienced vegetation degradation.

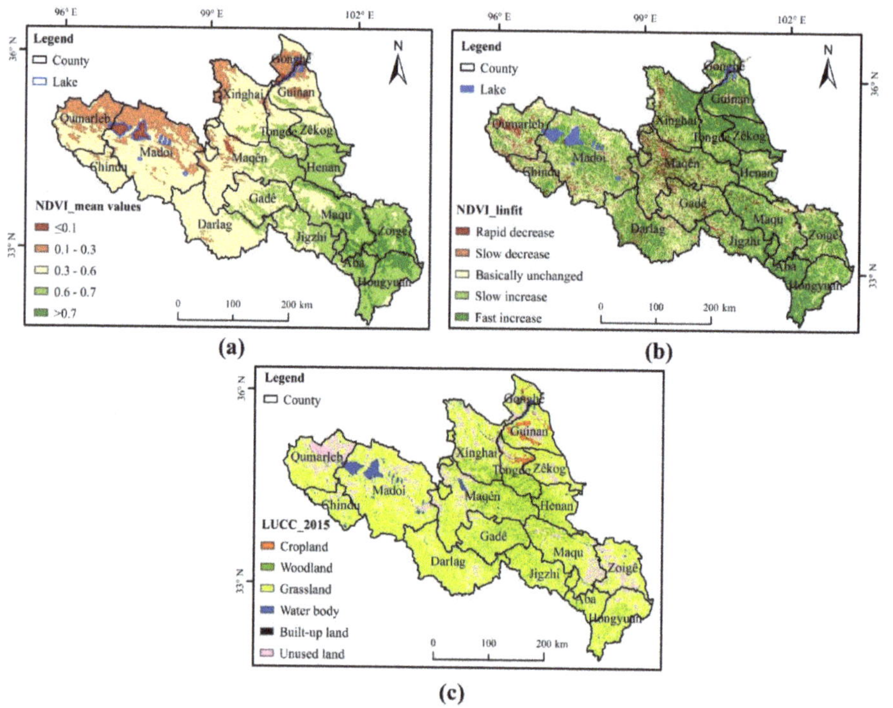

Figure 2. Spatial distribution in the SRYR from 1998 to 2016: (**a**) mean NDVI value; (**b**) NDVI change trend; (**c**) land-use types in 2015. In Figure 2b, rapid decrease represents a slope <–0.003; slow decrease represents –0.003 < slope < –0.001; basically unchanged represents –0.001 < slope < 0.001; slow increase represents 0.001 < slope < 0.003; rapid increase represents a slope <0.003.

As we can see in Figure 3, the NDVI in the SRYR increased slowly over the past 19 years, with a slope of 0.00204/a. Before 2005, the NDVI was lower than the multiyear average values, and then it fluctuated around the average, indicating that the vegetation coverage had improved since 2005. The State Council approved and launched the "master plan for ecological protection and construction of the Three-River-Source Nature Reserve in Qinghai" in 2005 and implemented a series of engineering measures. The results of this article showed that the implementation of these projects had a certain effect on vegetation restoration and protection. From the different land-use types, the trend of the grassland NDVI was the most consistent with that of the whole region. The NDVI values for different land-use types in the region showed an upward trend. The increasing trend of cropland NDVI was the most obvious, with a linear tendency of 0.00559/a, an average NDVI value of 0.46, and a change point that occurred in approximately 2004. Both the woodland NDVI and grassland NDVI showed slow trends, with linear tendencies of 0.00248/a and 0.00214/a, respectively. The average woodland NDVI value was 0.60, and the change point was between 2009 and 2011. The average grassland NDVI value was 0.50, and the change point was approximately 2006. The unused land NDVI showed slight increasing trends, with linear tendencies of 0.00147/a, an average NDVI value of 0.40, and the abrupt point occurred in approximately 2003. In the study area, the unused land in the west mainly included sandy land and the Gobi with low NDVI values; the eastern part, namely, the Zoigê wetland, was dominated by marshland with high NDVI values. In summary, the distribution of the NDVI values in the different land-use types was woodland > grassland > cropland >> unused land. In terms of trends, the grassland NDVI contributed significantly to the annual NDVI in the study area.

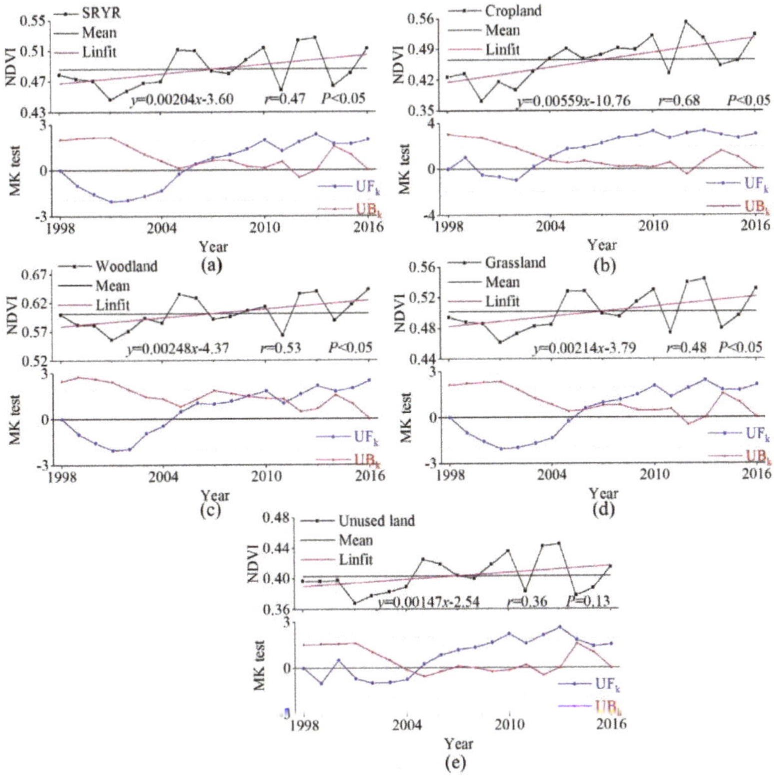

Figure 3. Temporal change of NDVI in different land-use types: (**a**) the SRYR; (**b**) cropland; (**c**) woodland; (**d**) grassland; (**e**) unused land.

The increasing trends in NDVI were obvious in many regions in China; however, the NDVI changes in the SRYR were relatively small, even though many protective measures were adopted by the government in the region at the same time, and increasing trends of climate change were faster than the average in China. Therefore, we discussed the main factors affecting NDVI in the subsequent analysis.

3.2. Impact of Meteorological Elements on the NDVI

Many studies have shown a clear response of NDVI to climate change. The impact of climate change on vegetation is mainly reflected in the hydrothermal conditions. Evapotranspiration data for long-term continuous observations are difficult to obtain. Therefore, climate change can be attributed as a cause of changes in precipitation and temperature. Temperature and precipitation are the most direct and important factors for plant growth [37,38]. Figure 4 shows that the precipitation in the growing season showed an upward trend, with a linear trend of 7.17/a and an average value of 449.52 mm. The average temperature linearly increased at 0.04/a, with an average value of 6.42 °C. The precipitation and temperature were mainly below average before 2005. The precipitation had the lowest value in 2001. After 2005, the climate elements fluctuated around the mean value. The consistency between NDVI and temperature was better than the consistency between NDVI and precipitation.

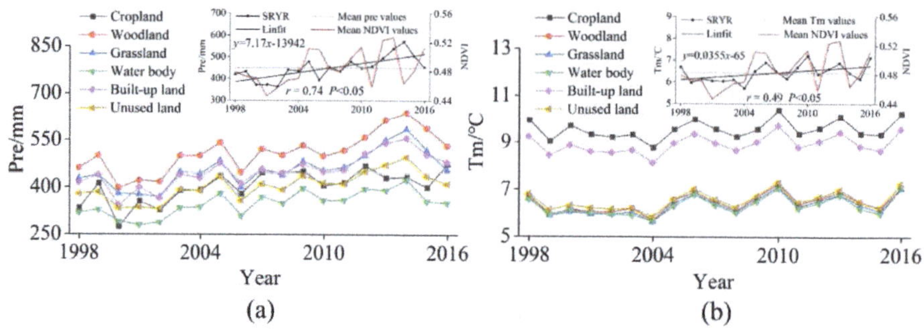

Figure 4. Temporal change in the climate elements from 1998 to 2016: (**a**) precipitation; (**b**) temperature.

Correlation coefficients between climate elements and NDVI in different land-use types are shown in Table 2. The partial correlation coefficient between the NDVI and precipitation was 0.221 ($P = 0.289$), and the coefficient between the NDVI and temperature was 0.467 ($P = 0.131$). Among the NDVI values for the different land-use types, precipitation exhibited the best correlation with cropland, and passed the 5% significance level test. Cropland mainly located in the central of Guinan and north of Tongde. The driving force of precipitation on cropland was stronger than temperature. Temperature had the best correlation with grassland, followed by woodland.

Table 2. Partial correlation coefficient.

Climate Elements	SRYR	Cropland	Woodland	Grassland	Unused Land
Pre	0.221	0.669 *	0.276	0.216	0.230
Tm	0.467	0.363	0.471	0.490	0.411

* indicates significance at the 5% level.

Table 3 shows the partial correlation between climate elements and NDVI in different counties. The correlation coefficient between precipitation and NDVI was positive in all counties. Among the different counties, at the significance level of 0.05, Guinan exhibited the highest significant correlation proportion that covered 71.08% of the total area, followed by Zêkog and Gonghe. The significant correlation proportion in these counties were all above 50%. The counties where the significant correlation proportion were between 25% and 50% were mainly distributed in Xinghai, Tongde, Madoi, Zoigê, Henan and Qumarleb. The counties where the significant correlation proportion were <25% mainly included Maqên, Hongyuan, Darlag, Maqu, Aba, Jigzhi, Gadê and Chindu.

Table 3. Climate elements and correlation coefficients in different counties.

Counties	Pre			Tm		
	Mean Value/mm	Correlation Coefficients	Significant Correlation Proportion/%	Mean Value/°C	Correlation Coefficients	Significant Correlation Proportion/%
SRYR	449.52	0.221	31.01	6.42	0.467	56.40
Zoigê	547.11	0.285	30.79	8.75	0.499	64.41
Hongyuan	623.22	0.257	24.30	7.96	0.556	75.39
Aba	609.28	0.206	19.24	8.27	0.588 *	83.15
Henan	512.34	0.332	28.92	7.19	0.524	72.32
Maqu	567.74	0.181	19.97	7.49	0.510	66.48
Jigzhi	633.57	0.066	17.13	6.00	0.541	74.71
Zêkog	459.47	0.519	62.79	7.50	0.550 *	81.04
Gadê	522.78	0.060	16.72	5.20	0.459	52.45
Tongde	447.81	0.443	45.50	7.95	0.442	47.50
Maqên	450.46	0.137	24.35	5.07	0.417	44.27
Darlag	494.39	0.084	24.02	5.14	0.535	70.44
Xinghai	376.10	0.372	46.62	6.76	0.292	16.94
Guinan	392.23	0.553	71.08	10.09	0.366	34.68
Chindu	350.83	0.036	11.72	4.29	0.560	72.07
Madoi	319.84	0.144	31.09	5.05	0.474	58.37
Qumarleb	267.72	0.228	27.97	4.87	0.456	53.48
Gonghe	379.30	0.439	54.92	11.00	0.252	10.80

* Indicates significance at the 5% level.

The counties where the significant correlation proportion between NDVI and temperature were above 75% mainly included Aba, Zêkog and Hongyuan. Aba exhibited the highest significant correlation proportion that covered 83.15% of the total area. Both Aba and Zêkog correlation coefficients passed the 5% significance level test. That is, the driving force of temperature on vegetation was stronger than precipitation in these areas. The counties where the significant correlation proportion were between 50% and 75% were mainly distributed in Jigzhi, Henan, Chindu, Darlag, Maqu, Zoigê, Madoi, Qumarleb and Gadê. The counties where the significant correlation proportion were between 25% and 50% were mainly included Tongde, Maqên and Guinan. The counties where the significant correlation proportion were <25% mainly included Xinghai and Gonghe. The contents of correlation coefficients are shown in Figure 5.

Figure 5 shows the relationship between NDVI, climatic elements and partial correlation coefficient in different counties. Precipitation and temperature were positively correlated with NDVI. That is, in different counties, the NDVI increased gradually with increasing precipitation and temperature. The higher the NDVI is, the weaker the correlation between precipitation and NDVI. The higher the NDVI is, the stronger the correlation between temperature and NDVI. Figure 5c,d show the partial correlation coefficient between climatic elements and NDVI. As shown in the figure, precipitation, temperature and the correlation coefficient were negatively correlated; that is, with the increase in precipitation and temperature, the correlation between NDVI and climatic factors weakened. In general, the lower the precipitation and temperature of the county were, the stronger the correlation between climate factors and NDVI. Generally, the effects on vegetation were more obvious under unfavourable climate conditions than under suitable ones.

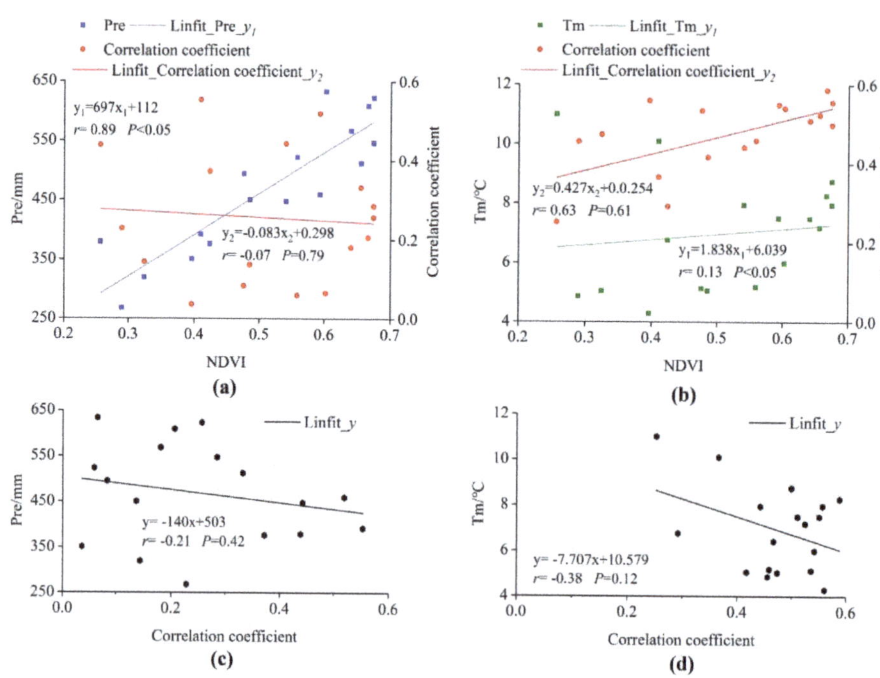

Figure 5. Correlation analysis in different counties for: (**a**) NDVI, precipitation, and correlation coefficient; (**b**) NDVI, temperature, and correlation coefficient; (**c**) precipitation and correlation coefficient; (**d**) temperature and correlation coefficient.

The spatial distribution of the partial correlation coefficient between the NDVI and climate elements in the growing seasons is shown in Figure 6. At the pixel scale, the partial correlation coefficient between the NDVI and precipitation showed a significant positive correlation with precipitation that covered 27.08% of the total area in Figure 6a. This correlation was mainly distributed in Guinan, which mainly includes grassland and sandy land, most of this region is located between 2559 and 4759 m above sea level, and the annual average precipitation is below 400 mm; significant positive correlations were also observed in Zêkog, Gonghe, Tongde, Xinghai, and northwestern Madoi. A total of 68.99% of the area was not significantly related. The areas with significant negative correlations covered 3.93% of the total area, and points were distributed in Darlag and Madoi. This region is located between 3787 and 5236 m above sea level, and the annual average temperature is below 5.5°C. There was an increase in precipitation and widespread melting of glaciers and snows, which fed glacial lakes and wetlands, reducing the vegetation coverage in glacial snow regions to a certain extent.

The area where there was a significantly positive correlation between NDVI and temperature covered 56.34% of the total area in Figure 6b, and was mainly distributed in Zêkog, the southern Madoi, Chindu, Darlag, and the southeastern SRYR. The area that was not significantly related covered 43.60%, and was mainly distributed in Gonghe, Xinghai, northern Guinan, Maqên and Gadê.

Overall, the NDVI exhibited a positive correlation with precipitation and temperature in the SRYR, and the correlation with temperature was higher than that with total precipitation. This result showed that the sensitivity of the NDVI to temperature was higher than that of precipitation, which showed that temperature had a greater impact on vegetation.

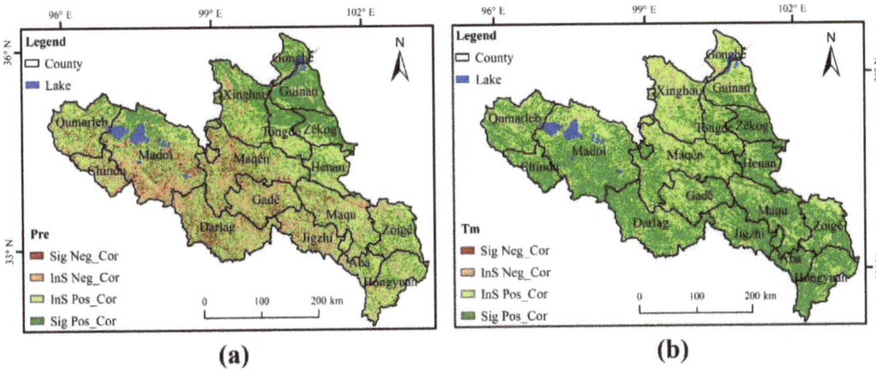

Figure 6. Spatial distribution of the correlation analysis for the NDVI and climate elements: (**a**) precipitation; (**b**) temperature. In the figure, Sig Neg_Cor represents a significant negative correlation; InS Neg_Cor represents a nonsignificant negative correlation; InS Pos_Cor represents a nonsignificant positive correlation; and Sig Pos_Cor represents a significant positive correlation.

3.3. Impact of Human Activities on NDVI

Numerous studies have shown that alpine vegetation, which is highly sensitive to global changes [39,40], has been severely affected by global climate change and human activities. The impact of human activities on vegetation changes mainly includes the promotion of increased vegetation cover (ecological engineering, etc.) and the destructive effect of reduced vegetation cover (grazing, urban expansion, etc.). Spatial distribution of the residual analysis for NDVI are shown in Figure 7. Eight typical areas were selected in the figure, and human activities information in these areas were collected to verify the residual results.

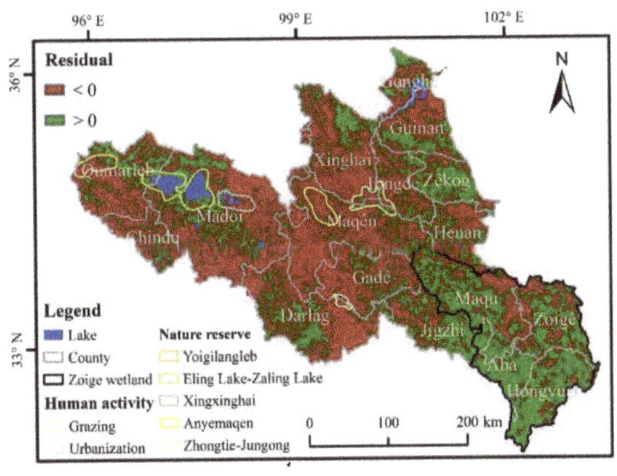

Figure 7. Spatial distribution of the residual analysis for NDVI.

Figure 7 shows that 53.58% of the residual values were negative, which mainly included the central and western regions in the SRYR, and Maqên accounted for the highest proportion. Human activities in these areas play a negative role in vegetation. The values in Maqên within the territory of the Anyemaqen Mountains, were sensitive and exhibited risk of change [41]. These areas are high-altitude regions with the following basic characteristics: poor water-heat conditions and strong solar radiation, which are not conducive to the implementation of ecological construction projects. The ecological environment continues to deteriorate. Second, at the border of Gadê and Darlag, human activity had a negative effect on vegetation. Over the last 19 years, the desertification and environmental degradation of this region have mainly been attributed to human activities such as overgrazing, under the background of regional climate changes. Liu reported that grassland degradation was the most important land-cover change in the SRYR [42]. Furthermore, in the central part of Gonghe, land-use changes were caused by the rapid expansion of built-up land and had a negative effect on local vegetation.

In addition, 46.42% of the area exhibited positive residuals, mainly in the Zoigê wetland and nature reserves. Human activities in these areas play a positive role in vegetation. The Zoigê wetland mainly includes Hongyuan, Aba, Zoigê, and Maqu. The residuals in core areas of nature reserves [43] were positive, which mainly including Yoigilangleb, Eling Lake–Zaling Lake, and Zhongtie-Jungong. The NDVI values in these areas showed an increasing trend, indicating that decreasing trends of vegetation and expanding desertification were restrained, and wetland expansion and increasing vegetation cover were obvious. To a certain extent, the effects of the establishment of the Three-River-Source Nature Reserve (2000) were confirmed, and the ecological protection construction project (2005) has already achieved initial results. The establishment of the Three-River-Source National Park in 2020 indicated that the ecological protection of the Yellow River source area had reached a new level.

3.4. Trend Prediction

Multivariate linear regression equations were used to obtain the regression coefficients of the observed NDVI values and the observed climate elements (precipitation and temperature) from 1998 to 2016. The regression coefficients were fitted based on the climate forecast data from RegCM4 during the same period to simulate the pixel-based change trend of the NDVI. The comparison in Figure 8b shows that the simulated NDVI tendency value with linear tendencies of 0.00207/a, was the same as the observed NDVI tendency value with linear tendencies of 0.00204/a. This result shows that the credibility of the simulated NDVI trend was high. For the simulated future time period, we chose 2020–2038, which was similar to the past time length and close to the present time. Based on the grid, using the established pixel-scale NDVI-climate model, NDVI change trend distribution from 2020 to 2038 was analysed at the pixel scale with MATLAB. The statistics were calculated with the equation:

NDVI (2020–2038) = Precipitation regression coefficient (1998–2016) * precipitation (2020–2038) + Temperature regression coefficient (1998–2016) * temperature (2020–2038).

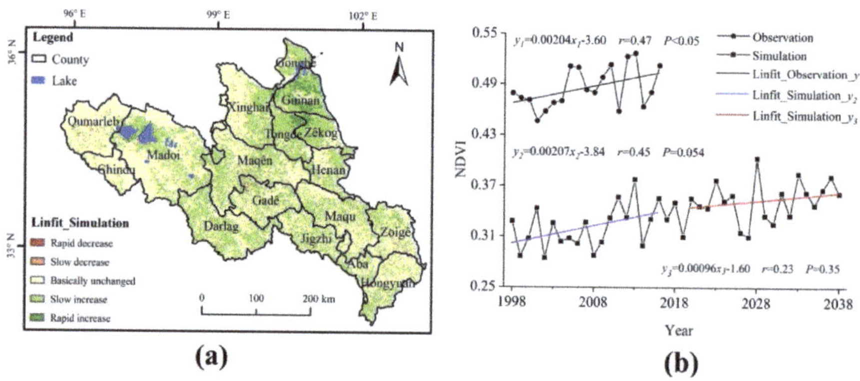

Figure 8. Prediction of the NDVI trend from 2020 to 2038: (**a**) spatial distribution; (**b**) temporal series. In the Figure 8a, rapid decrease represents a slope <–0.003; slow decrease represents –0.003 < slope < –0.001; basically unchanged represents –0.001 < slope < 0.001; slow increase represents 0.001 < slope < 0.003; rapid increase represents a slope <0.003.

According to Figure 8, the NDVI will show a slight upward trend over the next 19 years, with a slope of 0.00096/a. From 2020 to 2038, the areas where the NDVI will basically remain unchanged and slowly increase cover 54% and 42.43% of the total area, respectively. Among these areas, the basically unchanged areas are mainly distributed in Chindu and Qumarleb, the proportions of which are 91.16% and 86.15% in each county. The slowly increasing areas are mainly concentrated in Zêkog and Tongde, covering 70.76% and 69.83% of the county, respectively. In addition, NDVI has been increasing rapidly in the areas of Guinan, Zêkog and Tongde, where there is currently a large amount of cropland and a small amount of sandy land, following a similar trend over the past 19 years. The increasing NDVI trend in Guinan is the most obvious, with a rapid growth rate of 0.00267/a, covering 83.34% of the county.

The inputs to the prediction model are mainly precipitation and temperature, so the increase in NDVI is related to global warming. Rising temperatures, melting glaciers and increasing precipitation provide a good environment for vegetation cover. New studies have found that shrubs and grasses are springing up around Mount Everest [44], and the temperature in Antarctica exceeded 20 °C for the first time. These results suggest that Himalayan ecosystems are highly vulnerable to climate-induced shifts in vegetation, and the effects of global warming are spreading. Climate change affects vegetation growth, and vegetation change reflects climate variation. The SRYR is a sensitive area to climate, and the past and future trends of NDVI both demonstrate the warming and wetting trends of climate, which should arouse attention.

The climate simulation model was different from the weather forecast model and the short-term climate prediction model. The dates in the model were not equal to actual calendar dates. Therefore, the results of this study were only for the simulation of future NDVI trends and do not represent current NDVI values.

4. Discussion

The SRYR is an important water conservation and recharge area in the Yellow River basin due to its unique climatic characteristics and rich wetland system. We found that the NDVI increased over more than 70% of the study area, and the rate of increase ranged from 0 to 0.00559/a. Compared with that in 1998, the NDVI increased in the majority of the area in 2016. However, in considerable parts of Qumarleb and Maqên, the NDVI decreased, indicating that the natural ecological environment needs to be protected. He reported that grazing exclusion was an effective restoration approach to rehabilitate degraded alpine meadows in Maqin [45]. The annual precipitation in the study area increased from 1998 to 2016, with significant changes in different stages. The results were consistent with those of Li [46]. The temperature increased with a linear tendency of 0.0355/a. The annual average temperature increase over the past several years was mainly caused by the increase in the average annual minimum temperature [47]. Related studies have also confirmed that the climate in the SRYR is warming [26,47,48]. The temperature changes in the region over the past 19 years were consistent with those throughout China [49], and all regions showed increasing temperatures. However, the temperature increase was larger in the SRYR than the overall increases in China, which also confirmed the sensitivity of the alpine region to global changes [50]. This study showed that the temperature and precipitation of the SRYR have been increasing over the past 19 years. The climate of the region will enter a warmer and wetter period, which will be conducive to the restoration and establishment of vegetation. In addition, the NDVI was found to be more sensitive to temperature than precipitation.

In addition, the variations in grassland vegetation responded not only to long- and short-term changes in climate but also to the impact of human activities and their associated perturbations. State-approved ecological protection construction efforts involve major projects such as returning pastures to grasslands and ecological immigration. The implementation of continuous ecological restoration and ecological protection projects has increased the vegetation coverage to a certain extent, but the vegetation degradation caused by intensified human activities in local areas cannot be ignored.

The region has been overused and overgrazed for a long time throughout history. Overgrazing is the main factor causing the degradation of grassland ecosystems [41]. The residual results showed that human activities had a higher negative residual on the vegetation in the study area, indicating that the degradation of the alpine vegetation had not been effectively contained, which was consistent with the results of related research. The grassland NDVI was closely related to climate. Our research showed that the trend of the NDVI was most consistent with the grassland NDVI trend, and the partial correlation between the grassland NDVI and temperature was the best. Yang [49] selected the period from 1998–2007 to analyse the vegetation trends in the SRYR. The results showed that vegetation was improving [51], which is consistent with the results of this article. Ongoing climate change and human interference have greatly affected vegetation. Therefore, the wetting tendency of the climate and vegetation restoration projects might be the main reasons for vegetation improvements in the SRYR.

The prediction of the NDVI in the SRYR showed that it may increase over the next 19 years, which was consistent with the trends from 1998 to 2016. In addition to climate elements and human activities, NDVI is also affected by air pollution, soil degradation, slope and other factors. The quantification of the relationships between vegetation change and these factors has become difficult. The choice of data image resolution and time series and factors affecting NDVI will be the direction of our future research. Currently, the frequency of remote-sensing images has shifted from an annual scale to finer scales, such as monthly, ten-day, and daily periods. Methods for improving the spatial and temporal resolution quality of NDVI data through scientific data fusion methods by using high-resolution MODIS NDVI, high-resolution SPOT NDVI, and long time-series GIMMS NDVI data are worth exploring.

5. Conclusions

Based on the NDVI and climate data in the growing season from 1998 to 2016, this study analysed the spatial and temporal characteristics and impact mechanisms of the NDVI in the SRYR and predicted future NDVI trends. The results showed the following:

(1) The average NDVI in the growing season was 0.486, which decreased from northwest to southeast and showed obvious regional differences. The NDVI values were concentrated between 0.3 and 0.6 over 50.77% of the total area. The NDVI showed a trend of "increasing overall and decreasing locally", and 71.40% of the area showed an increasing trend. Among the different land-use types, woodland had the highest NDVI value, and the grassland NDVI trend coincided best with the overall NDVI trend.

(2) From 1998 to 2016, both precipitation and temperature showed an increasing trend. These conditions may be the main reason for the warm and humid climate in the SRYR in recent years. This trend was conducive to the improvement of vegetation. The sensitivity of vegetation and temperature was higher than that of precipitation. Among the different counties, the effects on vegetation were more obvious under unfavourable climate conditions than under suitable ones. The results of the residual analysis indicated that human activities had a positive impact on 46.42% of the SRYR. However, 53.58% of the area was still negatively affected by human activities, which proves that the trend of grassland degradation had not been effectively contained.

(3) The trend simulation results suggested that the NDVI showed a slight upward trend from 2020 to 2038. The NDVI has been increasing rapidly in the areas of Guinan, Zêkog and Tongde. The past and future NDVI trends in the SRYR both demonstrate climate warming and wetting trends, which should arouse attention.

Due to the limitations in data coverage for earlier years, this article analysed the spatiotemporal changes in the source area over only the last 19 years and simulated the trends for the next 19 years. Studies on long time-series data are the next research direction.

Author Contributions: Mingyue Wang and Jun'e Fu conceived the general idea and designed the experiments. Mingyue Wang and Zhitao Wu performed the experimens. Mingyue Wang, Jun'e Fu and Zhiguo Pang made the dataset. Mingyue Wang wrote the whole paper, and all authors participated in the analysis of results and edited the paper. All authors have read and agreed to the published version of the manuscript.

Funding: This work was supported by the National Key Point Research and Invention Program of the Thirteenth Five Year Plan (2017YFA0605003), and the Natural Science Foundation of China (51779269).

Acknowledgments: The authors would like to thank work fellows of Research Center on Flood & Drought Disaster Reduction of the Ministry of Water Resources for their support during this research. Meanwhile, thanks to the five anonymous reviewers and all the editors who contributed with their comments and suggestions to improve the quality of the manuscript in the process of revision.

Conflicts of Interest: The authors declare no conflicts of interest.

References

1. Hughes, T.P.; Kerry, J.T.; Connolly, S.R.; Baird, A.H.; Eakin, C.M.; Heron, S.F.; Hoey, A.S.; Hoogenboom, M.O.; Jacobson, M.; Liu, G.; et al. Ecological memory modifies the cumulative impact of recurrent climate extremes. *Nat. Clim. Chang.* **2019**, *9*, 40–43. [CrossRef]
2. Walther, G.R.; Post, E.; Convey, P.; Menzel, A.; Parmesan, C.; Beebee, J.C.; Fromentin, J.M.; Ove, H.G.; Bairlein, F. Ecological responses to recent climate change. *Nature* **2002**, *416*, 389–395. [CrossRef] [PubMed]
3. Wang, C.; Sun, Y.; Li, L.; Zhang, Q. Quantitative evaluation of regional vegetation ecological environment quality by using remotely sensed data over Qingjiang, Hubei. In Proceedings of the SPIE Second International Conference on Space Information Technology, Wuhan, China, 10 November 2007; Volume 6795.
4. Zhong, B.X.; Jiong, X.X.; Wei, Z. Spatiotemporal variations of vegetation cover on the Chinese Loess Plateau (1981–2006): Impacts of climate changes and human activities. *Sci. China Ser. D Earth Sci.* **2008**, *51*, 67–78.
5. Qin, Z.H.; Zhu, Y.X.; Li, W.J.; Xu, B. Mapping vegetation cover of grassland ecosystem for desertification monitoring in Hulun Buir of Inner Mongolia, China. In Proceedings of the Remote Sensing for Agriculture, Ecosystems, and Hydrology, Cardiff, Wales, UK, 16–18 September 2008; Volume 7104.
6. Xu, C.; Li, Y.T.; Hu, J.; Yang, X.J.; Sheng, S.; Liu, M.S. Evaluating the difference between the normalized difference vegetation index and net primary productivity as the indicators of vegetation vigor assessment at landscape scale. *Environ. Monit. Assess.* **2011**, *184*, 1275–1286. [CrossRef] [PubMed]
7. Kawabata, A.; Ichii, K.; Yamaguchi, Y. Global monitoring of interannual changes in vegetation activities using NDVI and its relationships to temperature and precipitation. *Int. J. Remote Sens.* **2001**, *22*, 1377–1382. [CrossRef]
8. Liu, X.F.; Zhu, X.F.; Pan, Y.Z.; Li, Y.Z.; Zhao, A.Z. Spatiotemporal changes in vegetation coverage in China during 1982–2012. *Acta Ecol. Sin.* **2015**, *35*, 5331–5342. (In Chinese)
9. Piao, S.L.; Fang, J.Y. Dynamic vegetation cover change over the last 18 years in China. *Quat. Sci.* **2001**, *4*, 294–302. (In Chinese)
10. Xu, G.C.; Zhang, J.X.; Li, P.; Li, Z.B.; Lu, K.X.; Wang, X.K.; Wang, F.C.; Cheng, Y.T.; Wang, B. Vegetation restoration projects and their influence on runoff and sediment in China. *Ecol. Indic.* **2018**, *95*, 233–241. [CrossRef]
11. Zhang, D.D.; Yan, D.H.; Wang, Y.C.; Lu, F.; Wu, D. Changes in extreme precipitation in the Huang-Huai-Hai River basin of China during 1960–2010. *Theor. Appl. Climatol.* **2015**, *120*, 195–209. [CrossRef]
12. Cui, L.F.; Wang, L.C.; Singh, R.P.; Lai, Z.P.; Jiang, L.L.; Yao, R. Association analysis between spatiotemporal variation of vegetation greenness and precipitation/temperature in the Yangtze River Basin (China). *Environ. Sci. Pollut. Res.* **2018**, *25*, 21867–21878. [CrossRef]
13. Xu, W.X.; Liu, X.D. Response of Vegetation in the Qinghai-Tibet Plateau to Global Warming. *Chin. Geogr. Sci.* **2007**, *17*, 151–159. [CrossRef]
14. Hou, W.J.; Gao, J.B.; Wu, S.H.; Dai, E.F. Interannual Variations in Growing-Season NDVI and Its Correlation with Climate Variables in the Southwestern Karst Region of China. *Remote Sens.* **2015**, *7*, 11105–11124. [CrossRef]
15. Jin, H.; He, R.; Cheng, G.; Wu, Q.; Wang, S.; Lü, L.; Chang, X. Changes in frozen ground in the Source Area of the Yellow River on the Qinghai–Tibet Plateau, China, and their eco-environmental impacts. *Environ. Res. Lett.* **2009**, *4*, 45206. [CrossRef]

16. Li, J.; Sheng, Y.; Wu, J.C.; Feng, Z.L.; Ning, Z.J.; Hu, X.Y.; Zhang, X.M. Landform-related permafrost characteristics in the source area of the Yellow River, eastern Qinghai-Tibet Plateau. *Geomorphology* **2016**, *269*, 104–111. [CrossRef]
17. Hu, G.Y.; Jin, H.J.; Dong, Z.B.; Lu, J.F.; Yan, C.Z. Driving forces of aeolian desertification in the source region of the Yellow River: 1975–2005. *Environ. Earth Sci.* **2013**, *70*, 3245–3254. [CrossRef]
18. Zhang, H.W. Ecological protectionand high-quality development in the yellow river basin are guaranteed by scientific management methods. *Yellow River* **2020**, *42*, 148–155. (In Chinese)
19. Qin, Y.; Yang, D.; Gao, B.; Wang, T.; Chen, J.; Chen, Y.; Zheng, G. Impacts of climate warming on the frozen ground and eco-hydrology in the Yellow River source region, China. *Sci. Total Environ.* **2017**, *605*, 830–841. [CrossRef]
20. Mudassar, I.; Wen, J.; Wang, S.P.; Tian, H.; Muhammad, A. Variations of precipitation characteristics during the period 1960–2014 in the Source Region of the Yellow River, China. *J. Arid Land* **2018**, *10*, 388–401.
21. Yang, J.P. Studies on eco-environmental change in source regions of the Yangtze and Yellow Rivers of China: Present and future. *Sci. Cold Arid Reg.* **2019**, *11*, 173–183.
22. Jiang, C.; Zhang, L.B. Climate Change and Its Impact on the Eco-Environment of the Three-Rivers Headwater Region on the Tibetan Plateau, China. *Int. J. Environ. Res. Public Health* **2015**, *12*, 12057–12081. [CrossRef]
23. Guo, W.Q.; Yang, T.B.; Dai, J.G.; Shi, L.; Lu, Z.Y. Vegetation cover changes and their relationship to climate variation in the source region of the Yellow River, China, 1990–2000. *Int. J. Remote Sens.* **2008**, *29*, 2085–2103. [CrossRef]
24. Liang, S.H.; Ge, S.M.; Wan, L.; Xu, D.W. Characteristics and causes of vegetation variation in the source regions of the Yellow River, China. *Int. J. Remote Sens.* **2012**, *33*, 1529–1542. [CrossRef]
25. Adnani, A.E.; Habib, A.; Khalidi, K.E.; Zourarah, B. Spatio-Temporal Dynamics and Evolution of Land Use Land Cover Using Remote Sensing and GIS in Sebou Estuary, Morocco. *J. Geogr. Inf. Syst.* **2019**, *11*, 551–566. [CrossRef]
26. Iqbal, M.; Wen, J.; Lan, Y.C.; Anjum, M.N.; Adnan, M.; Wang, X.; Tian, H. Assessment of Air Temperature Trends in the Source Region of Yellow River and Its Sub-Basins, China. *Asia-Pac. J. Atmos. Sci.* **2018**, *54*, 111–123. [CrossRef]
27. Hu, Y.R.; Maskey, S.; Uhlenbrook, S.; Zhao, H.L. Streamflow trends and climate linkages in the source region of the Yellow River, China. *Hydrol. Process.* **2011**, *25*, 3399–3411. [CrossRef]
28. Li, L.; Shen, H.Y.; Dai, S.; Xiao, J.S.; Shi, X.H. Response of runoff to climate change and its future tendency in the source region of Yellow River. *J. Geogr. Sci.* **2012**, *22*, 431–440. [CrossRef]
29. Xu, X.L. China Monthly Vegetation Index (NDVI) spatial distribution data set. In *Data Registration and Publishing System of the Resource and Environmental Data Cloud Platform Centre of the Chinese Academy of Sciences*; Resource and Environmental Data Cloud Platform Centre of the Chinese Academy of Sciences: Beijing, China, 2020. (In Chinese)
30. Xu, Y.; Gao, X.J.; Shen, Y.; Xu, C.H.; Shi, Y.; Giorgi, F. A Daily Temperature Dataset over China and Its Application in Validating a RCM Simulation. *Adv. Atmos. Sci.* **2009**, *26*, 763–772. [CrossRef]
31. Wu, J.; Gao, X.J. Simulation of tropical cyclones over the western north pacific and landfalling in China by REGCM4. *J. Trop. Meteorol.* **2019**, *25*, 437–447.
32. Chu, L.; Huang, C.; Liu, G.H.; Liu, Q.S.; Zhao, J. Analysis on vegetation changes of Maqu alpine wetlands in the Yellow River source region. In Proceedings of the Land Surface Remote Sensing II, Beijing, China, 13–16 October 2014; Volume 9260.
33. Pirnia, A.; Darabi, H.; Choubin, B.; Omidvar, E.; Onyutha, C.; Haghighi, A.T. Contribution of climatic variability and human activities to stream flow changes in the Haraz River basin, northern Iran. *J. Hydro-Environ. Res.* **2019**, *25*, 12–24. [CrossRef]
34. Evans, J.; Geerken, R. Discrimination between climate and human-induced dryland degradation. *J. Arid Environ.* **2004**, *57*, 535–554. [CrossRef]
35. Chen, H.; Liu, X.N.; Ding, C.; Huang, F. Phenology-Based Residual Trend Analysis of MODIS-NDVI Time Series for Assessing Human-Induced Land Degradation. *Sensors* **2018**, *18*, 3676. [CrossRef] [PubMed]
36. Yuan, L.H.; Chen, X.Q.; Wang, X.Y.; Xiong, Z.; Song, C.Q. Spatial associations between NDVI and environmental factors in the Heihe River Basin. *J. Geogr. Sci.* **2019**, *29*, 1548–1564. [CrossRef]
37. Morecroft, M.D.; Paterson, J.S. Effects of temperature and precipitation changes on plant communities. *Plant Growth Clim. Chang.* **2006**, *16*, 146–164.

38. Nemani, R.; Keeling, C.; Hashimoto, H.; Jolly, W.; Piper, S.; Tucker, C.; Myneni, R.; Running, S. Climate-driven increases in global terrestrial net primary production from1982 to 1999. *Science* **2003**, *300*, 1560–1563. [CrossRef]
39. Liu, S.L.; Zhao, H.D.; Su, X.K.; Deng, L.; Dong, S.K.; Zhang, X. Spatio-temporal variability in rangeland conditions associated with climate change in the Altun Mountain National Nature Reserve on the Qinghai-Tibet Plateau over the past 15 years. *Rangel. J.* **2015**, *37*, 67. [CrossRef]
40. McGuire, A.D.; Sturm, M.; Chapin, F.S., III. Arctic Transitions in the Land–Atmosphere System (ATLAS): Background, objectives, results, and future directions. *J. Geophys. Res.-Atmos.* **2003**, *108*. [CrossRef]
41. Peng, J.F.; Gou, X.H.; Chen, F.H.; Li, J.B.; Liu, P.B.; Zhang, Y. Altitudinal variability of climate–tree growth relationships along a consistent slope of Anyemaqen Mountains, northeastern Tibetan Plateau. *Dendrochronologia* **2008**, *26*, 87–96. [CrossRef]
42. Liu, L.S.; Zhang, Y.L.; Bai, W.Q.; Yan, J.Z.; Ding, M.J.; Shen, Z.X.; Li, S.C.; Zheng, D. Characteristics of grassland degradation and driving forces in the source region of the Yellow River from 1985 to 2000. *J. Geogr. Sci.* **2006**, *16*, 131–142. [CrossRef]
43. Shao, Q.Q.; Liu, J.Y.; Huang, L.; Fan, J.W.; Xu, X.L.; Wang, J.B. Integrated assessment on the effectiveness of ecological conservation in Sanjiangyuan National Nature Reserve. *Geogr. Res.* **2013**, *32*, 1645–1656. (In Chinese)
44. Anderson, K.; Fawcett, D.; Cugulliere, A.; Benford, S.; Jones, D.; Leng, R. Vegetation expansion in the subnival Hindu Kush Himalaya. *Glob. Chang. Biol.* **2020**, *26*, 1608–1625. [CrossRef]
45. He, H.; Li, H.; Zhu, J.; Mao, S.; Li, Y.; Yang, Y.; Zhang, F. Effects of Grazing Exclusion on Soil Properties in Maqin Alpine Meadow, Tibetan Plateau, China. *Pol. J. Environ. Stud.* **2016**, *25*, 1583–1587.
46. Li, Q.; Yang, M.X.; Wan, G.N.; Wang, X.J. Spatial and temporal precipitation variability in the source region of the Yellow River. *Environ. Earth Sci.* **2016**, *75*, 594. [CrossRef]
47. Hu, Y.; Maskey, S.; Uhlenbrook, S. Trends in temperature and rainfall extremes in the Yellow River source region, China. *Clim. Chang.* **2012**, *110*, 403–429. [CrossRef]
48. Chen, L.; Chang, J.X.; Wang, Y.M. Assessing runoff sensitivities to precipitation and temperature changes under global climate-change scenarios. *Hydrol. Res.* **2018**, *50*, 24–42. [CrossRef]
49. Du, Q.Q.; Zhang, M.J.; Wang, S.J. Changes in air temperature over China in response to the recent global warming hiatus. *Acta Geogr. Sin.* **2019**, *29*, 496–516. [CrossRef]
50. Chen, F.; Zhang, Y.; Shao, X.M.; Li, M.Q.; Yin, Z.Y. A 2000-year temperature reconstruction in the Animaqin Mountains of the Tibet Plateau, China. *Holocene* **2016**, *26*, 1904–1913. [CrossRef]
51. Yang, Z.P.; Gao, J.X.; Zhou, C.P.; Shi, P.L.; Zhao, L.; Shen, W.S.; Ouyang, H. Spatio-temporal changes of NDVI and its relation with climatic variables in the source regions of the Yangtze and Yellow rivers. *J. Geogr. Sci.* **2011**, *21*, 979–993. [CrossRef]

International Journal of
Geo-Information

MDPI

Article

Identification of Poverty Areas by Remote Sensing and Machine Learning: A Case Study in Guizhou, Southwest China

Jian Yin [1,2,*], Yuanhong Qiu [1,2] and Bin Zhang [1,2]

1 Center for China Western Modernization, Guizhou University of Finance and Economics, University City, Huaxi District, Guiyang 550025, China; yuanhongq@mail.gufe.edu.cn (Y.Q.); gzcdzbin@mail.gufe.edu.cn (B.Z.)
2 College of Big Data Application and Economic, Guizhou University of Finance and Economics, University City, Huaxi District, Guiyang 550025, China
* Correspondence: jiany@mail.gufe.edu.cn

Abstract: As an objective social phenomenon, poverty has accompanied the vicissitudes of human society, which is a chronic dilemma hindering human civilization. Remote sensing data, such as nighttime lights imagery, provides abundant poverty-related information that can be related to poverty. However, it may be insufficient to rely merely on nighttime lights data, because poverty is a comprehensive problem, and poverty identification may be affected by topography, especially in some developing countries or regions where agriculture accounts for a large proportion. Therefore, some geographical features may be necessary for supplements. With the support of the random forest machine learning method, we extracted 23 spatial features base on remote sensing including nighttime lights data and geographical data, and carried out the poverty identification in Guizhou Province, China, since 2012. Compared with the identifications using support vector machines and the artificial neural network, random forest showed a better accuracy. The results supported that nighttime lights and geographical features are better than those only by nighttime lights features. From 2012 to 2019, the identified poor counties in Guizhou Province showed obvious dynamic spatiotemporal characteristics. The number of poor counties has decreased consistently and contiguous poverty-stricken areas have fragmented; the number of poor counties in the northeast and southwest regions decreased faster than other areas. The reduction in poverty probability exhibited a pattern of spreading from the central and northern regions to the periphery parts. The poverty reduction was relatively slow in areas with large slope and large topographic relief. When poor counties are adjacent to more non-poor counties, they can get rid of poverty easier. This study provides a method for feature selection and recognition of poor counties by remote sensing images and offers new insights into poverty identification and regional sustainable development for other developing countries and areas.

Keywords: poverty probability; random forest; nighttime lights; spatiotemporal characteristics

Citation: Yin, J.; Qiu, Y.; Zhang, B. Identification of Poverty Areas by Remote Sensing and Machine Learning: A Case Study in Guizhou, Southwest China. *ISPRS Int. J. Geo-Inf.* **2021**, *10*, 11. https://doi.org/10.3390/ijgi10010011

Received: 11 November 2020
Accepted: 27 December 2020
Published: 30 December 2020

1. Introduction

As an objective social phenomenon, poverty has accompanied the vicissitudes of human society, which is a chronic dilemma hindering human civilization [1]. China, the world's largest developing country, has been undergoing rapid economic development [2]. In the past years, China has taken a large number of comprehensive poverty-alleviation work and has achieved remarkable success in poverty reduction since the beginning of the economic reforms [3]. China was therefore the first country in the world to successfully achieve the target of Millennium Development Goals of having extreme poverty in 2012, which was ahead of schedule [4]. However, there are still a large number of poor people, and, at the same time, new aspects of poverty have emerged in China. Poverty remains a serious issue for China's modernization [5]. In Southwest China, the high altitude and mountainous terrain lead to the low efficiency of land use and the underdeveloped transportation, which are the important causes of poverty [6]. Guizhou, a relatively poor province in Southwest China, has more mountainous areas, less flat land, and backward economic development [7]. Coordinating regional development, reducing poverty, and preventing the reoccurrence of poverty are the major challenges of its social and economic development [8].

Being able to accurately identify poor areas is the underlying premise of poverty reduction [8]. The spatiotemporal analysis of poverty is conducive to the formulation of regional policies for the purpose of poverty-alleviation [9]. Traditionally, the data sources for poverty assessment are the census or household surveys collected by local governments or national organizations [10,11]. Although the quality and quantity of economic data for developing countries have improved recently, a lack of data remains a major obstacle for relevant assessment [12]. In particular, the data related to poverty are usually scarce and inadequate in coverage. As the poverty research involves multi-dimensional assessment currently, more accurate indicators are needed to support poverty identification [13]. The substantial monetary cost, time and effort required to perform a census or survey means that developing countries can only conduct them periodically [14]. Moreover, different calibers of data sources make it extremely difficult to have a dynamic spatiotemporal analysis of poverty [9]. In addition, not all countries or organizations conduct periodical in-depth surveys [6]. The limitations of traditional data sources have posed great challenges to the identification of regional poverty in developing countries and areas [9].

Compared to traditional data, remote sensing data are unique, objective, consistent, and valuable resources that can provide spatial information for a variety of research purposes [16]. However, remote sensing cannot completely replace the traditional poverty identification. For example, the survey of household income and subjective feelings of poverty cannot be obtained by remote sensing. However, remote sensing technology has certain advantages in the spatial identification of poverty, which can reduce the workload of surveys and statistics [11]. Many scholars have demonstrated that poverty can be investigated using remote sensing [9,11,13,17–24]. The research areas mainly include regional poverty identification [8,9], slum mapping [17–22], and the link of the morphological structure to social poverty [23–25]. The first one is mainly used to solve the problem of identifying contiguous poverty areas at a large scale [9]. The latter two have advantages in solving slums investigation and urban problems [16,19,21], mostly based on high-resolution remote sensing data [23,24]. Rural poverty is the main form of poverty in China [26]. According to the government survey, by the end of 2018, there were 16.60 million rural poor people in China, most of whom lived in mountainous areas [27,28]. The poor counties are usually obviously contiguous in space, while the farmers' houses are scattered in vast rural areas [29]. The high-spatial-resolution recognition of poor residential areas is affected by the complexity of underlying surface [17]. Using morphological structure recognition or slum mapping to study this kind of regional poverty problem has certain challenges.

Nighttime light imagery, which is a common remote sensing data source, has been successfully used to explore changes in human activities. As one of the few reproducible and objective observation sources on a global scale, it has been used for research on regional issues including ecological civilization, social economy, urbanization, and energy-related issues, with a great reduction of statistical costs [30–32]. Recently, there have been promising signs that nighttime light imagery can provide accurate and up-to-date indications of regional poverty [11,13,15,33–35]. However, mostly studies used imagery in combination with statistical data [36]; thus, they were restricted by the statistical data. The advantages of remote sensing data including high efficiency and low cost have not been given full play. Therefore, there is a need to explore a more achievable approach to evaluate poverty using night-time light imagery. With the help of machine learning, slum mapping can be achieved with high precision based on remote sensing data [2,19–21]. Inspired by this, some scholars have tried to study regional poverty based on nighttime light imagery and machine learning. Jean et al. [11] combined survey data and satellite imagery to predict poverty using a machine learning approach. Li et al. [9,36] employed several typical machine learning approaches to identify high-poor counties based on nighttime light imagery. However, without consideration of geographical indicators, it may result in a certain degree of uncertainty in poverty identification in special topography areas [9]. Guizhou is one of the provinces with the largest proportion of rural poverty in China [37], whose special geographical environment has a great potential impact on the occurrence of poverty [26–29]. Therefore, it is necessary to consider the geographical indicators when identifying poverty.

Against this backdrop, we integrate remote sensing data of nighttime lights and geographical environment, use machine learning algorithms to identify the poor counties, and investigate the spatiotemporal dynamics of regional poverty in Guizhou from 2012 to 2019. The second part of the study introduces the materials and methods. The third part describes the dynamic spatiotemporal characteristics of poor counties in Guizhou. The fourth part provides a discussion and the conclusion.

2. Materials and Methods

2.1. Data and Area

The county is the most basic administrative unit in China. The economic development level and spatial distribution pattern of this unit are the visual performances of the status quo of the regional economic development in China [27]. Therefore, we take the county as the basic unit to identify regional poverty. Figure 1 provides a map of administrative division at the county-level in Guizhou and shows its fifty national-level poor counties released by the state council leading group office of poverty-alleviation and development (http://www.cpad.gov.cn) in 2012. The data of administrative boundaries are from National Geomatics Center of China (http://www.resdc.cn). Guizhou Province is located in Southwest China, the eastern part of Yungui Plateau. Mountain and hilly areas account for 90% of the total area, of which 70% are karst landforms. The topography impacts the regional poverty significantly [27]. The poverty line set by the Chinese government is based on the constant price of RMB 2300 per capita in 2011. According to the minimum annual income standard, the incidence of poverty in Guizhou Province reached 26.80% in 2012. Guizhou is faced with an arduous task of poverty reduction.

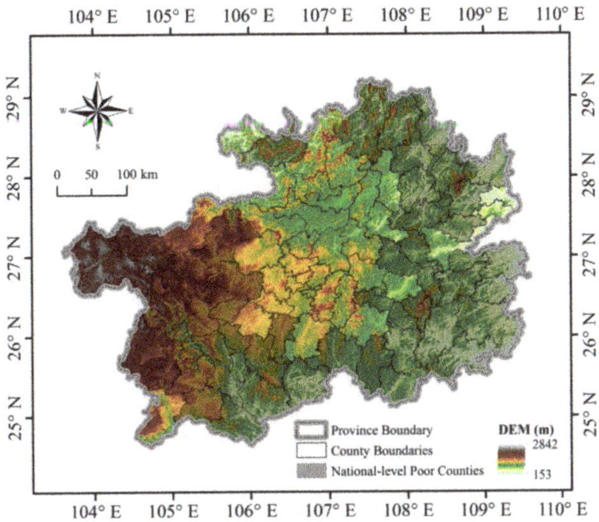

Figure 1. Distribution of the national-level poor counties of Guizhou in 2012.

The study conducts an analysis of poverty identification on a yearly basis starting from 2012. The nighttime light imagery data were provided by the Visible Infrared Imaging

Radiometer Suite (VIIRS). VIIRS is one of five instruments onboard the Suomi National Polar-orbiting Partnership (SNPP) satellite platform (https://www.ngdc.noaa.gov/eog/viirs). The spatial resolution and the illumination resolution of VIIRS are 6 times and 250 times those of Defense Meteorological Satellite Program/Operational Linescan System (DMSP/OLS) detector, respectively, and VIIRS has solved the problem of overflow due to over-saturation of the brightness value, giving the captured night images higher resolution and a greater value for a broader scale of research.

Natural topography has been regarded as one of the most important factors that controls the economic development of a county in China [38]. The studies [26,27,38,39] showed that the complex conditions of the geographical environment have a positive driving effect on the spatial distribution of the poverty-stricken counties in China. Therefore, the Digital Elevation Model (DEM) by Shuttle Radar Topography Mission (SRTM) (http://gdex.cr.usgs.gov/gdex), land use coverage, street data were also used. Vegetation fraction, water coverage, and building coverage were selected for land-use dynamic analysis. The vegetation fraction was obtained based on normalized difference vegetation index [40] that was obtained in the Moderate Resolution Imaging Spectroradiometer (MODIS) band analysis (https://modis.gsfc.nasa.gov/data/dataprod/mod13.php). Water coverage and building coverage were extracted based on the albedo of near infrared and visible light bands of Landsat Image (http://earthexplorer.usgs.gov/). The street network information was collected from Open Street Map (OSM) platform (https://www.openstreetmap.org).

2.2. Methodology

2.2.1. SNPP-VIIR Data Processing

In 2012, the Day-Night Band (DNB) of VIIRS sensor mounted on the SNPP satellite began to provide nighttime lights data with higher spatial resolution and better data quality. Compared with the DMSP/OLS data, it represents great improvements in many aspects and is unprecedentedly powerful in nocturnal observation. The drawback of VIIRS is that some of the noise is not filtered. The SNPP-VIIRS sensor has 22 bands in total, and DNB is one of its bands with the wavelengths from 500 nm to 900 nm and a spatial resolution of about 750 m. DNB features high accuracy in radiation measurement and provides on-board calibration to ensure the accuracy and stability of the data. NOAA provides two forms of SNPP-VIIRS DNB data: daily data and synthetic data. Since only the synthetic data in 2015 is available, extra work needs to be done to get the data of other years. The 2015 annual average data has been officially processed, it can be used as masking data to eliminate light anomalies and background noise. We reduced the noise by a combination of median filtering and low threshold denoising [41]. We corrected the geometric errors in the nighttime light imagery [42] and used the maximum threshold method [43] to remove the abnormal values caused by transient light. Finally, we synthesized the processed data into the annual average data.

2.2.2. Identification Features Selection

Given the applications of nighttime lights data and the identification features used in related fields [9,36,44–47], we adopted 12 statistical and spatial features to extract meaningful information from nighttime light imagery and identify poor counties. The selected features reveal the differences in the quantity, complexity, diversity, and variability of nighttime lights intensities between counties. Three aspects of statistical features were selected to describe the characteristics of the nighttime light distribution in each county: central tendency, degree of dispersion, and distribution features. The central tendency reflects the general data patterns. The dispersion degree reflects the representation of the minority data. The general distribution characteristics of night-time lights in each county can be used to demonstrate the statistical discrepancies between different counties. We also used the identification features that reflect the geographical environment. Topography is an important factor restricting the development of rural economics in China. Seventy percent of the poor counties in China are characterized by poor topographic condition [27].

By contrast, non-poor counties are mainly located in areas with good topographic conditions [38]. Natural topography determines land availability and regional accessibility and further influences the objective environment of wealth creation [27]. Therefore, the topography and reachability were considered as the features. In addition, China is a country undergoing rapid urbanization. The process of urbanization also reflects the regional development [39]. The land use change can describe the process of urbanization. So, we also chose the land use coverage as the identification features. So, the geographical features include the following easily remote-sensed spatial variables: topography, surface coverage, and reachability based on the SRTM DEM, resource satellite remote sensing image, and OSM data. Table 1 summarizes a total of 23 features for poverty identification and their extraction methods.

Table 1. Description of feature variables.

Feature Types	Aspects	Descriptions and Statistical Methods	Data Sources
Nighttime lights	Central tendency	Average of all pixels of the nighttime lights within the county's boundary Median value of all pixels within the county's boundary Average light index of all pixels within the county's boundary	SNPP-VIIRS
	Dispersion degree	Variance of all pixels within the county's boundary Standard deviation of all pixels within the county's boundary Sum of squares of deviation of all pixels within the county's boundary	
	Distribution characteristics	Total value of all pixels within the county's boundary Number of pixels within the county's boundary Number of pixels greater than zero within the county's boundary Largest value of all pixels within the county's boundary Smallest value of all pixels within the county's boundary Range between the largest and smallest value of all pixels within the county's boundary	
Topography	Elevation	Standard deviation of all pixels of the elevation within the county's boundary Average of all pixels of the elevation within the county's boundary	SRTM DEM
	Slope	Average of all pixels of the slope data within the county's boundary Percentage of pixels with slope greater than 20 degree within the county's boundary	
Surface coverage	Vegetation coverage water coverage Building coverage	Average of all pixels of vegetation fraction within the county's boundary Percentage of water coverage within the county's boundary Percentage of building coverage within the county's boundary	MODIS/ Landsat
Reachability	Total length of road	Total length of OSM primary and secondary roads within the county's boundary	OSM
	Road network density	The ratio of the total length of all roads to the total area within the county's boundary	
	Distance to county capital	Average distance of all pixels to the nearest county capital within the county's boundary	

2.2.3. Machine Learning Method

Classification is an important direction of research on data mining. At present, many machine-learning methods, including mainly single classification algorithms and ensemble learning algorithms, can be used for classification [11]. As an ensemble learning algorithm, Random Forest (RF) has better performance in classification than some other classification algorithms such as Support Vector Machines (SVM), Artificial Neural Network (ANN), and K-Nearest Neighbors (KNN) [48–51]. RF deals very well with the problems of missing data, non-equilibrium and multi-collinearity in the data set [52]. Currently, it is one of the algorithms with better results in classification and prediction of multi-variate data [53]. This paper used the RF to predict the poverty probability at the county level in Guizhou from 2012 to 2019 based on the 23 classification features. The classification approaches were conducted using the caret and random forest packages in R statistical software.

A RF model is constructed based on randomly generated training sets and multiple decision trees. The classification results of the test sample set are selected based on votes. The specific steps are described as follows:

1. The 23 identification features were calculated in each county.
2. The bootstrap sampling method [54] was used to randomly sample with replacement from the collected data of poor counties to construct a poverty training sample set with the same number of samples as the original data set. The unselected samples formed a poverty test sample set to measure the recognition error of the decision tree formed by the poverty training set. Two thirds of the samples were used to build the model, while the remaining one third are used as the test set. The sample sets were based on the national-level poor counties of Guizhou in 2012.
3. The multiple poverty features were selected as the basis for construction of the decision tree, and the multiple decision trees were built based on multiple training sample sets constructed. According to the principle of the Classification and Regression Trees (CART) algorithm [55], the classification feature with the smallest Gini coefficient was selected from the m features as the branch node, and the optimal cut-point was determined based on the Gini coefficient after the classification feature was split to complete the construction of the CART tree.
4. Starting from the root node, following the procedures in Step (3), the greedy algorithm [56] was employed to select the classification features from top to bottom, until the node cannot be split any further. Thus, the decision tree was constructed. The stopping condition was that the remaining sample number of new leaf node was less than 3.
5. The above steps were repeated many times to construct multiple decision trees to form a RF.
6. When there was a need to classify the sample in the poverty test set, multiple recognition results of the sample were obtained through RF, the conditional probability of recognition result of the sample was calculated, and the Boyer–Moore majority vote algorithm [55] was used to determine the result with the highest probability as the poverty identification result of the sample.

In the RF, the number of decision tree split attributes (mtry) and the number of decision trees (ntree) are two important attributes, which have an important impact on the performance of poverty identification. It is very important for the construction of the model to determine the appropriate values for ntree and mtry. Therefore, the results of the algorithms with different ntree and mtry were compared and analyzed. When mtry was 4, the accurate poverty identification model showed the lowest error rate. When ntree was greater than 300, the increase in the number of decision trees cannot reduce the error rate significantly but would instead reduce the operating efficiency of the model. Based on the above analysis, mtry and ntree are set to 4 and 300, respectively.

The choice of training sets is the key to the accuracy of classification. This paper selected poor counties from the national-level poor counties identified by the Chinese government in 2012, and non-poor counties from other counties identified by the government as non-poverty-stricken counties. During the adjustment and optimization of training sets, the training sets were subjected to 10-fold cross-validation, and the training sets with high accuracy were selected.

The classification result was denoted by the median value of probability, which can reveal the poverty level of each county and reflect the characteristics of relative poverty. Specifically, the closer the probability value is to 1, the greater the probability of the county to be a poor county. To a certain extent, the choice of threshold will have a significant impact on the accuracy and the prediction error of the model. We set the counties with poverty probability over 0.8 as poor counties. Accuracy is assessed by calculating the coincidence rate in space between the identification results and the 2012 government designation. On the premise that better identification results can be obtained for 2012, we used the RF model derived for 2012 to predict poverty probability at the county level of all the nighttime light imagery from 2012 to 2019 and to investigate the relative spatiotemporal patterns of poor counties in Guizhou.

2.2.4. Spatial and Temporal Analysis of Poverty Probability

In the study, Global Moran's I and Local Indicators of Spatial Association (LISA) are employed to investigate the spatiotemporal dynamics of poor counties. The Global Moran's Index is calculated by Equation (1), and its values ranges from -1 to 1 [57,58]. The closer it is to 1, the stronger the positive correlation is, while the closer it is to -1, the stronger the negative correlation; if it is close to 0, the correlation is not significant. LISA analysis uses five attributes (high-high, low-low, high-low, low-high, no significant) to describe the correlation of spatial units [59].

$$\text{Moran/s I} = \frac{n\sum_{i=1}^{n}\sum_{j=1}^{n} w_{ij}(x_i-\overline{x})(x_i-\overline{x})}{\sum_{i=1}^{n}\sum_{j=1}^{n} w_{ij}\sum_{i=1}^{n}(x_i-x)^2} \quad (1)$$

where n is the number of calculation units (such as the number of counties), x_i is the poverty probability of the ith county, the upper horizontal line represents the mean value, and w_{ij} is the spatial symmetric weight.

Spatial clustering of poverty probability at county level is calculated with the Getis-Ord G_i* statistic [60]. The Getis-Ord G_i* statistic is used to identify significant spatial clusters of high (hot spots) and low values (cold spots). High values surrounded by high values are considered as hot spots, and low values surrounded by low values are considered as cold spots. The G_i* is calculated by Equation (2), where the variables represented by letters were the same as those in Equation (1).

$$G_i^* = \frac{\sum_{j=1}^{n} w_{ij}x_j - \frac{1}{n}\left(\sum_{j=1}^{n} x_j\right)\left(\sum_{j=1}^{n} w_{ij}\right)}{\sqrt{\frac{n\sum_{j=1}^{n} x_j{}^2-\left(\sum_{j=1}^{n} x_j\right)^2}{n^2}}\sqrt{\frac{n\sum_{j=1}^{n} w_{ij}{}^2-\left(\sum_{j=1}^{n} w_{ij}\right)^2}{n-1}}} \quad (2)$$

3. Results

3.1. Performance of the Poverty Identification

In order to make a comparison between the RF and other machine learning models, we used two typical classification algorithms of SVM and ANN to identify the poverty probability in 2012, and adopted four evaluation indicators, namely, accuracy, precision, recall, and F-value to evaluate the effects of these three models in poverty identification. Table 2 shows the accuracy, precision, recall and F-value of the SVM, ANN, and RF in poverty identification. It also shows the identification results using only nighttime lights features as a comparison. Compared with ANN and SVM, the RF model reaches higher values in its accuracy, precision, recall and F-value, indicating the RF-based identification had a better performance in poverty identification. The performance of the above three methods based on comprehensive features are all higher than that using only nighttime lights features. Figure 2 shows the Receiver Operating Characteristic (ROC) curves of the identification by the four methods. The Area Under Curve (AUC) values from large to small are RF, SVM, ANN, and RF using only nighttime lights features. Therefore, RF is the most reliable among the several schemes tested. The identification results of the above four patterns were shown in Figures 3–6.

Table 2. Evaluation indicators in poverty identification by the typical algorithms.

Algorithm	Evaluation Indicator			
	Accuracy	Precision	Recall	F-Value
RF	0.9432	0.9592	0.9400	0.9495
ANN	0.8636	0.8958	0.8600	0.8776
SVM	0.8750	0.8980	0.9167	0.9072
RF only nighttime lights features	0.7614	0.8085	0.7917	0.8000

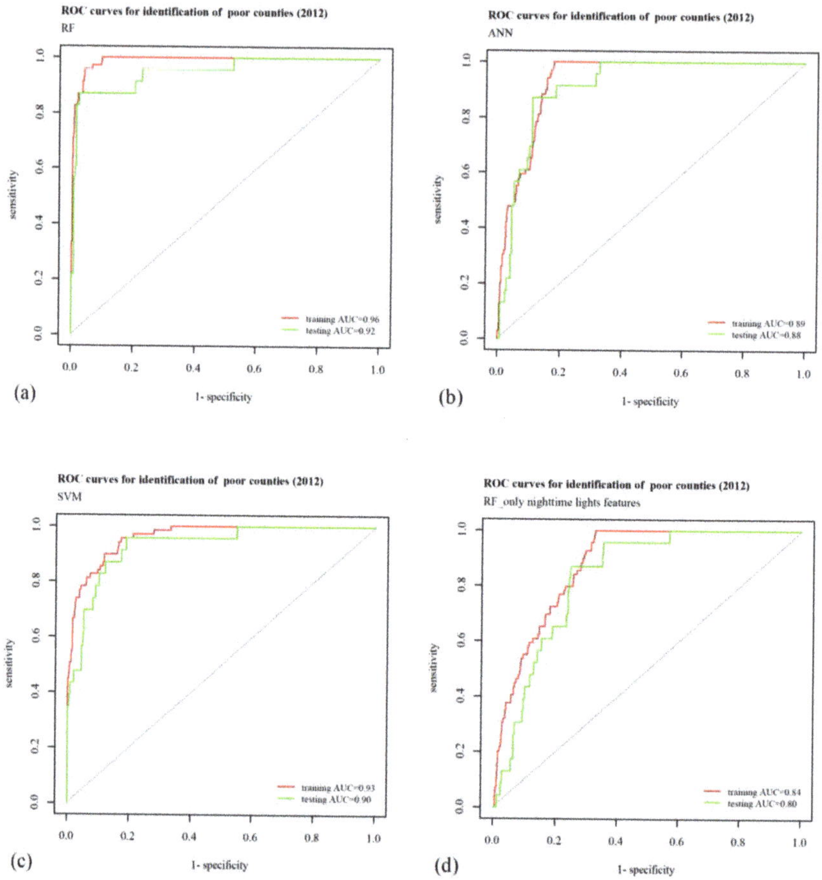

Figure 2. The Receiver Operating Characteristic (ROC) curves of poor counties identification in 2012. (**a**) ROC curves of poor counties identification by RF, (**b**) ROC curves of poor counties identification by ANN, (**c**) ROC curves of poor counties identifi-cation by SVM, (**d**) ROC curves of poor counties identifi-cation by RF_only nighttime lights fea-tures.

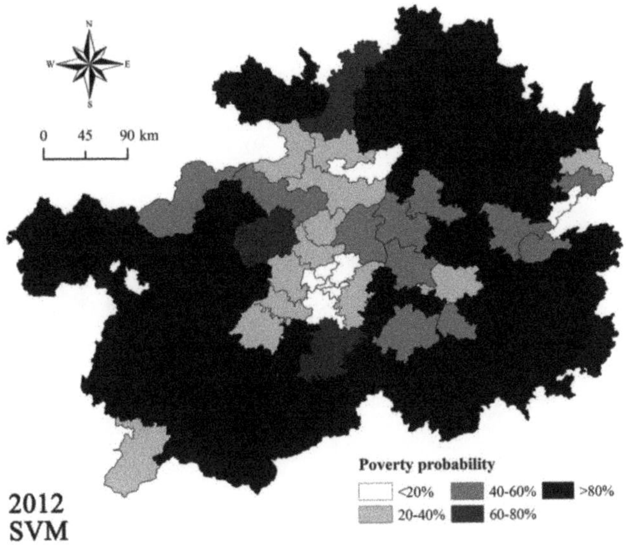

Figure 3. Distribution of poverty probability at the county level by Support Vector Machines (SVM) in 2012.

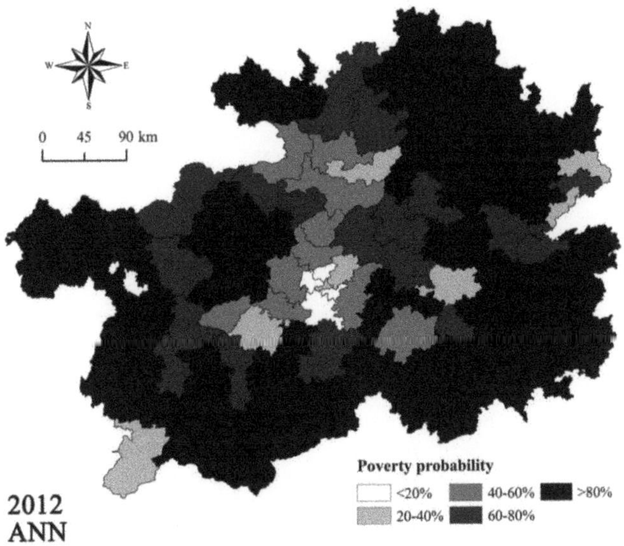

Figure 4. Distribution of poverty probability at the county level by the Artificial Neural Network (ANN) in 2012.

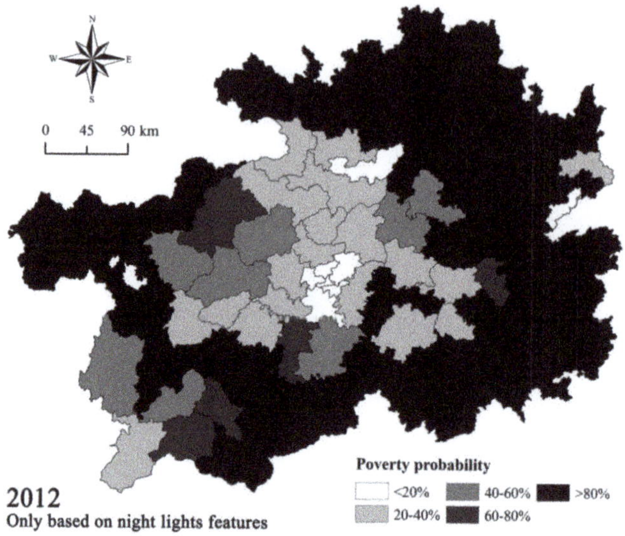

Figure 5. Distribution of poverty probability at the county level by Random Forest (RF) only based on nighttime lights features in 2012.

The poverty probability at the county level in 2012 is obtained and compared with the 50 national-level poor counties designated by the Chinese government. Using the RF model, there are 47 counties with a poverty probability greater than 0.8, which is in good agreement with the national-level poor counties. Figure 6 shows the spatiotemporal pattern of poverty probability at the county level in Guizhou from 2012 to 2019. As shown in the figure, there are evident characteristics and dynamics in the spatial distribution of poor counties over time. The poor counties are mainly distributed in the eastern, southern and high-altitude regions of Guizhou. It is obvious that the poor counties of Guizhou are contiguous in distribution. With the support of the poverty-alleviation work, the poor counties have been decreasing. The number of poor counties in the northeast and southwest of Guizhou has declined faster than the number of poor counties in other areas. In terms of the spatial distribution of the poor counties becoming non-poor counties, poor counties that are adjacent to non-poor counties are more easily transformed into non-poor counties. The poverty probability of each county in Guizhou Province changed in fluctuations. The poverty probability of its western region fluctuated greatly, and its eastern and southern regions were consistently recognized as areas with higher poverty probability.

The distribution of poor counties in Guizhou is contiguous, and the spatiotemporal characteristics of contiguous poverty-stricken areas from 2012 to 2019 are evident in Guizhou. The changes in the boundaries of impoverished areas indicate that the spatial distribution of impoverished areas has become more complex, showing obvious regional characteristics. We extracted the spatial distribution characteristics of the topography at the county level (Figures 7 and 8). It is found that the boundary change in the poverty-stricken areas is highly related to the terrain features, and that poverty-alleviation is easier to implement in areas with a high proportion of plains. The plain area in central Guizhou has the lowest incidence of poverty. The area with high altitude and high topographic relief usually has a high incidence of poverty. In addition, the internal fragmentation of the contiguous poverty-stricken areas is intensifying.

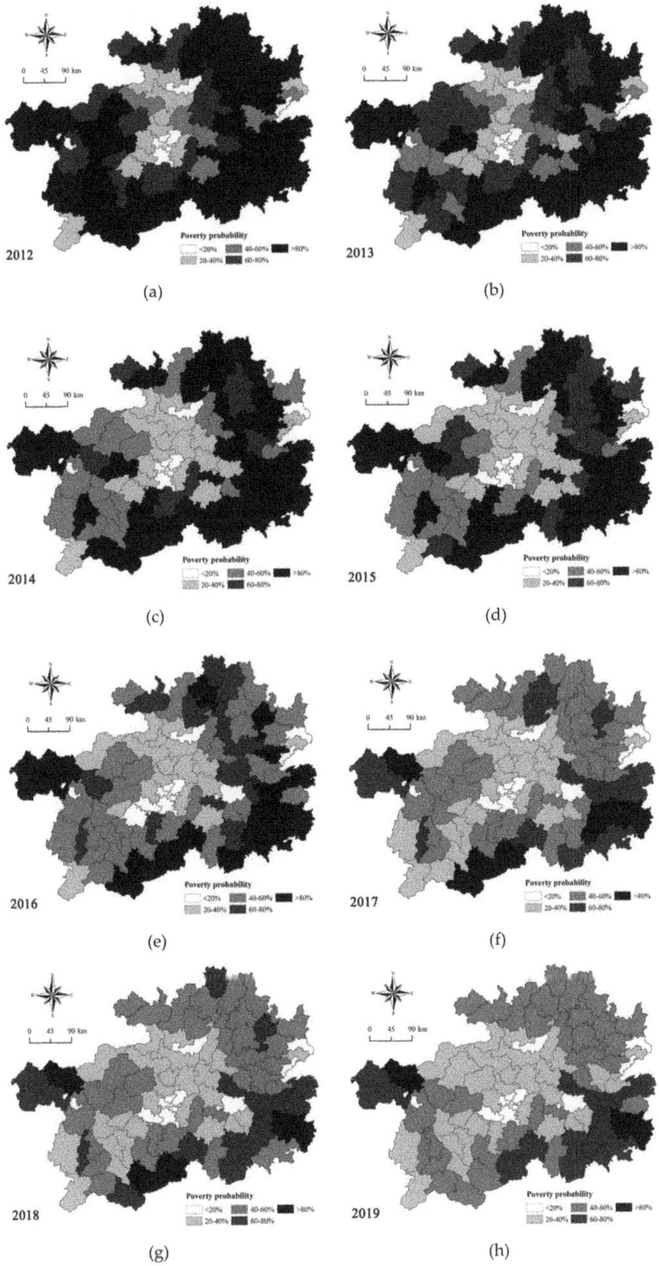

Figure 6. Distribution of poverty probability at the county level by RF from 2012 to 2019, (a) poverty probability at the county level by RF in 2012, (**b**) poverty probability at the county level by RF in 2013, (**c**) poverty probability at the county level by RF in 2014, (**d**) poverty probability at the county level by RF in 2015, (**e**) poverty probability at the county level by RF in 2016, (**f**) poverty probability at the county level by RF in 2017, (**g**) poverty probability at the county level by RF in 2018, (**h**) poverty probability at the county level by RF in 2019.

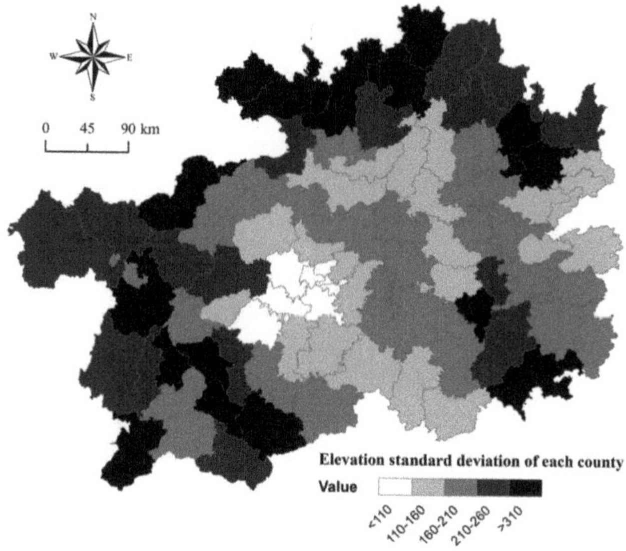

Figure 7. Distribution of elevation standard deviation at the county level.

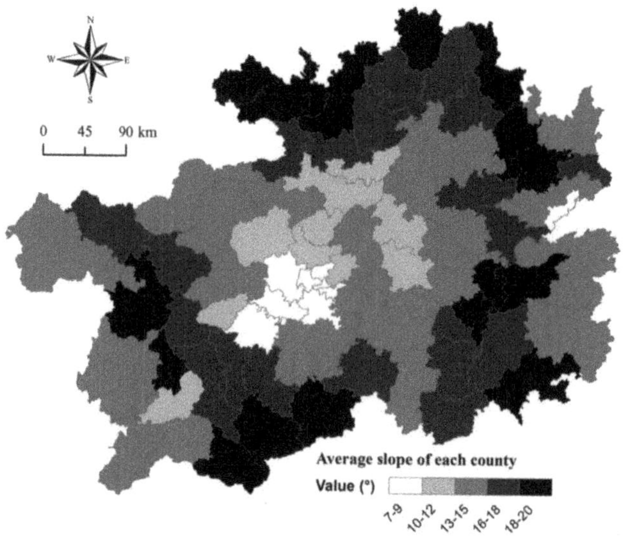

Figure 8. Distribution of average slope at the county level.

3.2. Spatiotemporal Dynamics of Poverty Probability

Figure 9 shows the Global Moran's I of poverty probability at the county level from 2012 to 2019. The results reveal that the estimates of the 8 years are all above 0, indicating that the poverty probability at the county level in Guizhou has a positive spatial autocorrelation. The values are greater than 0.7 from 2012 to 2015, indicating that this spatial autocorrelation is relatively strong during the years. The Global Moran's I has relatively small fluctuations and a decreasing trend, indicating that the poverty probability at the county level in Guizhou has been relatively stable since 2012 but with a trend of spatial dispersion.

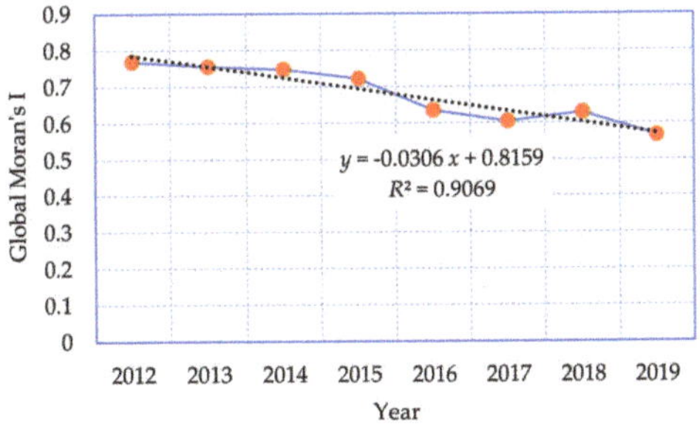

Figure 9. Global Moran's I of poverty probability from 2012 to 2019 in Guizhou Province.

Figure 10 shows the local autocorrelation of poverty probability at the county level in 2012. As shown in the figure, the high-high clusters are primarily located in the eastern and southwestern regions of Guizhou. The central part is identified as having low-low clusters. These results indicate that relatively poor and non-poor counties have obvious regional distribution features. The high-low distribution is mainly found around the low-low clusters, which is related to the rapid poverty-alleviation in this area. The low-high areas were scattered, which were the areas with high possibility of returning to poverty and were also the key areas of poverty-alleviation.

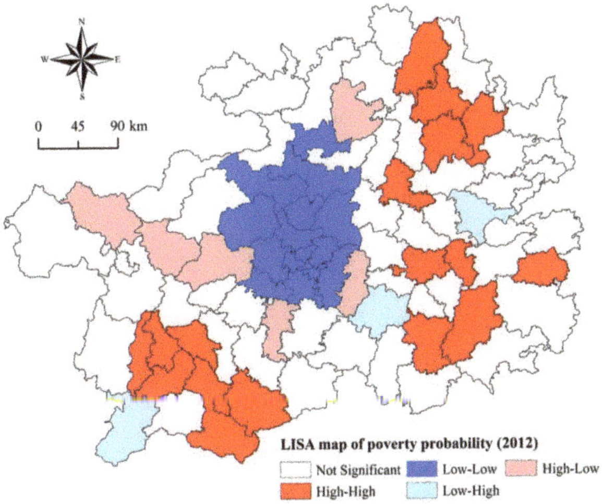

Figure 10. Local Indicators of Spatial Association (LISA) map of poverty probability at the county level for 2012.

Figure 11 shows the results of the hot spot analysis of poverty in Guizhou. As shown in the figure, the hot spots are mainly distributed in the southeast, southwest, and northeast of Guizhou, while the cold spots are mainly distributed in the central region. After 2016, the poverty hot spots in northeast of Guizhou disappear, indicating that there are no obvious contiguous poverty-stricken areas in the region. Within the whole province, the number of hot spots is decreasing, indicating that the contiguous poverty-stricken areas are gradually shrinking. However, there are still hot spots in southeast and south of Guizhou characterized with spatial agglomeration in 2019. The cold spots gather in the central. It indicates that the southeastern and southwestern parts of Guizhou are the areas with high poverty probability, while the central part of Guizhou has a low poverty probability.

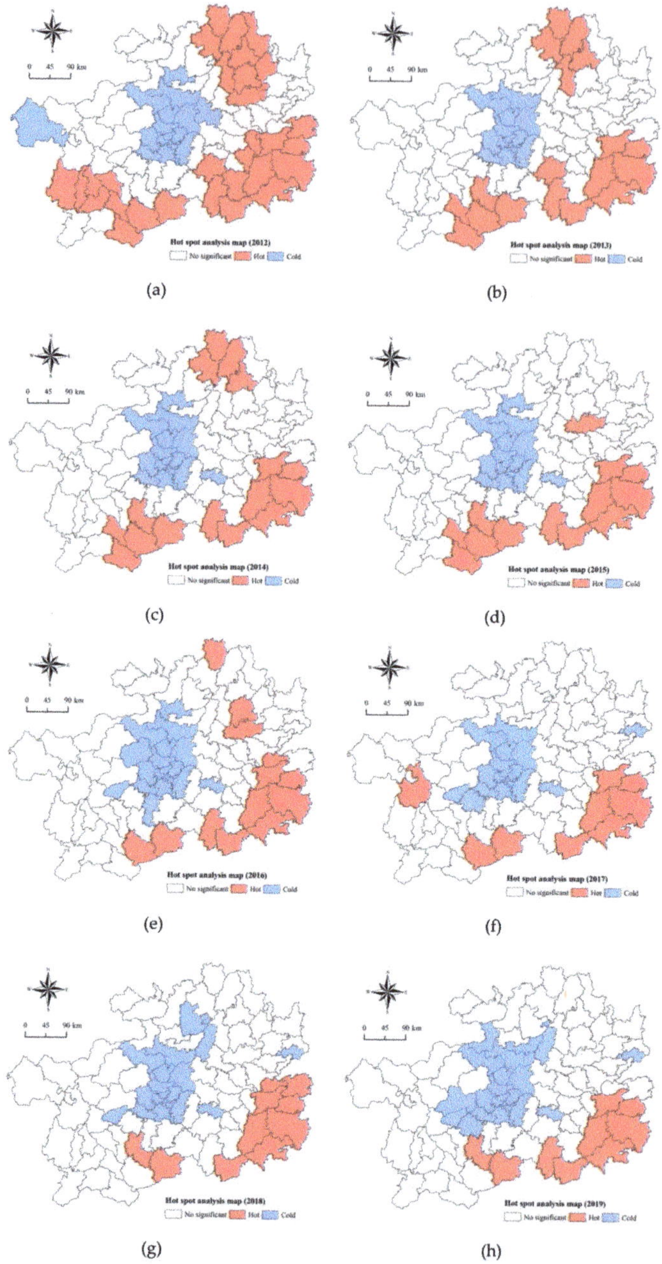

Figure 11. Hot-cold spot map of poverty probability at the county level; (**a**) hot-cold spot analysis map of poverty probability at the county level in 2012, (**b**) hot-cold spot analysis map of poverty probability at the county level in 2013, (**c**) hot-cold spot analysis map of poverty probability at the county level in 2014, (**d**) hot-cold spot analysis map of poverty probability at the county level in 2015, (**e**) hot-cold spot analysis map of poverty probability at the county level in 2016, (**f**) hot-cold spot analysis map of poverty probability at the county level in 2017, (**g**) hot-cold spot analysis map of poverty probability at the county level in 2018, (**h**) hot-cold spot analysis map of poverty probability at the county level in 2019.

There are three national contiguous poverty-stricken areas (Figure 12) that located in Guizhou partly. They are Wuling Mountain Area (WLMA), Wumeng Mountain Area (WMMA), and Rocky Desertification Area in Yunnan, Guizhou, and Guangxi (RDAYGG). Figure 13 shows the average poverty probability of counties at various regional scales since 2012. The average value of national-level poor counties in Guizhou is greater than that of the WLMA, WMMA, RDAYGG, and the whole province; the average poverty probability of national-level poor counties has approximated that of the whole province; the average value of national-level poor counties is falling faster than that of the whole province. The average value of RDAYGG is far greater than that of the whole province. Since 2016, the average poverty probability of the WLMA has been less than that of the whole province. It indicates the effect of poverty reduction has been remarkable in the region.

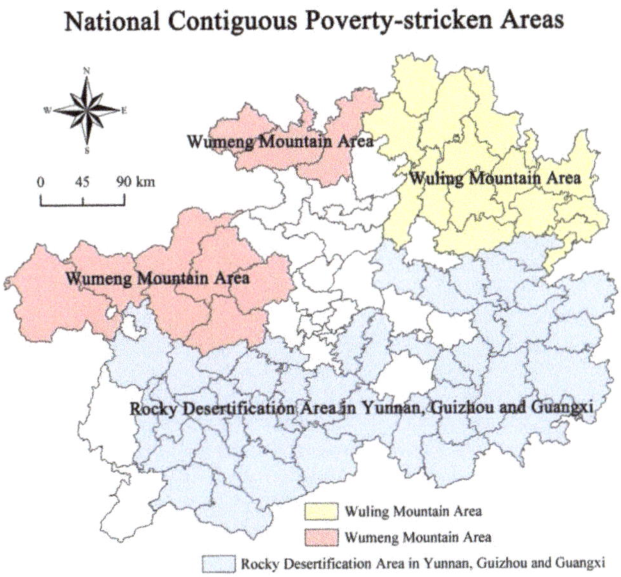

Figure 12. Distribution of national contiguous poverty-stricken areas of Guizhou Province.

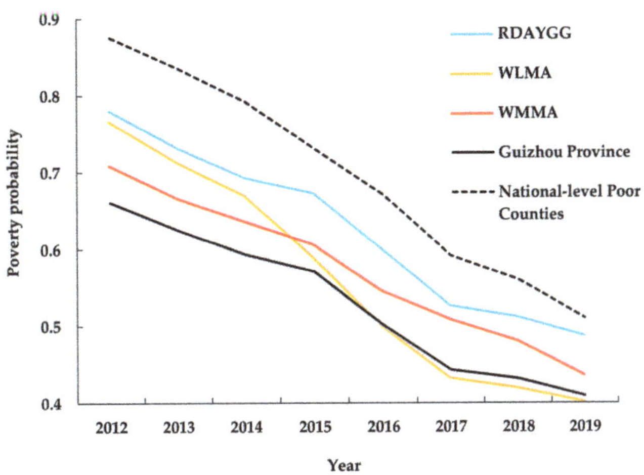

Figure 13. Average poverty probability from 2012 to 2019 at different spatial scales.

4. Discussion and Conclusions

4.1. Poverty Measurement

Poverty is a complex issue that has inextricable relationship with society, economy, environment and so on. It is a global problem and the primary obstacle for the realization of sustainable development. Most of the existing research works rely on statistics data from the government or other organizations, to a certain extent, which has limited the authenticity, validity, and timeliness of poverty-related research, making it difficult to accurately recognize regional poverty and study its dynamic spatiotemporal characteristics. Nighttime light imagery data provides abundant poverty-related information that improves the efficiency of poverty identification. However, it is also insufficient to rely merely on nighttime lights remote sensing data. Poverty is a comprehensive problem, and poverty identification may be affected by indicators such as topography and environment. Considering that the geographical conditions of Guizhou are the important reason for poverty, this paper extracted 23 spatial features including nighttime light imagery data and geographical data and carried out the poverty identification at the county level in Guizhou since 2012. Compared with the results of the study using only nighttime lights remote sensing data, the accuracy has been improved. Therefore, it is necessary to introduce geographical data into poverty identification. Our results provide an approximate description of the variation of poor counties in Guizhou from 2012 to 2019 and identify the regions that were impoverished in these years. As an exploratory study, our research represents the attempt to estimate poverty only using remote-sensed data. We hope that our research can serve as a reference for future researchers and facilitate more accurately targeted antipoverty strategies.

4.2. Spatiotemporal Dynamics of Poverty

Poor counties present obvious spatiotemporal dynamics in Guizhou from 2012 to 2019. The number of poor counties fluctuates but overall decreases over time. The non-poor counties are mainly distributed in the central part of Guizhou, while the poor counties are mainly distributed in the southwestern, southeastern and northwestern regions. Geographical barriers hinder or limit regional development and are an important cause of poverty. The reduction of poor counties started from the central and northern regions in Guizhou and spread to the periphery. Affected by the topography, poverty reduction is relatively slow in areas with large slope and large topographic relief. When poor counties are adjacent to more non poor counties, it is easier for them to get rid of poverty. The results are consistent with the law of regional development and practical conditions.

In order to study the dynamic changes in the poverty-alleviation phases of Guizhou and the impact of the phased policies on the distribution of poor counties, we further analyzed the poor counties at some important phases. In the year of 2016, China over fulfilled the goal of reducing rural poverty by 10 million with the support of national policies, which was a milestone in the history of poverty alleviation in China. Figure 6 shows the number of counties with a poverty probability greater than 80% decrease significantly in 2016, reflecting the poverty-alleviation achievements. The Chinese government pointed out that 2019 was a crucial year to win the fight against poverty. As shown in Figure 6, the poverty probability in Guizhou is greatly reduced in 2019, and there are only three counties with a poverty probability greater than 80%. These findings reveal that the regional economic development and national policy implementation are highly related to the spatiotemporal dynamics of poverty, and the key to achieving regional poverty reduction is the development of the regional economy and the implementation of national macrolevel policies. Therefore, the results of the spatiotemporal dynamic of poverty can be used to make targeted policies for poverty reduction in impoverished areas, to achieve more sustainable and effective poverty-alleviation.

4.3. Applications and Implications

Compared with research that relies on census and commercial data sets, machine learning method based on remote sensing data can effectively improve the efficiency of poverty identification while allowing for real-time monitoring. As nighttime light imagery data and other remote sensing data are accessible at any time for free, the cost of poverty identification can be greatly reduced. This method can cope with the problems of lack of statistical information and high statistical costs and can serve as a reference for poverty surveys and targeted poverty alleviation in regional and even global underdeveloped areas.

The county is the most basic administrative unit in China and around the world. The economic development and spatial distribution pattern of this unit are the visual performances of the status quo of the regional economic development in China [27]. Therefore, the poverty analysis at the county level could support the regional coordinated development and the national macrolevel formulation of policies.

4.4. Limitations and Prospects

Nighttime light imagery data is a kind of comprehensive information. In the future, the exploration could be made to build a correlation model between geographical characteristics and nighttime lights data, and analyze whether certain features of nighttime lights data can replace the geographical data. In addition, limited by the available verification data, this study only verified the results in 2012. The RF with higher accuracy than ANN and SVM was used to identify the poverty probability after 2012. It needs to consider that the best RF for poverty probability identification may change over time. Therefore, the results may not be the most accurate identification by using the RF of 2012 in the next few years. Finally, this study only qualitatively discusses the correlation between geographical factors and poverty. Future research will focus on geostatistical analysis to examine multiple factors affecting poverty identification quantitatively.

Author Contributions: Conceptualization, Jian Yin; methodology, Jian Yin; validation, Jian Yin, Yuanhong Qiu and Bin Zhang; formal analysis, Jian Yin; investigation, Jian Yin; resources, Bin Zhang; data curation, Jian Yin; writing—original draft preparation, Jian Yin; writing—review and editing, Jian Yin and Bin Zhang; visualization, Jian Yin and Yuanhong Qiu. All authors have read and agreed to the published version of the manuscript.

Funding: This work was partially supported by the MOE (Ministry of Education in China) Liberal arts and Social Sciences Foundation (Grant No. 19YJCZH228), and the University Project of Guizhou University of Finance and Economics (2019XJC03). The authors are grateful to the reviewers for the help and thought-provoking comments.

Informed Consent Statement: Not applicable.

Data Availability Statement: The data presented in this study are available on request from the corresponding author.

Acknowledgments: We thank the editors and the anonymous reviewers for their valuable comments and suggestions.

Conflicts of Interest: The authors declare no conflict of interest.

References

1. Zhao, X.; Yu, B.; Liu, Y.; Chen, Z.; Li, Q.; Wang, C.; Wu, J. Estimation of poverty using random forest regression with multi-source data: A case study in Bangladesh. *Remote Sens.* **2019**, *11*, 375. [CrossRef]
2. Lo, K.; Wang, M. How voluntary is poverty-alleviation resettlement in China? *Habitat Int.* **2018**, *73*, 34–42. [CrossRef]
3. Sun, J.; Xia, T. China's Anti-poverty strategy and post-2020 relative poverty line. *China Econ.* **2020**, *15*, 62–75. [CrossRef]
4. Guo, Y.; Zhou, Y.; Cao, Z. Geographical patterns and anti-poverty targeting post-2020 in China. *J. Geogr. Sci.* **2018**, *28*, 1810–1824. [CrossRef]
5. Wu, Y.; Qi, D. A gender-based analysis of multidimensional poverty in China. *Asian J. Womens Stud.* **2017**, *23*, 66–88. [CrossRef]
6. Luo, G.; Wang, B.; Luo, D.; Wei, C. Spatial agglomeration characteristics of rural settlements in poor mountainous areas of Southwest China. *Sustainability* **2020**, *12*, 1818. [CrossRef]

7. Yang, L.; Jiang, C.; Ren, X.; Walker, R.; Xie, J.; Zhao, Y. Determining Dimensions of Poverty Applicable in China: A Qualitative Study in Guizhou. *J. Soc. Serv. Res.* **2020**, 1–18. [CrossRef]
8. Xu, Z.; Cai, Z.; Wu, S.; Huang, X.; Liu, J.; Sun, J.; Su, S.; Weng, M. Identifying the geographic indicators of poverty using geographically weighted rgression: A case study from Qiandongnan Miao and Dong Autonomous Prefecture, Guizhou, China. *Soc. Indic. Res.* **2019**, *142*, 947–970. [CrossRef]
9. Li, G.; Chang, L.; Liu, X.; Su, S.; Cai, Z.; Huang, X.; Li, B. Monitoring the spatiotemporal dynamics of poor counties in China: Implications for global sustainable development goals. *J. Clean. Prod.* **2019**, *227*, 392–404. [CrossRef]
10. Labar, K.; Bresson, F. A multidimensional analysis of poverty in China from 1991 to 2006. *China Econ. Rev.* **2011**, *22*, 646–668. [CrossRef]
11. Jean, N.; Burke, M.; Xie, M.; Davis, W.M.; Lobell, D.B.; Ermon, S. Combining satellite imagery and machine learning to predict poverty. *Science* **2016**, *353*, 790–794. [CrossRef] [PubMed]
12. Hick, R. Material poverty and multiple deprivation in Britain: The distinctiveness of multidimensional assessment. *J. Public Policy* **2016**, *36*, 277–308. [CrossRef]
13. Yu, B.; Shi, K.; Hu, Y.; Huang, C.; Chen, Z.; Wu, J. Poverty evaluation using NPP-VIIRS nighttime light composite data at the county level in China. *IEEE J. Sel. Top. Appl. Earth Obs. Remote Sens.* **2017**, *8*, 1217–1229. [CrossRef]
14. Njuguna, C.; McSharry, P. Constructing spatiotemporal poverty indices from big data. *J. Bus. Res.* **2017**, *70*, 318–327. [CrossRef]
15. Elvidge, C.D.; Ziskin, D.; Baugh, K.E.; Tuttle, B.T.; Ghosh, T.; Pack, D.W.; Erwin, E.H.; Zhizhin, M. A fifteen year record of global natural gas flaring derived from satellite data. *Energies* **2009**, *2*, 595–622. [CrossRef]
16. Bunte, J.B.; Desai, H.; Gbala, K.; Parks, B.; Runfola, D.M. Natural resource sector FDI, government policy, and economic growth: Quasi-experimental evidence from Liberia. *World Dev.* **2018**, *107*, 151–162. [CrossRef]
17. Kuffer, M.; Pfeffer, K.; Sliuzas, R. Slums from space—15 years of slum mapping using remote sensing. *Remote Sens.* **2016**, *8*, 455. [CrossRef]
18. Mahabir, R.; Croitoru, A.; Crooks, A.T.; Agouris, P.; Stefanidis, A. A critical review of high and very high-resolution remote sensing approaches for detecting and mapping slums: Trends, challenges and emerging opportunities. *Urban Sci.* **2018**, *2*, 8. [CrossRef]
19. Wurm, M.; Stark, T.; Zhu, X.X.; Weigand, M.; Taubenboeck, H. Semantic segmentation of slums in satellite images using transfer learning on fully convolutional neural networks. *ISPRS J. Photogramm. Remote Sens.* **2019**, *150*, 59–69. [CrossRef]
20. Wurm, M.; Taubenbck, H.; Weigand, M.; Schmitt, A. Slum mapping in polarimetric SAR data using spatial features. *Remote Sens. Environ.* **2017**, *194*, 190–204. [CrossRef]
21. Mast, J.; Wei, C.; Wurm, M. Mapping urban villages using fully convolutional neural networks. *Remote Sens. Lett.* **2020**, *11*, 630–639. [CrossRef]
22. Engstrom, R.; Hersh, J.; Newhouse, D. Poverty from Space: Using High-Resolution Satellite Imagery for Estimating Economic Well-Being. *World Bank Policy Res. Work. Pap.* **2017**, 8284. [CrossRef]
23. Wurm, M.; Taubenböck, H. Detecting social groups from space—Assessment of remote sensing-based mapped morphological slums using income data. *Remote Sens. Lett.* **2018**, *9*, 41–50. [CrossRef]
24. Hannes, T.; Jeroen, S.; Xiao, Z.; Christian, G.; Stefan, D.; Michael, W. Are the poor digitally left behind? Indications of urban divides based on remote sensing and twitter data. *ISPRS Int. J. Geo Inf.* **2018**, *7*, 304. [CrossRef]
25. Niu, T.; Chen, Y.; Yuan, Y. Measuring urban poverty using multi -source data and a random forest algorithm: A case study in Guangzhou. *Sustain. Cities Soc.* **2020**, *54*, 102014. [CrossRef]
26. Liu, Y.; Liu, J.; Zhou, Y. Spatio-temporal patterns of rural poverty in China and targeted poverty-alleviation strategies. *J. Rural Stud.* **2017**, *52*, 66–75. [CrossRef]
27. Zhou, L.; Xiong, L. Natural topographic controls on the spatial distribution of poverty-stricken counties in China. *Appl. Geogr.* **2018**, *90*, 282–292. [CrossRef]
28. National Bureau of Statistics. Available online: http://www.stats.gov.cn/tjsj/zxfb/201908/t20190829_1694202.html (accessed on 10 December 2020).
29. Ren, Q.; Huang, Q.; He, C.; Tu, M.; Liang, X. The poverty dynamics in rural china during 2000–2014: A multi-scale analysis based on the poverty gap index. *J. Geogr. Sci.* **2018**, *28*, 1427–1443. [CrossRef]
30. Huang, Q.; Yang, X.; Gao, B.; Wang, L.; Hu, Y.; Wang, J.; Huang, W. Application of DMSP/OLS nighttime light images: A meta-analysis and a systematic literature review. *Remote Sens.* **2014**, *6*, 6844–6866. [CrossRef]
31. Keola, S.; Andersson, M.; Hall, O. Monitoring economic development from space: Using nighttime light and land cover data to measure economic growth. *World Dev.* **2015**, *66*, 322–334. [CrossRef]
32. Shao, S.; Tian, Z.; Fan, M. Do the rich have stronger willingness to pay for environmental protection? New evidence from a survey in China. *World Dev.* **2018**, *105*, 83–94. [CrossRef]
33. Wang, W.; Cheng, H.; Zhang, L. Poverty assessment using DMSP/OLS night-time light satellite imagery at a provincial scale in China. *Adv. Space Res.* **2012**, *49*, 1253–1264. [CrossRef]
34. Pan, W.; Fu, H.; Zheng, P. Regional poverty and inequality in the Xiamen-Zhangzhou-Quanzhou city cluster in China based on NPP/VIIRS night-time light imagery. *Sustainability* **2020**, *12*, 2547. [CrossRef]
35. Shi, K.; Chang, Z.; Chen, Z.; Wu, J.; Yu, B. Identifying and evaluating poverty using multisource remote sensing and point of interest (POI) data: A case study of Chongqing, China. *J. Clean Prod.* **2020**, *255*, 120245. [CrossRef]

36. Li, G.; Cai, Z.; Liu, X.; Liu, J.; Su, S. A comparison of machine learning approaches for identifying high-poor counties: Robust features of DMSP/OLS night-time light imagery. *Int. J. Remote Sens.* **2019**, *40*, 5716–5736. [CrossRef]
37. Xu, J.; Song, J.; Cao, X.; Sun, F. Spatial pattern of poverty and its influencing factors Based on CART Model in Guizhou Province. *Econ. Geogr.* **2020**, *40*, 166–173. [CrossRef]
38. Ward, P.S. Transient poverty, poverty dynamics, and vulnerability to Poverty: An empirical analysis using a balanced panel from rural China. *World Dev.* **2016**, *78*, 541–553. [CrossRef]
39. Zhang, K.; Dearing, J.A.; Dawson, T.P.; Dong, X.; Yang, X.; Zhang, W. Poverty-alleviation strategies in eastern china lead to critical ecological dynamics. *Sci. Total Environ.* **2015**, *506–507*, 164–181. [CrossRef]
40. Gong, Z.; Zhao, S.; Gu, J. Correlation analysis between vegetation coverage and climate drought conditions in North China during 2001–2013. *J. Geogr. Sci.* **2017**, *27*, 143–160. [CrossRef]
41. Zhong, L.; Liu, X.; Yang, P. Method for SNPP-VIIRS nighttime lights images denoising. *Bull. Surv. Mapp.* **2019**, *3*, 21–26. [CrossRef]
42. Wang, W.; Cao, C.; Bai, Y.; Blonski, S.; Schull, M.A. Assessment of the NOAA S-NPP VIIRS geolocation reprocessing improvements. *Remote Sens.* **2017**, *9*, 974. [CrossRef]
43. Chen, Z.; Yu, B.; Hu, Y.; Huang, C.; Shi, K.; Wu, J. Estimating house vacancy rate in metropolitan areas using NPP-VIIRS nighttime light composite data. *IEEE J. Sel. Top. Appl. Earth Obs. Remote Sens.* **2015**, *8*, 2188–2197. [CrossRef]
44. Li, X.; Ge, L.; Chen, X. Detecting Zimbabwe's decadal economic decline using nighttime light imagery. *Remote Sens.* **2013**, *5*, 4551–4570. [CrossRef]
45. Small, C.; Elvidge, C.D. Night on earth: Mapping decadal changes of anthropogenic night light in Asia. *Int. J. Appl. Earth Obs. Geo Inf.* **2013**, *22*, 40–52. [CrossRef]
46. Wu, J.; Wang, Z.; Li, W.; Peng, J. Exploring factors affecting the relationship between light consumption and GDP based on DMSP/OLS nighttime satellite imagery. *Remote Sens. Environ.* **2013**, *134*, 111–119. [CrossRef]
47. Ma, T.; Zhou, Y.; Zhou, C.; Haynie, S.; Pei, T.; Xu, T. Night-time light derived estimation of spatio-temporal characteristics of urbanization dynamics using DMSP/OLS satellite data. *Remote Sens. Environ.* **2015**, *158*, 453–464. [CrossRef]
48. You, H.; Ma, Z.; Tang, Y.; Wang, Y.; Yan, J.; Ni, M.; Cen, K.; Huang, Q. Comparison of ANN (MLP), ANFIS, SVM, and RF models for the online classification of heating value of burning municipal solid waste in circulating fluidized bed incinerators. *Waste Manag.* **2017**, *68*, 186. [CrossRef]
49. Yuan, H.; Yang, G.; Li, C.; Wang, Y.; Liu, J.; Yu, H.; Feng, H.; Xu, B.; Zhao, X.; Yang, X. Retrieving soybean leaf area index from unmanned aerial vehicle hyperspectral remote sensing: Analysis of RF, ANN, and SVM regression models. *Remote Sens.* **2017**, *9*, 309. [CrossRef]
50. Sun, T.; Chen, F.; Zhong, L.; Liu, W.; Wang, Y. GIS-based mineral prospectivity mapping using machine learning methods: A case study from Tongling ore district, eastern China. *Ore Geol. Rev.* **2019**, *109*, 26–49. [CrossRef]
51. Luo, L. Reserch on targeted poverty indentification model based on random forest algorithms. *J. Huazhong Agric. Univ.* **2019**, *144*, 21–29. [CrossRef]
52. Mutanga, O.; Adam, E.; Cho, M.A. High density biomass estimation for wetland vegetation using WorldView-2 imagery and random forest regression algorithm. *Int. J. Appl. Earth Obs. Geoinf.* **2012**, *18*, 399–406. [CrossRef]
53. Stumpf, A.; Kerle, N. Object-oriented mapping of landslides using random forests. *Remote Sens. Environ.* **2011**, *115*, 2564–2577. [CrossRef]
54. Rodriguez-Galiano, V.F.; Ghimire, B.; Rogan, J.; Chica-Olmo, M.; Rigol-Sanchez, J.P. An assessment of the effectiveness of a random forest classifier for land-cover classification. *ISPRS J. Photogramm. Remote Sens.* **2012**, *67*, 93–104. [CrossRef]
55. Breiman, L. Random forests. *Mach. Learn.* **2001**, *45*, 5–32. [CrossRef]
56. Halstead, J.B. Recruiter selection model and implementation within the United States Army. *IEEE Trans. Syst. Man Cybern. Part C* **2009**, *39*, 93–100. [CrossRef]
57. Moran, P.A.P. The interpretation of statistical maps. *J. R. Stat. Soc. Ser. B Stat. Methodol.* **1948**, *10*, 243–251. [CrossRef]
58. Su, S.; Zhou, H.; Xu, M.; Ru, H.; Wang, W.; Weng, M. Auditing street walkability and associated social inequalities for planning implications. *J. Transp. Geogr.* **2019**, *74*, 62–76. [CrossRef]
59. Su, S.; Pi, J.; Xie, H.; Cai, Z.; Weng, M. Community deprivation, walkability, and public health: Highlighting the social inequalities in land use planning for health promotion. *Land Use Policy* **2017**, *67*, 315–326. [CrossRef]
60. Songchitruksa, P.; Zeng, X. Getis–Ord spatial statistics to identify hot spots by using incident management data. *Transp. Res. Rec.* **2010**, *2165*, 42–51. [CrossRef]

International Journal of *Geo-Information*

MDPI

Article

Mapping Landslide Hazard Risk Using Random Forest Algorithm in Guixi, Jiangxi, China

Yang Zhang [1], Weicheng Wu [1,*], Yaozu Qin [1], Ziyu Lin [1], Guiliang Zhang [2], Renxiang Chen [2], Yong Song [2], Tao Lang [2], Xiaoting Zhou [1], Wenchao Huangfu [1], Penghui Ou [1], Lifeng Xie [1], Xiaolan Huang [1], Shanling Peng [1] and Chongjian Shao [1]

1 Key Laboratory of Digital Lands and Resources and Faculty of Earth Sciences, East China University of Technology, Nanchang 330013, China; 201810818002@ecut.edu.cn (Y.Z.); qyz60010@ecut.edu.cn (Y.Q.); zylin@ecut.edu.cn (Z.L.); 201900818004@ecut.edu.cn (X.Z.); 201810818004@ecut.edu.cn (W.H.); 201810818013@ecut.edu.cn (P.O.); 201810705007@ecut.edu.cn (L.X.); 201810705009@ecut.edu.cn (X.H.); pshanling@ecut.edu.cn (S.P.); scj350936@ecut.edu.cn (C.S.)

2 264 Geological Team of Jiangxi Nuclear Industry, Ganzhou 341000, China; zgl-63@163.com (G.Z.); crxkcy@163.com (R.C.); songy_6611@163.com (Y.S.); ecitlangtao@163.com (T.L.)

* Correspondence: wuwch@ecut.edu.cn or wuwc030903@sina.com

Received: 10 October 2020; Accepted: 13 November 2020; Published: 23 November 2020

Abstract: Landslide hazards affect the security of human life and property. Mapping the spatial distribution of landslide hazard risk is critical for decision-makers to implement disaster prevention measures. This study aimed to predict and zone landslide hazard risk, using Guixi County in eastern Jiangxi, China, as an example. An integrated dataset composed of 21 geo-information layers, including lithology, rainfall, altitude, slope, distances to faults, roads and rivers, and thickness of the weathering crust, was used to achieve the aim. Non-digital layers were digitized and assigned weights based on their landslide propensity. Landslide locations and non-risk zones (flat areas) were both vectorized as polygons and randomly divided into two groups to create a training set (70%) and a validation set (30%). Using this training set, the Random Forests (RF) algorithm, which is known for its accurate prediction, was applied to the integrated dataset for risk modeling. The results were assessed against the validation set. Overall accuracy of 91.23% and Kappa Coefficient of 0.82 were obtained. The calculated probability for each pixel was consequently graded into different zones for risk mapping. Hence, we conclude that landslide risk zoning using the RF algorithm can serve as a pertinent reference for local government in their disaster prevention and early warning measures.

Keywords: Random Forest; landslide hazard risk; integrated multisource dataset; field sample rasterization; weight assignment

1. Introduction

Landslides are a major natural hazard and can be defined as phenomena in which a rock and soil body on a slope slides down a certain interface under the action of gravity, rainfall, and groundwater. Landslides are one of the most frequent geological disasters in China. In 2019, 6181 geohazards were recorded in China, including 4220 landslides, accounting for 68.27% of the total hazards. These geohazards resulted in 211 deaths, 13 missing persons, and 75 injuries, and a direct economic loss of 2.77 billion yuan (China Geological Survey, 2020) [1]. According to the nationwide distribution of landslides in 2019, Jiangxi province ranks number two. The Ministry of Natural Resources of the People's Republic of China announced that 1747 geological hazards occurred in the first half of 2020, with a direct economic loss of 1.01 billion yuan. It was predicted that the situation will remain severe in the second half of the year (http://www.mnr.gov.cn/).

Due to complex natural conditions, spatiotemporal differences, and uncertainties of the landslide mechanism, it is difficult to accurately predict the occurrence time, scale, and impact range of landslides. However, based on the existing geohazard research and available technologies, it is possible to conduct effective landslide risk prediction and mapping. This will assist local authorities to take preventative and early warning measures to reduce damage and loss of life and property.

Recently, machine learning approaches have been applied in risk prediction and mapping. The advantage of the machine learning methods lies in its capacity to deal with a large amount of geospatial data within multi-dimensional and even hyper-dimensional space, and in its ability to achieve accurate prediction and classification (Wu et al. 2016, and 2018) [2,3]. These learning algorithms may provide the probability of the spatial occurrence of a landslide and identify the importance of different geo-environmental causal factors that play a potential role in these landslide events [4]. Several machine learning approaches have been utilized for landslide assessment in the past decade, such as Support Vector Machines (SVMs) [5,6], Artificial Neural Networks (ANNs) [7,8], Deep Learning Neural Networks (DLNNs) [9,10], Convolutional Neural Networks (CNNs) [11], Boosted Regression Trees (BRTs) [12,13], and Random Forests (RFs) [13]. A number of case studies show that these algorithms have a good prediction performance [14,15]; in particular, the RF has gained a high reputation for its outperformance in both classification and prediction compared to other approaches. Therefore, we adopted this algorithm for landslide risk mapping in our study.

Preprocessing of different geo-environmental causal factors is essential to geohazard risk prediction and assessment. At present, no standard exists to address this issue. Some methods use each causal factor as a categorical variable, e.g., a range of values from the same variable. However, prediction is not based on the accurate value of each pixel, which may affect the prediction efficiency of the model. When dealing with linear factors such as faults, roads, and rivers, these factors are not processed in a hierarchical manner, but in buffer zones with a different order of propensity weight in terms of proximity. Furthermore, the range of the buffer zones can be too large to be efficient for accurate prediction in space. Rainfall is a fundamental factor triggering landslides [16,17]; nevertheless, few studies included seasonal rainfall in landslide risk assessment.

For the reasons outlined above, the objectives of this research were to identify a relevant digitization approach to quantify the causal factors for landslide risk prediction and zoning using the RF algorithm for Guixi, Jiangxi Province, and to produce a risk map of landslides, and thus provide support and advice for local governments and decision-makers to implement landslide hazard prevention and early warning measures.

2. Materials and Methods

2.1. Study Area

Guixi is located in the northeast of Jiangxi Province, China, in the middle reaches of the Xinjiang River, and is bordered by the Wuyishan Mountains (Mts) on the south. The study area lies between 27°50′53″ N and 28°37′33″ N in latitude and between 116°57′43″ E and 117°28′06″ E in longitude, covering an area of about 2292 km^2. Topographically, Guixi is generally characterized by high mountains in the south and low hills in the north, cut by the Xinjiang River running west in the central region. The elevation of the study area varies from 20 to 1504 m above sea level (Figure 1). About 65.3% of the study area has a slope gradient <15°, whereas areas with gradients of 15–25°, 25–35°, 35–45° and >45° account for 19.8%, 10.6%, 3.7%, and 0.7%, respectively.

As a part of the subtropical monsoon climatic zone, Guixi receives an annual rainfall of 1789.3 mm with 163 annual rainfall days on average during the period 1958–2017. The rainy season occurs in March to July, with a mean accumulated rainfall of 1227.1 mm, or 68.6% of the annual rainfall (Figure 2). The annual mean temperature is 18.2 °C. Guixi is one of the major forest resource counties in Jiangxi, with a forest cover rate of 56%. The Guixi National Forest Park, situated in the south of the county, is covered with 2929.93 ha of forests, occupying 98.2% of its total area.

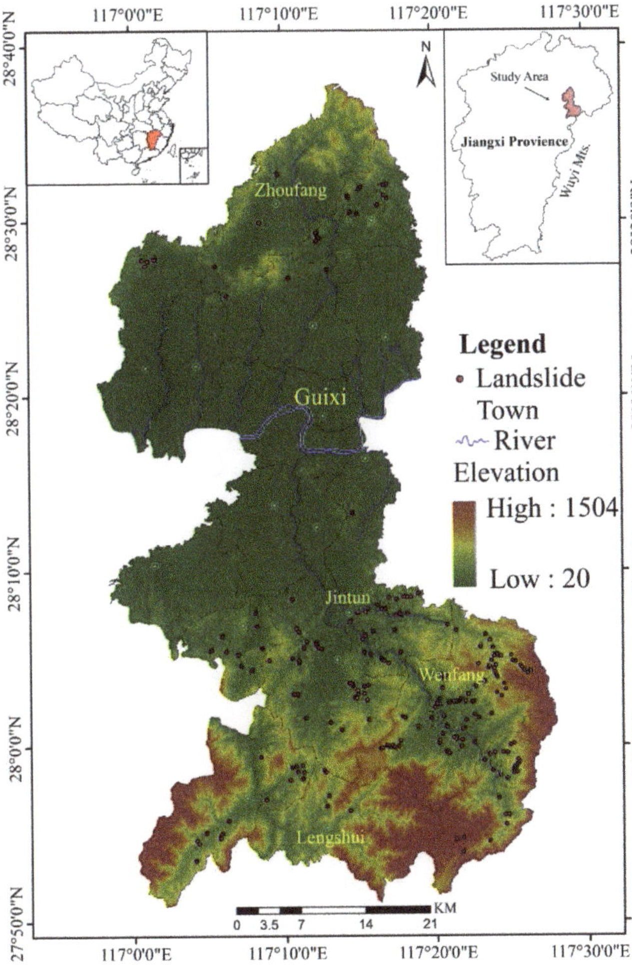

Figure 1. Location of the study area and distribution of the landslides.

Geologically, the formations in the study area include the strata from the Mesoproterozoic to the Cenozoic, with a stratigraphic sequence of Qingbaikou-Cambrian, Carboniferous-Permian, and Triassic-Quaternary. Magmatic rocks are well exposed and mainly distributed in the south, comprising the northern region of the Wuyishan Mts with multiple lithologies, such as basic, acidic, and neutral rocks. In terms of formation time, magmatic activities occurred partly in the Caledonian (around 508–408 ma) and predominantly in the Yanshanian periods (208–65 ma) [18].

According to the historical records, eight earthquakes in total had occurred in Guixi since AD 445, all with a magnitude below 5, and hence, this factor was not considered in our study.

Field investigation revealed that a total of 428 houses, 568 m of highway, 50 m of water channel and 5.1 ha of farmland have been destroyed by landslides in the last ten years in the study area. The damages to social properties were estimated about 3.54 million yuan. However, few efforts have been taken to predict the occurrence of these landslides for disaster reduction purpose.

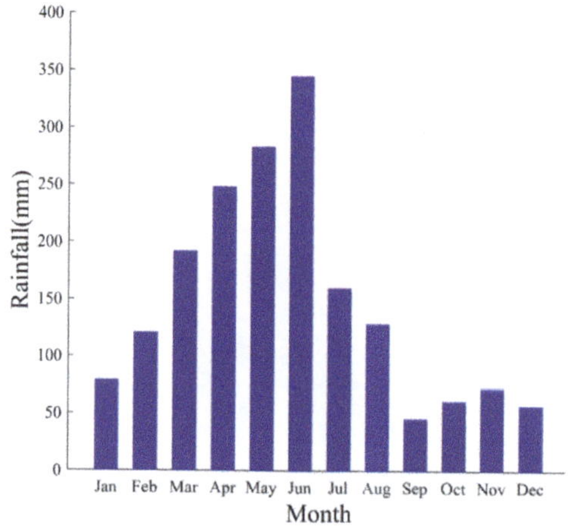

Figure 2. Diagram showing the averaged monthly rainfall in the study area.

2.2. Data and Preparation

2.2.1. Field Data, Training and Validation Sets

The first landslide inventory was conducted in the period September 2014–December 2015 and compiled into the Geological Hazard Survey Report (1:50,000) of Guixi by the 264 Geological Team of Jiangxi Nuclear Industry. The second survey was undertaken by ourselves in July and October 2019, and August 2020.

In this study, a total of 273 landslides that had taken places in the past ten years were identified. They are all small in volume, in which the smallest one is about 5 m^3, and the largest one around 20,000 m^3 with an average of about 533 m^3. As more than 88% of the landslides are less than 900 m^3, the landslide sites (points) have been repalced with polygons of 30 m $\times$ 30 m in size to facilitate the successive analysis. This field landslide dataset was divided randomly into two groups: training set and validation set, which took up respectively 70% and 30% of the total samples. Against the landslide events, 380 no-risk stable points (defined in the same size of polygons) in lowlands, croplands and urban where slope is $< 3°$ were selected. These no-risk polygons were also separated stochastically into two groups, 70% and 30%, and then incorporated respectively into the training set (191 landslides and 266 non-landslides) and validation set (82 landslides and 114 non-landslides).

As the successive landslide risk mapping was based on a binary classification using RF algorithm, these two classes of samples in both training and validation sets were assigned with a probability value of 1.0 for the occurred landslides, and 0.0 for no-risk samples. And then, these two sets were converted into raster of 30 m size according to the approach proposed by Wu et al. (2018) [3].

2.2.2. Landslide Causal Factors and Integrated Hyper-Dimensional Dataset

Identification, selection and preprocessing of landslide causal factors is a key procedure for risk modeling and zoning. Previous studies have utilized various factors and attempted to reveal their potential roles in landslide events [13]. Based on this and our field knowledge, 21 landslide-related factors such as geological formations, elevation, slope, aspect, plan curvature, profile curvature, thickness of the weathering crust, soil type and texture (clay, sand and silt contents), land use, the normalized difference vegetation index (NDVI), average annual rainfall, March–July rainfall, May–July rainfall, distance to the geological boundaries, distance to faults, distance to roads, and distance to rivers were identified (Table 1). These factors were processed in GIS and all converted to raster with a cell size of 30 m after weight assignment for the non-digital ones.

Table 1. Geo-information layers used as landslide causal factors.

No	Causal Factors	Resolution	Sources
1	Elevation	30 m	GDEMV 3 NASA (https://earthdata.nasa.gov/)
2	Slope		DEM-derived
3	Aspect		
4	Plan curvature		
5	Profile curvature		
6	Depth of the weathering crust (soil thickness)		Kriging interpolation
7	Soil type	1 km	Data Center for Resources and Environmental Sciences, Chinese Academy of Sciences (RESDC) (http://www.resdc.cn/)
8	Soil texture (sand content)		
9	Land use	30 m	Landsat 5 TM
10	NDVI		Landsat 5 TM
11	Average annual rainfall		264 Geological Team of Jiangxi Nuclear Industry
12	March-June rainfall		
13	March-July rainfall		
14	May-July rainfall		
15	June-July rainfall		
16	June-August rainfall		
17	Lithology	1:50,000	264 Geological Team of Jiangxi Nuclear Industry
18	Distance to geological boundaries		
19	Distance to faults		
20	Distance to roads	1:5000	Google Earth
21	Distance to rivers		

Quantification and Weight Assignment

1. Topographic features are critical for landslide hazard risk assessments [19]. A digital elevation model (DEM), ASTER GDEM V003 product of 30 m in resolution, was obtained from the NASA (https://earthdata.nasa.gov/) for the study area. This DEM was further used to derive elevation (Figure 1), slope (Figure 3a), aspect, plan curvature and profile curvature.

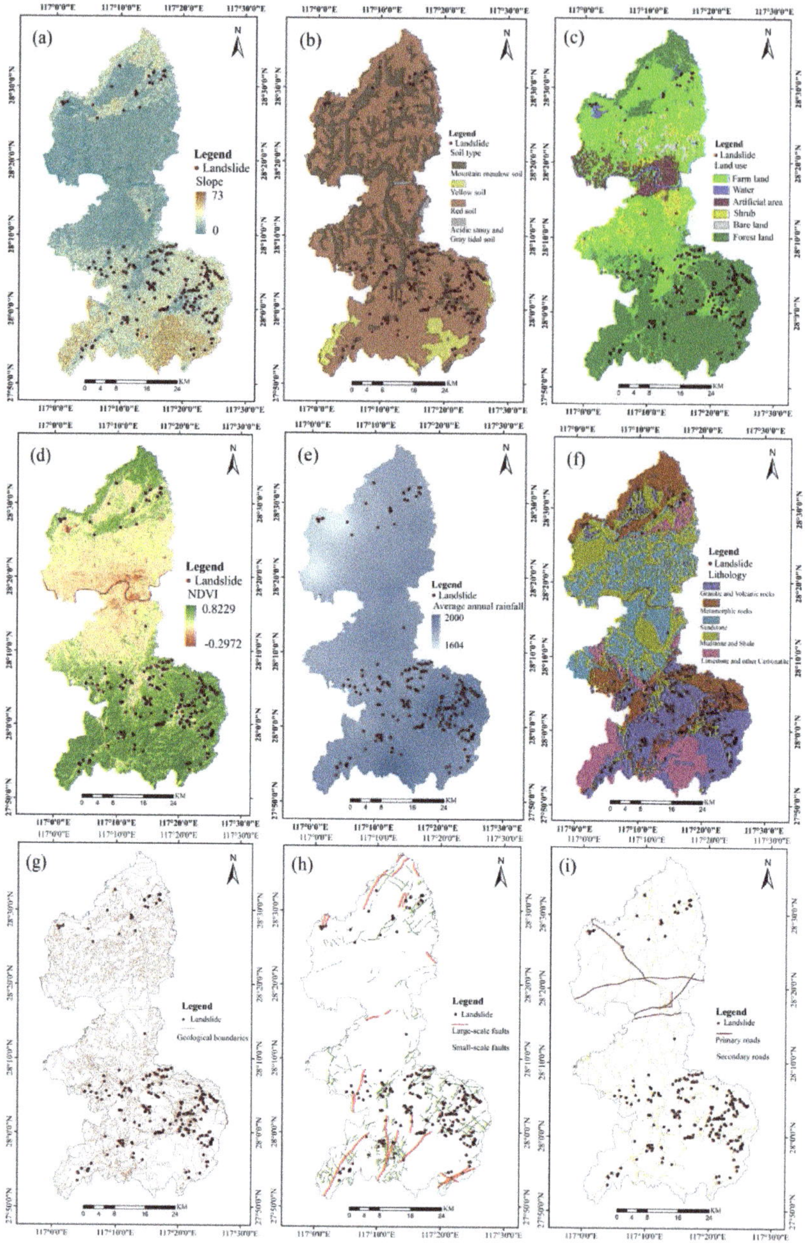

Figure 3. Landslide causal factors used in the study taking the following factors as an example: (**a**) slope, (**b**) soil type, (**c**) land use, (**d**) NDVI, (**e**) average annual rainfall, (**f**) lithology, (**g**) geological boundaries, (**h**) faults, (**i**) roads.

2. The weathering crust provides materials and sites for landslides and is the hoster of the latter, and can be considered as an important controlling factor of landslide event [20–22]. The interaction between this crust and rainfall causes the occurrence of landslides. The survey on the thickness of

the weathering crust in Guixi lacks detailed data except for some landslide profiles. We extracted the ridge and valley lines based on the DEM data and assumed that the lowland plain and valleys had a thick crust of about 10 m, and it decreased as the slope and altitude increased, and at the ridge, it was about 0.5 m. A Kriging interpolation approach was employed to produce the thickness map of the weathering crust.

3. Edaphic factor is also necessary for risk modeling and prediction as it influences the occurrence of landslides [23,24]. Soil type and texture data of the study area were obtained from the Data Center for Resources and Environmental Sciences, CAS (RESDC: http://www.resdc.cn/) (Figure 3b). No matter which type of soil, the important feature is the soil texture, i.e., percentage of sands, which influences greatly the soil property. For example, the higher percentage of sands, the higher porosity for rainwater permeation, leading to a higher risk of landslide where clay on the interface may play a role as lubricant. Thus, soil with sand percentage of > 40%, 20–40%, 10–20% and 0–10% were respectively assigned a propensity value of 10, 7, 4, and 1. Then soil map was resampled to pixels with size of 30 m to match the other data.
4. Land use (Figure 3c) is an indicator of human activity that illustrates the relationship between man and environment. The exploitation of land resources has been regarded as an unignorable factor that may affect negatively our environment and the occurrence of landslides [25,26]. The land use map of the study area was produced by using Landsat 5 TM images acquired on May 31 and November 07, 2010, obtained from the Geospatial Data Cloud (http://www.gscloud.cn), using the approaches proposed by Wu et al. (2016) [2,27]. With a mapping accuracy of 91.44%, six land use classes were identified, namely artificial area (urban, rural village and infrastructure), farm land, forests, shrubs, bare land, and water bodies, and were assigned respectively a proneness weight of 0, 0, 1, 4, 10 and 0. Here low slope urban and farmland have lowest proneness, and forest cover has also low propensity while bare land without vegetation protection is the most vulnerable category given the same natural conditions.
5. Vegetation condition and abundance, which can be represented by vegetation index, e.g., the normalized difference vegetation index (NDVI), have been reported of a high correlation with the occurrence of landslides [12]. As a complement to land cover, NDVI (Figure 3d) was selected and included in the analysis of this study. We obtained the late autumn (October 24–November 07, when herbaceous vegetation became withered and most crops were harvested) Landsat 5 TM images of the period 2005–2010, from the same data server as mentioned above. The TM images were atmospherically corrected using the COST model (Chavez 1996; Wu 2003; Wu et al. 2013) [28–30] in which both additive scattering and multiplicative path transmission effects were minimized. NDVI was calculated using the formula (NIR-R)/(NIR+R) [31] from each scene and then averaged to get the multiyear mean NDVI.
6. Rainfall is often considered as a triggering factor of landslide events [32,33]. In this study, the average annual rainfall (Figure 3e), March–June, March–July, May–July, June–July and June–August rainfall were taken into account as hazard-causative factors. Our purpose was to investigate which months' rainfall or combined accumulation is the most important for assessing the landslide risk. The daily rainfall data of the period 2008–2017 from 104 ground stations were acquired and used to create different accumulative monthly rainfall combinations, which were further gridded into raster layers using Inverse Distance Weighted (IDW) interpolation approach.
7. Geological strata, especially, their lithologies (Figure 3f) and bedding, can play different roles in the occurrence of landslides because of their different resistance to weathering and bedding structure, in particular, together with joints and fractures, which may serve as rainfall permeation pathways and slippery interface. The lithological data of the study area were digitized from the Geological Map on a scale of 1/50,000 [18]. The hazard-causative propensity weight of each formation lithology was assigned in terms of its resistance to landslide, e.g., higher resistance formation was assigned with lower weight value or vice versa. More concretely, granitic and volcanic rocks were assigned a weight value of 1, metamorphic rocks 5, sandstone 7, limestone

and other carbonatite 8, and mudstone and shale 10. The higher value assigned, the higher proneness of the factor tends to contribute to landslides [34].

8. Geological boundaries (Figure 3g) are the connection belts of inhomogeneous geological formations and usually fragile zones that are susceptible to weathering and fracturing. It is thus considered a potential factor influencing the slope stability. Actually, the closer to the boundary the higher risk may exist [9]. The geological boundaries were extracted from the above-mentioned geological map and buffered into different zones as 0–30 m, 30–60 m, 60–90 m and 90–120 m, which were assigned a weight value of 5, 3, 2 and 1, respectively.
9. Linear features: Faults (Figure 3h) often play an active role in landslide events as they are fractures and subject to water permeation and extensive weathering. This tends to increase the vulnerability of geological bodies and slope instability. Road construction (Figure 3i) is a direct human action on the slope resulting in an instability of the latter. The change in landform and the loss of support from the underlying massif lead to the increase of tension on the upper slope that promotes the development of cracks [4]. Rivers are usually an active factor in modification of landscape by cutting the different geological formations and making their adjacent massif fragile through liquidization. A number of studies revealed that not only rivers but also reservoirs influence the stability of slope [35,36].

In this study, faults, roads and rivers were buffered in line with their scales. For example, small faults, roads and rivers were buffered with distance intervals of 0–30, 30–60, 60–90 and 90–120 m; large-scale faults, main river and reservoirs were buffered with distances of 0–60, 60–120, 120–180 and 180–240 m from the borders. Each buffer zone was assigned a weight in terms of its potential proneness to landslide, e.g., zones 0–30, 30–60, 60–90 and 90–120 m were assigned respectively 10, 7, 4, and 1, and zones 0–60, 60–120, 120–180 and 180–240 m respectively 20, 15, 10, and 5. This assignment was based on the rule that the closer to the linear features, the higher propensity of landslide. These buffers of different linear factors were converted into raster layers of 30 m in cell size.

Integrated Hyper-dimensional Geo-information Set

The above rasterized 21 hazard-causative factors including elevation, slope, aspect, plan curvature, profile curvature, thickness of the weathering crust, soil texture (especially, sand %), land use, NDVI, average annual rainfall, March–July rainfall, May–July rainfall, lithology, distance to faults, distance to roads, and distance to rivers, etc., were stacked together to compose a 21-layer geo-information dataset. Specifically, this is an integrated dataset with 21 dimensions, a realistic hyper-dimensional data space.

2.3. Risk Prediction and Modeling

2.3.1. RF Algorithm

As one of the machine learning approaches, the RF algorithm achieves learning and prediction using an ensemble of growing decision-trees, or rather, of classification and regression trees (CARTs) and their majority voting (Breiman 2001) [37]. One critical technique of this algorithm lies in its bootstrap sampling from the training set to build trees followed with a randomized selection of the input variables to determine the best split for each node. In the meantime, the out-of-bag (OOB) estimates are applied within the RF algorithm to determine the generalization error and the importance of each predictive variable (Breiman, 2001) [37]. Moreover, there shall not be the overfit problem with RF if the number of decision trees (NT) is large enough. In other words, the RF algorithm makes use of the strong law of large numbers, i.e., the more features employed, the less error generated (Breiman 2001; Wu et al. 2018) [3,37]. Thus, NT should be large enough so as to minimize the OOB error of classification or regression to a stable level during the training procedure. Another advantage of the RF algorithm is its capacity to deal with hyper-dimensional data using limited training samples but achieving results of high accuracy. Instead of classification of land cover types, we employed this RF algorithm here to classify the probability of risk and no-risk for each pixel in the whole study area.

2.3.2. Risk Prediction and Modeling

RF modeling was conducted within EnMap-Box, an image processing package developed by DLR (German Aerospace Center) [38]. Using the RF Classification (RFC) function, the integrated 21-layer geo-information dataset was input as predictive variables with the training set for training.

Before risk modeling, a set of parameters have to be set up, for example, NT, the number of randomly selected features at each node, and the stop criteria (for node splitting). How to set up these parameters can be referred to Wu et al. (2018) [3].

Risk modeling was actually a parameterization procedure versus the training set with an internal validation. The generated RF risk model was applied back to the integrated dataset to perform a binary classification of risk probability to derive the probability map.

2.3.3. Verification and Reliability Analysis

To assess the performance of landslide risk modeling, the predicted results were verified against the independent validation set [2,3,39] rather than the training set. Two metrics were used, i.e., overall accuracy (OA) and Kappa Coefficient (KC), which were calculated based on the confusion matrix of the trained landslide risk models versus the validation set. Here KC is a direct indicator of reliability of the risk modeling and prediction [2,3,39,40]. For a value of 0, it indicates a poor consistency between the prediction and the observation, whereas, a value of 1 implies a perfect agreement between the two. The KC-based agreement levels proposed by Landis and Koch [40] were followed: poor (0–0.2), fair (0.21–0.40), moderate (0.41–0.60), substantial (0.61–0.80) and almost perfect (0.81–1.0).

2.3.4. Assessment of the Importance of Landslide Causal Factors

It is necessary to understand that the role of each hazard-causative factor may differ from one place to another, depending on the assessment model and landslide mechanism in different geo-environments. This implies that a geo-environmental factor may take an active part in landslide prediction in one model in one place but play a tiny role in another elsewhere. Therefore, the contribution of a causative factor is conditional and various. The importance of each factor for the landslide events in this study was evaluated using the OOB ranking procedure of the RF classifier.

3. Results

3.1. *NT within the RF Algorithm*

The NT affected the predication results when the RF risk modeling was conducted (Table 2). In spite of its capacity to deliver rather accurate prediction when NT was set to 100, the prediction was more robust with higher OA and KC when it was set to 300 and 500. As a confirmation to other authors [2], OA and KC declined slightly when NT was 1000 (Table 2). Hence, 300-500, especially, 300 would be advised to use for NT when tackling landslide risk prediction and zoning.

Table 2. Performance of the RF algorithm with different Number of Trees (NT).

Number of Trees	Overall Accuracy	Kappa Coefficient
100	90.75	81.08
300	91.23	82.02
500	91.07	81.70
1000	90.75	81.04

3.2. *Landslide Hazard Risk Map*

As shown above, the RF algorithm performed best when NT was set to 300, and thus the modeling results of this case were selected for landslide hazard risk mapping. The computed risk probability, ranging from 0 to 1.0 in each pixel, was classified into five levels, i.e., No risk (0–0.2), Low risk (0.2–0.4),

Median risk (0.4–0.6), High risk (0.6–0.8), and Extremely high risk (0.8–1.0). Thence, the landslide risk zonation map was produced and presented in Figure 4.

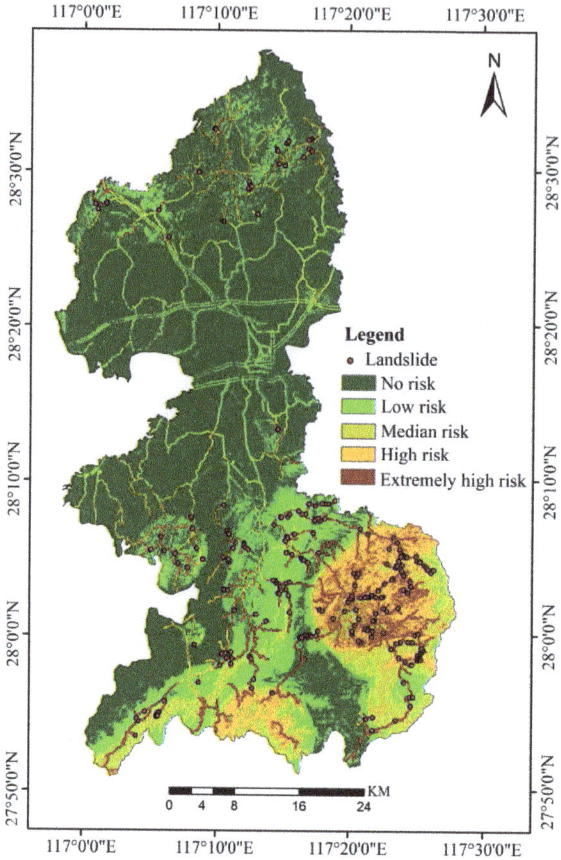

Figure 4. Landslide risk map.

This map shows that the landslide-prone areas are mainly distributed along the roads and in the high slope hilly and mountainous areas in the north and south of Guixi where the predicted High risk and Extremely high risk zones are distributed. There are also Median and High risk zones in the southwest where there is abundant rainfall. Nevertheless, the risk is relatively low in the central part, a plain with gentle topographic relief in Guixi.

It is seen in Figure 4 that two types of landslide risks were predicted, i.e., one is man-made landslides distributed along the roads or cut slopes as a consequence of road construction and housing development, and the other is natural ones distributed in the mountainous slopes in the south and southeast of the study area (Figure 4).

For the modeling result obtained when NT was 300, the OA of this risk map is 91.23%, and KC 0.82 versus the validation set, reaching the "almost perfect (0.81–1.0)" level. As statistics revealed, the number of the observed landslides falling in the zones No risk, Low, Medium, High and Extremely high risk, accounted for, respectively, 0.73%, 1.10%, 2.93%, 2.57% and 92.67% of the total.

3.3. Importance of the Hazard-Causative Factors

As seen in Figure 5, all 21 hazard-causative factors have contributed to the landslide events but the first five, i.e., distance to road, slope, May–July rainfall, average annual rainfall and elevation, comprise 65.45% of the total contribution to the occurrence of landslide disasters. That is to say, they have played a more important role in landslide events than other factors. The contribution of soil type and faults is relatively low.

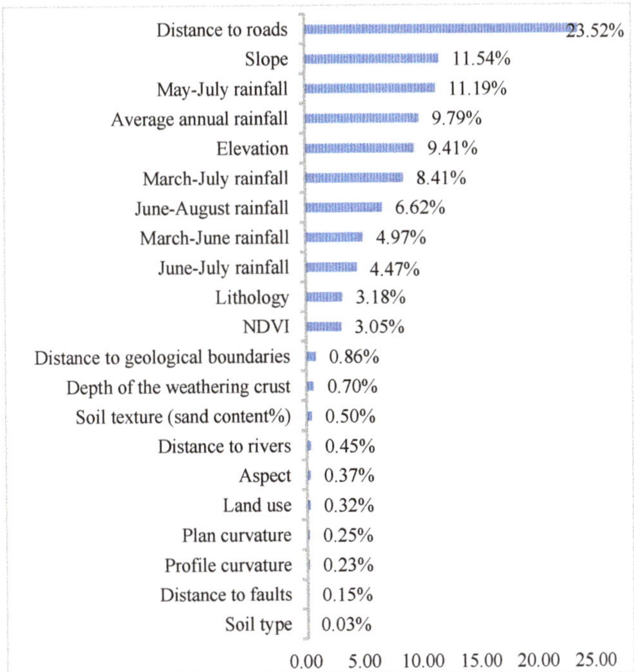

Figure 5. Importance of hazard-causative factors.

Field survey revealed that these landslides were neither triggered by earthquake nor by active tectonic movements but by human activity, e.g., road construction and housing development, and rainfall. Actually, anthropogenic landslides accounted for 98.9%, where the slope ranges from 8° to 25°, much lower than the threshold, 28–35°, of the natural landslides as proposed by Fan et al. (2016) [23]. In addition, 209 landslides, 76.3% of the total, occurred in the talus accumulation with a thickness of about 0.5–10 m. The occurrence time of landslides is mainly in March–July, in particular, in June–July. The number of landslides of these two months accounts for more than 50% of the total.

4. Discussion

4.1. Algorithm for Landslide Risk Assessment

As previously mentioned, a number of data-driven approaches have been applied for landslide risk prediction. Xiong et al. (2020) [13] noted that among the machine learning algorithms, BRTs performed best in debris flow susceptibility assessment in Sichuan Province whereas Chen et al. [15] concluded that RF achieved the best prediction in Chongren, Jiangxi. Actually, one of our parallel research (Zhou et al. under review [34]) conducted in Ruijin, Jiangxi and that of Sun et al. (2020) [19] in Fengjie, Chongqing, both pointed out that RF is capable for providing accurate landslide risk prediction. This study, using the RF algorithm to fulfill the task with a high satisfactory level, "almost perfect", confirmed their conclusion. However, care has to be taken while employing different geo-environmental data for RF-based modeling as the landslides used as training samples were mostly small in scale, i.e., less than one Landsat pixel in surface area. It is hence necessary to use high resolution data to highlight such disaster risk while modeling and mapping are conducted, and data with resolution of coarser than 30 m will not be recommended.

4.2. The Different Importance of the Causal Factors

As revealed in Figure 5, distance to roads, slope, rainfall and elevation are the most important factors in landslide events in Guixi. The order of importance of the geo-environmental factors may be different from one site to another, e.g., slope, rock type, distance to river and NDVI [5], slope and distance to roads [6], lithological formation, distance from roads, and NDVI [12], and elevation and annual rainfall [19]. But all these studies point at a fact, that is, slope, distance to road, elevation and rainfall are the commonly important factors causing landslide hazards, and that is what we have uncovered in this study.

Since the road construction constitutes the most active human factor leading to such geohazard, it shall be necessary to design the road system by avoiding the most risky area and taking the geological strata bedding and slope stability into account.

It is worth mentioning that the importance ranking of all factors related to rainfall accounts for 45.45%, which shows a clear relationship of rainfall with the landslide events, especially, the accumulated rainfall of May–July (Figure 5). However, this importance weight seems still underestimated. Theoretically, rainfall is the triggering factor of the most landslides and should have more importance. The exploration on this topic seems not possible until we have grasped the exact occurrence time of these landslides. Only with such information, can we decide how to combine more reasonably the daily or monthly rainfall for risk modeling.

Some factors, such as faults, edaphic features and geological boundaries, used to be considered as necessary. Nevertheless, the factor importance analysis revealed that they were not as significant as expected in this study. Hence, it is possible to optimize the selection of the causal factors in landslide risk modeling and mapping in the similar geo-environment, in particular, when computing capacity is low.

4.3. Landslide Types

In terms of our field survey, the majority of landslides observed is small in volume, provoked by concentrated rainfall superimposed on the road construction and slope cutting for housing development. Rainfall is able to infiltrate into subsurface along the fractures to reach and liquidize the sliding interface, causing landslides.

There exist relatively big and deep landslides with a volume of about 20,000 m^3 but driven by different occurrence mechanisms: (1) landslide occurs after the action of the accumulated rainfall, especially, when Quaternary sediment (talus) has a clear interface with the underlying rocks in which the unconformity serves as slide surface after infiltration of rainfall; (2) multicycle landslides at the same place, they begin with small volumes of slide after rainfall but little by little extending out and deepening after the repeated rainfall events, and finally these slides become a big one; (3) big landslide within downhill strata bedding, which does not take place in the talus or weathered crust but inside the geological strata after the bedding surface has been lubricated by the penetrated rainfall when there are faults and joints. The rapidity of the big landslide relies on the dip of the strata bedding. For high dip bedding, rocky landslide may happen quickly as long as rock mass gravity exceeds the resistant friction of the underlying formation. For low dip bedding, the overlying strata and weathered crust do not constitute a rapid slide but a creep moving downward gradually. When there is a slope cutting, the downward movement becomes faster. This was also clearly observed in Ruijin, another city in southern Jiangxi, constituting a threat to the newly established Longzhu Temple and the No 6

Middle School of Ruijin [37]. It is hence essential to take measure to prevent the huge loss and damage before such big hazard happens.

5. Conclusions

This study made use of the RF algorithm for landslide risk modeling, prediction and mapping in Guixi with an integrated 21 geo-environmental factors and field data. During the modeling, we employed machine learning technique to determine which hazard-prone factors are the most important in provoking landslides. We also demonstrated the procedure on how to digitize non-digital geo-environmental factors and assign a weight value in terms of their proximity or propensity so that quantitative analysis and modeling were made possible. This study provides not only key information on how to research landslide mechanism but also operational approach on how to investigate hazard risk to prevent our society from further damage. In particular, our risk zone map with high reliability may serve as a reference for the local governments and decision-makers of Guixi to implement landslide prevention and early warning measures in the landslide-prone areas.

One key finding of this research is the surprising importance of road construction and housing development in landslide events. This reveals the role of human activity in provoking such geo-disaster, and suggests that when designing road systems, more comprehensive slope protection measures and more profound geological investigation on downhill strata bedding should be necessary so that man-made landslides can be minimized.

Author Contributions: Conceptualization, Yang Zhang, Weicheng Wu, Yaozu Qin and Ziyu Lin; methodology, Weicheng Wu and Yang Zhang; software, Weicheng Wu; validation, Yang Zhang, Yaozu Qin, Wenchao Huangfu, and Penghui Ou; formal analysis, Yang Zhang, Xiaoting Zhou, Wenchao Huangfu, and Penghui Ou; investigation, Yang Zhang, Weicheng Wu, Shanling Peng, Chongjian Shao, Penghui Ou, Wenchao Huangfu, Xiaoting Zhou, Lifeng Xie and Xiaolan Huang; resources, Weicheng Wu, Guiliang Zhang, Guiliang Zhang, Yong Song, Zhiling Wang and Tao Lang; data curation, Yang Zhang, Xiaolan Huang, Xiaoting Zhou Renxiang Chen and Penghui Ou; writing—original draft preparation, Yang Zhang and Weicheng Wu; writing—review and editing, Weicheng Wu; visualization, Wenchao Huangfu, Lifeng Xie and Xiaolan Huang; project administration, Weicheng Wu and Guiliang Zhang; funding acquisition, Weicheng Wu. All authors have read and agreed to the published version of the manuscript.

Funding: This research was financially supported by the Jiangxi Talent Program (Grant No: 900/2120800004) and the Start-up Fund for Scientific Research of the East China University of Technology (Grant No. DHTP2018001) which were both granted to Weicheng Wu.

Acknowledgments: A part of the first-hand landslide observation data and the Geological Map of Guixi on the Scale of 1/50,000 were provided by the 264 Geological Team of Jiangxi Nuclear Industry.

Conflicts of Interest: The authors declare no conflict of interest.

References

1. China Geological Survey. Notification on National Geological Hazard in 2019. Available online: http://www.cgs.gov.cn/gzdt/zsdw/202003/t20200331_504559.html (accessed on 15 August 2020).
2. Wu, W.; Zucca, C.; Karam, F.; Liu, G. Enhancing the performance of regional land cover mapping. *Int. J. Appl. Earth Obs. Geoinf.* **2016**, *52*, 422–432. [CrossRef]
3. Wu, W.; Zucca, C.; Muhaimeed, A.S.; Al-Shafie, W.M.; Al-Quraishi, A.M.F.; Vinay, N.; Zhu, M.; Liu, G. Soil salinity prediction and mapping by machine learning regression in Central Mesopotamia, Iraq. *Land Degrad. Dev.* **2018**, *29*, 4005–4014. [CrossRef]
4. Zhao, Y.; Wang, R.; Jiang, Y.; Liu, H.; Wei, Z. GIS-based logistic regression for rainfall-induced landslide susceptibility mapping under different grid sizes in Yueqing, Southeastern China. *Eng. Geol.* **2019**, *259*, 105147. [CrossRef]
5. Huang, F.; Cao, Z.; Guo, J.; Jiang, S.-H.; Li, S.; Guo, Z. Comparisons of heuristic, general statistical and machine learning models for landslide susceptibility prediction and mapping. *Catena* **2020**, *191*, 104580. [CrossRef]

6. Hong, H.; Pradhan, B.; Xu, C.; Bui, D. Spatial prediction of landslide hazard at the Yihuang area (China) using two-class kernel logistic regression, alternating decision tree and support vector machines. *Catena* **2015**, *133*, 266–281. [CrossRef]
7. Tan, Q.; Huang, Y.; Hu, J.; Zhou, P.; Hu, J. Application of artificial neural network model based on GIS in geological hazard zoning. *Neural Comput. Appl.* **2020**, *1*, 1–12. [CrossRef]
8. Wang, Y.; Feng, L.; Li, S.; Ren, F.; Du, Q. A hybrid model considering spatial heterogeneity for landslide susceptibility mapping in Zhejiang Province, China. *Catena* **2020**, *188*, 104425. [CrossRef]
9. Bui, D.T.; Tsangaratos, P.; Nguyen, V.T.; Liem, N.V.; Trinh, P.T. Comparing the prediction performance of a Deep Learning Neural Network model with conventional machine learning models in landslide susceptibility assessment. *Catena* **2020**, *188*, 104426. [CrossRef]
10. Hua, Y.; Wang, X.; Li, Y.; Xu, P.; Xia, W. Dynamic development of landslide susceptibility based on slope unit and deep neural networks. *Landslides* **2020**, *1*, 1–22. [CrossRef]
11. Fang, Z.; Wang, Y.; Ling, P.; Hong, H. Integration of convolutional neural network and conventional machine learning classifiers for landslide susceptibility mapping. *Comput. Geoences* **2020**, *139*, 104470. [CrossRef]
12. Reza, P.H.; Aiding, K.; Norman, K.; Farzin, S. Investigating the effects of different landslide positioning techniques, landslide partitioning approaches, and presence-absence balances on landslide susceptibility mapping. *Catena* **2020**, *187*, 104364. [CrossRef]
13. Xiong, K.; Adhikari, B.R.; Stamatopoulos, C.A.; Zhan, Y.; Wu, S.; Dong, Z.; Di, B. Comparison of Different Machine Learning Methods for Debris Flow Susceptibility Mapping: A Case Study in the Sichuan Province, China. *Remote Sens.* **2020**, *12*, 295. [CrossRef]
14. Wu, X.; Ren, F.; Niu, R. Landslide susceptibility assessment using object mapping units, decision tree, and support vector machine models in the Three Gorges of China. *Environ. Earth Ences* **2014**, *71*, 4725–4738. [CrossRef]
15. Chen, W.; Xie, X.; Peng, J.; Wang, J.; Duan, Z.; Hong, H. GIS-based landslide susceptibility modelling: A comparative assessment of kernel logistic regression, Naïve-Bayes tree, and alternating decision tree models. *Geomat. Nat. Hazards Risk* **2017**, *8*, 950–973. [CrossRef]
16. Segoni, S.; Piciullo, L.; Gariano, S.L. A review of the recent literature on rainfall thresholds for landslide occurrence. *Landslides* **2018**, *15*, 1483–1501. [CrossRef]
17. Huang, R.; Li, W. Formation, distribution and risk control of landslides in China. *J. Rock Mech. Geotech. Eng.* **2011**, *3*, 97–116. [CrossRef]
18. 264 Brigade of the Jiangxi Nuclear Industry Geological Bureau. The Guixi Geological Hazard Survey Project Implemented by Our Team Successfully Passed the Field Acceptance by the Expert Group. Available online: http://www.hgy264.com/show-27-6127-1.html (accessed on 27 October 2020).
19. Sun, D.; Wen, H.; Wang, D.; Xu, J. A random forest model of landslide susceptibility mapping based on hyperparameter optimization using Bayes algorithm. *Geomorphology* **2020**, *362*, 107201. [CrossRef]
20. Hung, L.Q.; Van, N.T.H.; Duc, D.M.; Ha, L.T.C.; Son, P.V.; Khanh, N.H.; Binh, L.T. Landslide susceptibility mapping by combining the analytical hierarchy process and weighted linear combination methods: A case study in the upper Lo River catchment (Vietnam). *Landslides* **2016**, *13*, 1285–1301. [CrossRef]
21. Xi, C. On the red weathering crusts of southern China. Quaternary Sciences. *Quat. Sci. (Chin. Engl. Abstr.)* **1991**, *1*, 1–8.
22. Zhu, X. Red clay and red weathered crust in southern China. *Res. Soil Water Conserv. (Chin. Engl. Abstr.)* **1995**, *4*, 94–101.
23. Fan, L.; Lehmann, P.; Or, D. Effects of soil spatial variability at the hillslope and catchment scales on characteristics of rainfall-induced landslides. *Water Resour. Res.* **2016**, *52*, 1781–1799. [CrossRef]
24. Kitutu, M.G.; Muwanga, A.; Poesen, J.; Deckers, J.A. Influence of soil properties on landslide occurrences in Bududa district, Eastern Uganda. *Afr. J. Agric. Res.* **2009**, *4*, 611–620. [CrossRef]
25. Hong, H.; Liu, J.; Zhu, A.-X. Modeling landslide susceptibility using LogitBoost alternating decision trees and forest by penalizing attributes with the bagging ensemble. *Sci. Total Environ.* **2020**, *718*, 137231. [CrossRef] [PubMed]
26. Napoli, M.D.; Carotenuto, F.; Cevasco, A.; Confuorto, P.; Martire, D.D.; Firpo, M.; Pepe, G.; Raso, E.; Calcaterra, D. Machine learning ensemble modelling as a tool to improve landslide susceptibility mapping reliability. *Landslides* **2020**, *17*, 1897–1914. [CrossRef]

27. Wu, W.; Mhaimeed, A.S.; Al-Shafie, W.M.; Al-Quraishi, A.M.F. Using Radar and Optical Data for Soil Salinity Modeling and Mapping in Central Iraq. In *Environmental Remote Sensing in Iraq*; Fadhil, A.M., Negm, A., Eds.; Springer: Cham, Switzerland, 2019; Chapter 2; pp. 19–40. [CrossRef]
28. Chavez, P.S. Image-Based Atmospheric Corrections-Revisited and Improved. *Photogramm. Eng. Remote Sens.* **1996**, *62*, 1025–1036.
29. Wu, W. Application de la Geomatique au Suivi de la Dynamique Environnementale en Zones Arides. Ph.D. Thesis, Université Paris 1, Paris, France, 2003.
30. Wu, W.; De Pauw, E.; Zucca, C. Use remote sensing to assess impacts of land management policies in the Ordos rangelands in China. *Int. J. Digit. Earth* **2013**, *6* (Suppl. 2), 81–102. [CrossRef]
31. Wu, W. The generalized difference vegetation index (GDVI) for dryland characterization. *Remote Sens.* **2014**, *6*, 1211–1233. [CrossRef]
32. Bordoni, M.; Galanti, Y.; Bartelletti, C.; Persichillo, M.G.; Barsanti, M.; Giannecchini, R.; Avanzi, G.D.A.; Cevasco, A.; Brandolini, P.; Galve, J.P. The influence of the inventory on the determination of the rainfall-induced shallow landslides susceptibility using generalized additive models. *Catena* **2020**, *193*, 104630. [CrossRef]
33. Chikalamo, E.E.; Mavrouli, O.C.; Ettema, J.; Westen, C.J.V.; Mustofa, A. Satellite-derived rainfall thresholds for landslide early warning in Bogowonto Catchment, Central Java, Indonesia. *Int. J. Appl. Earth Obs. Geoinf.* **2020**, *89*, 102093. [CrossRef]
34. Zhou, X.; Wu, W.; Lin, Z.; Zhang, G.; Chen, R.; Song, Y.; Wang, Z.; Lang, T.; Qin, Y.; Ou, P.; et al. Landslide risk zoning in Ruijin, Jiangxi, China. *Nat. Hazards Earth Syst. Sci. Discuss* **2020**, in press. [CrossRef]
35. Luo, X.; Li, J. Analysis on the Influence of Reservoir Impoundment on the Bank Landslide. *Des. Hydroelectr. Power Stn. (Chin. Engl. Abstr.)* **2003**, *3*, 61–64.
36. Wang, M.; Yan, E. Study on influence of reservoir water impounding on reservoir landslide. *Rock Soil Mech. (Chin. Engl. Abstr.)* **2007**, *12*, 2722–2725. [CrossRef]
37. Breiman, L. Random Forests. *Mach. Learn.* **2001**, *45*, 5–32. [CrossRef]
38. Waske, B.; van der Linden, S.; Oldenburg, C.; Jakimow, B.; Rabe, A.; Hostert, P. imageRF—A user-oriented implementation for remote sensing image analysis with Random Forests. *Environ. Model. Softw.* **2012**, *35*, 192–193. [CrossRef]
39. Jakimow, B.; Oldenburg, C.; Rabe, A. *Manual for Application: ImageRF (1.1)*; Universität Bonn and Humboldt Universität zu Berlin: Berlin, Germany, 2012.
40. Landis, J.R.; Koch, G.G. The Measurement of Observer Agreement for Categorical Data. *Biometrics* **1977**, *33*, 159–174. [CrossRef] [PubMed]

International Journal of
Geo-Information

MDPI

Article

Damage Signature Generation of Revetment Surface along Urban Rivers Using UAV-Based Mapping

Ting Chen [1], Haiqing He [2,3], Dajun Li [2,*], Puyang An [2] and Zhenyang Hui [2]

1 School of Water Resources & Environmental Engineering, East China University of Technology, Nanchang 330013, China; ct_201607@ecut.edu.cn
2 School of Geomatics, East China University of Technology, Nanchang 330013, China; hehaiqing@ecut.edu.cn (H.H.); zhouq@ecut.edu.cn (P.A.); huizhenyang2008@ecut.edu.cn (Z.H.)
3 Key Laboratory of Watershed Ecology and Geographical Environment Monitoring, National Administration of Surveying, Mapping and Geoinformation, Nanchang 330013, China
* Correspondence: djli@ecut.edu.cn; Tel.: +86-139-7041-8700

Received: 2 March 2020; Accepted: 22 April 2020; Published: 24 April 2020

Abstract: The all-embracing inspection of geometry structures of revetments along urban rivers using the conventional field visual inspection is technically complex and time-consuming. In this study, an approach using dense point clouds derived from low-cost unmanned aerial vehicle (UAV) photogrammetry is proposed to automatically and efficiently recognize the signatures of revetment damage. To quickly and accurately recover the finely detailed surface of a revetment, an object space-based dense matching approach, that is, region growing coupled with semi-global matching, is exploited to generate pixel-by-pixel dense point clouds for characterizing the signatures of revetment damage. Then, damage recognition is conducted using a proposed operator, that is, a self-adaptive and multiscale gradient operator, which is designed to extract the damaged regions with different sizes in the slope intensity image of the revetment. A revetment with slope protection along urban rivers is selected to evaluate the performance of damage recognition. Results indicate that the proposed approach can be considered an effective alternative to field visual inspection for revetment damage recognition along urban rivers because our method not only recovers the finely detailed surface of the revetment but also remarkably improves the accuracy of revetment damage recognition.

Keywords: revetment; damage signature; dense point clouds; unmanned aerial vehicle (UAV); gradient operator

1. Introduction

Revetment systems in urban rivers are constructed to protect riverbanks, infrastructures, and people, in an effort to control floods. Revetments are usually designed as slope protection and covered in concrete [1]. Floods can trigger revetment erosion to weaken revetments continuously and cause damage. In addition, revetments are damaged by complex factors, such as land subsidence, ground collapse, erosion, vegetation presence, riverbed degradation, and human interference [2,3]. Therefore, monitoring the condition of revetments is an essential task in the management of flood defense infrastructure and important in providing evidence for maintenance or improvements [4].

At present, some studies have been done to conduct an assessment of the condition of revetments by the use of remotely sensed data in countries such as England [4] and France [5], but these methods are not applicable to urban revetment monitoring, and the assessment of the condition of revetments is visually inspected by the Municipal Engineering Management Agency in China. However, field visual inspection is time-consuming and technically complex in obtaining complete information on revetments. Additionally, the assessment of the subsurface condition of revetments is difficult because visual inspection is preferred in investigating important signs, such as surface collapse. Notably, early damage recognition is highly beneficial in enabling maintenance and improvements in advance before further deterioration occurs [4].

Apart from field visual inspection, remote sensing technologies, such as unmanned aerial vehicle (UAV)-based photogrammetry, have been useful techniques for the creation of digital surface models (DSMs) and also widely used in obtaining revetment information due to their advantages in high-precision three-dimensional (3D) geometry reconstruction [6–8]. Terrestrial laser scanning is usually used to monitor revetment damage caused by revetment erosion in small areas rather in large-scale areas [9–11]. This method requires frequent measurements that usually involve expensive sensors and field logistics when monitoring large areas. For instance, airborne laser scanning was used to estimate the volume change in river valley walls caused by revetment erosion [12], and point clouds were used to analyze the protection of a revetment rock beach [13]. Pye et al. [14] assessed beach and dune erosion and accretion for coastal management. Ternate et al. [15] modeled water-related structures to assist the design of revetments. Although sensors enable the generation of dense 3D points for good reconstruction of the geometry structure for revetment monitoring, point clouds cannot directly provide the color texture of the revetment and are less intuitive in the damage interpretation of revetments. As a result, the noise in point clouds is difficult to remove. A portion of the revetment surface may typically be covered with vegetation (e.g., grass), which appears as 3D points in the fluctuating height values within dense point clouds. Other platforms have been used for revetment monitoring [16], but these platforms are unsuitable in certain areas with shallow water, such as urban rivers. Meanwhile, 3D point clouds derived from these sensors may be much more expensive than their image-derived counterpart, such as low-cost consumer-grade UAV-based photogrammetry [17–19]. In addition, image-derived 3D point clouds from UAV photogrammetry can capture the spatially detailed structure of the ground surface and offer more competitive accuracy compared with laser scanning-based products [20]. Compared with laser scanning, ground control points (GCPs) are needed in aerial photogrammetry and GCP measurement is a time-consuming task. Fortunately, several GCPs can be marked and measured once in advance, that is, multiple measurements are not required to collect GCPs for absolute orientation. Therefore, although laser scanning can produce high-resolution and dense 3D point clouds, this technology requires more complex operations and has a higher cost when collecting revetment information on urban rivers than the low-cost UAV-based mapping. In this study, a low-cost UAV platform equipped with a consumer-grade onboard camera (e.g., DJI Phantom quadcopters) is used to prove that it is suitable for recognizing the damage signatures with respect to finely detailed revetment surfaces.

In recent years, research on UAV has focused on understanding and modeling revetments, and high-resolution 3D data derived from low-cost UAV mapping has been widely used in the efficient and accurate monitoring of revetments for the implementation of relevant maintenance management strategies [19,21–25]. Hallermann et al. [21] and Kubota et al. [22] used the dense point clouds derived from low-cost UAV photogrammetry to visualize the deformation of revetments in the assessment of structural stability. Pitman et al. [19] obtained high-resolution and competitive accuracy of DSM of revetments derived from UAV-based mapping and compared the results with those derived from real-time kinematic global positioning systems (RTK GPS) and offered new possibilities (i.e., using UAVs) for measuring, monitoring, and understanding the deformation of revetments against the approaches of traditional geomorphology observation. This method achieves high-accuracy DSM, which are approximately equal to those obtained via airborne laser scanning. Although their research can reconstruct a good 3D geometry structure of a revetment using UAV-based mapping for monitoring, they ignored the automatic damage recognition from the image-derived point clouds. Moreover, photogrammetric surveying using UAV has been often used to monitor the changes in revetments for river management. Pires et al. [26] combined mapping and photogrammetric surveying in the revetment model to investigate coastal dynamics and shoreline evolution and contributed to coastal management. Jayson et al. [25] used UAV photogrammetry to reconstruct the delta revetment topography to analyze changes in beach sediments. Although many applications are effective in revetment monitoring using low-cost UAV photogrammetry, studies on the use of UAV-based mapping for revetment damage recognition along urban rivers have been rarely reported. Most importantly, the effectiveness and efficiency of UAV-based damage recognition are two indicators that determine whether this approach can be applied. Furthermore, the quality and efficiency of point cloud generation are critical to accurately characterize the surface of a revetment, and the reliability of damage signature generation from the derived point clouds is also equally important to damage recognition in revetments.

Revetments along urban rivers are usually designed as a relatively flat slope or curved surface, that is, the revetment surface is generally a simple irregular surface that can be modeled using a mathematical function. On this basis, this study proposes a dense point cloud-based approach derived from low-cost photogrammetry to extract the signatures of revetment damage from a slope intensity image instead of the prerequisite multitemporal data. For revetment damage recognition along urban rivers, information on damaged and nondamaged revetment surfaces is generally needed for comparison and analysis. In many cases, prior information on nondamaged surfaces is not typically obtained or finely reconstructed in municipal engineering management. Failure to accumulate historical data related to the surface of a revetment may result in poor revetment management due to unclear understanding of damage signatures. As an alternative to applications dependent on multitemporal data [20,25], we exploit an approach for revetment damage recognition that does not require nondamaged surface reconstruction or prior information. On the basis of the assumption that the surface of the revetment has roughly the same slope, dense point clouds are first transformed into a slope intensity image, in which feature extraction is then performed to generate the features of revetment damage. A self-adaptive and multiscale gradient operator (SMGO) is proposed for collecting damage information by using the omnidirectional (horizontal, vertical, and diagonal) operation, especially in feature extraction. SMGO is used to ensure that damage of different scales can be accurately extracted.

This study aims to exploit the workflow of revetment damage recognition along urban rivers through the dense point clouds derived from low-cost UAV photogrammetry, and the proposed point cloud and damage signature generation are both introduced to address damage recognition using UAV-based mapping. The main contribution of this study is the proposed approach based on photogrammetric point clouds, which offers new possibilities in revetment damage recognition. In our approach, pixel-by-pixel dense matching is simultaneously used with the combination of region growing and semi-global matching (SGM), which can reconstruct a finely detailed surface of a revetment by considering the contributions of adjacent 3D object points. In particular, feature image generation based on the proposed SMGO is suitable for recognizing the damage signatures on the surface of a revetment designed with slope protection under the assumption that the majority of the 3D points on the revetment surface remain unchanged and prior information is unnecessary.

2. Study Area and Materials

2.1. Test Site

Nanchang City (28°42′29″N, 115°48′58″E) in Jiangxi Province, China (Figure 1a,b), is the study area of this work. This study used a low-cost quadcopter UAV (i.e., DJI Mavic Air; DJI; Shenzhen, China) to investigate the revetments along urban rivers, and two parts (with lengths of 450 and 570 m, respectively) of the concrete revetment located in the west of Nanchang City were selected to test (Figure 1c) for the following reasons: different types of riverbank defense structures have been constructed along the different portions of the bank to manage the impact of lateral fluvial erosion, among such structures, the revetment is typically designed with a slope angle to protect the riverbanks and infrastructures in urban rivers [25]. The waterway is often covered with silt and gravel materials, and then a large number of sediments may mobilize and cause erosion to the revetment along with intense rainfall events. In addition to the presence of mass movements, complex external factors, such as groundwater penetration, also remarkably contribute to revetment erosion and damage. Revetments are characterized morphologically by using a slope approximately equal to 40°. The waterway basement geologically comprised unconsolidated sediments of clay, loose sand, and gravel deposits. Revetments are usually covered with weeds and continuously affected by lateral fluvial erosion, ground collapse, and riverbed degradation.

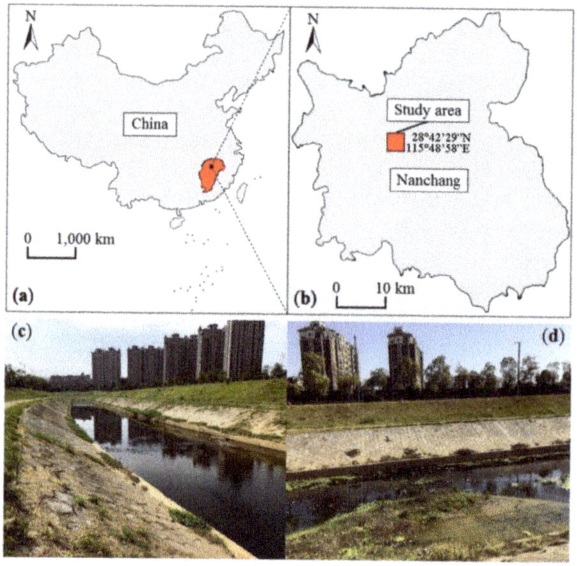

Figure 1. (**a**) Nanchang City located in Jiangxi Province in southeast China, (**b**) location of the study area in Nanchang City, and (**c**,**d**) landscape of the study area.

2.2. Acquisition of UAV Remote-Sensed Images and Measurement of Ground Control Points (GCPs)

Given its low-cost and flexible operation, a consumer-grade quadcopter DJI Mavic Air (DJI; Shenzhen, China) [27,28] (Figure 2a) was selected to capture high-resolution true-color remote sensing images in August 2019. In addition, the DJI Mavic Air is easily carried with its 430 g weight and folding design, it does not require a professional take-off and landing site, and the aircraft is simple to operate, allowing flexible flight plans for a variety of missions. The operator can safely monitor the revetment even under ultralow-altitude photogrammetry in urban areas through signature DJI technologies, such as obstacle avoidance and intelligent flight modes [27]. The DJI platform is inexpensive, efficient, and requires minimal expertise. Several user-friendly applications provided by DJI, including Ground Station Pro and a mission planning software package, were used to conduct autonomous flights with waypoints and nadir orientation of the consumer-grade camera during the acquisition of UAV-based images of stereo remote sensing [28]. Figure 2b (Part 1) shows an example of the extent of the survey and some survey parameters in the graphical user interface of this UAV survey application. Parameters, such as flight altitude, flight speed, and image overlap, can be obtained on the basis of the survey mission. To acquire high-resolution and non-blurry remote sensing images, a low-altitude flight with an above-ground level of 30 m and 2.8 m/s flight speed was conducted to reduce the atmospheric and environmental limitations. Thus, the ground sample distance was approximately equal to 2.0 cm/pix. To ensure the reliability of image matching with large overlaps, the front and side image overlaps were set to 80% and 60%, respectively. Once the flight parameters were set, the UAV was largely automated with the operator acquiring remote sensing images under a wind speed <10 m/s and non-rainy conditions. The total surveying flight time of the UAV in the two parts (450 and 570 m) of the concrete revetment along the urban rivers was around 10 and 14 min (less than the maximum flight time of 21 min) and was achievable in one battery charge. The UAV took approximately 232 and 287 images for the two parts to cover the study area, which also includes a buffer extent of approximately 15 m near the revetment. Moreover, to improve the performance of image matching, the system errors and interior orientation of the consumer-grade camera were eliminated by using the methods in a previous study [28].

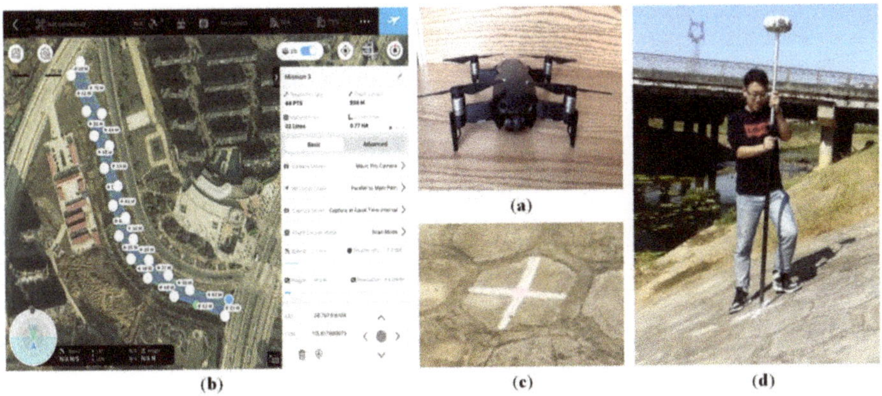

Figure 2. Unmanned aerial vehicle (UAV) data acquisition and ground control point measurement. (**a**) DJI Mavic Air, (**b**) DJI graphical user interface for mission planning, (**c**) ground control point (GCP) marked by a white cross with a pink center, and (**d**) field measurement of GCPs.

Additionally, 40 GCPs were evenly distributed on the revetment. The GCPs were placed across the study area and measured to validate the accuracy of the image-based DSM using RTK GPS. The GCPs were marked on the site, as shown in Figure 2c. Pixel-by-pixel dense point clouds were georeferenced with 5 and 7 GCPs for Parts 1 and 2, respectively. The other 28 GCPs (13 and 15 GCPs for Parts 1 and 2, respectively) were selected as check points (CPs), which were used to evaluate the accuracy of the surface reconstruction of the revetment.

3. Method

This study aims to exploit the workflow of revetment damage recognition along urban rivers through dense point clouds derived from low-cost UAV photogrammetry, and the proposed point cloud and damage signature generation are both introduced to address damage recognition using UAV-based mapping. The proposed approach demonstrated in Figure 3 mainly includes the following stages:

(1) Photogrammetric technologies are used to generate high-precision pixel-by-pixel dense point clouds for surface reconstruction of the revetment through a series of steps, that is, feature extraction and matching, incremental structure-from-motion (SfM), bundle adjustment, and region growing coupled with SGM.

(2) The slope intensity map of revetment is calculated and generated in terms of the height of the dense point clouds. The areas of revetment on both sides along the urban river are then extracted by segmenting and merging the superpixels, which are generated on the slope intensity map by using a simple linear iterative clustering (SLIC)-based algorithm.

(3) The signature of revetment damage is generated from the slope intensity image through vegetation removal, omnidirectional gradient operation and nonmaximum suppression, and denoising.

(4) Accuracy assessment is performed to validate the accuracy of the dense point clouds derived from the algorithm (i.e., region growing coupled with SGM) and evaluate the performance of revetment damage recognition along the urban rivers with quantitative analysis (e.g., indicators such as *Precision*, *Recall*, and *F1_score*) and visual assessment (i.e., ground field observation).

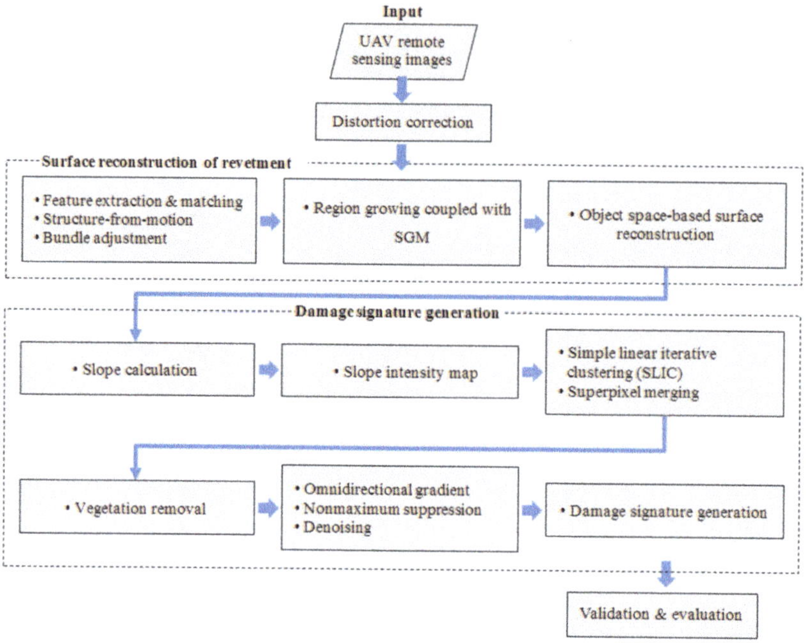

Figure 3. Workflow of the revetment damage recognition along urban rivers through dense point clouds derived from low-cost UAV photogrammetry.

3.1. Surface Reconstruction of Revetment

The camera mounted on the low-cost UAVs (e.g., consumer-grade DJI Phantom quadcopters) has large perspective distortions and poor camera geometry [28,29], which may cause systematic errors that need to be eliminated by using distortion correction for each UAV remote sensing image. Similar to previous studies [18], the digital camera should be calibrated strictly before the operation of aerial photography. Distortion correction is then performed by using the camera parameters and two radial and two tangential distortion coefficients, which are calculated from several views of a two-dimensional (2D) calibration pattern. These parameters will be further optimized by the following self-calibrating bundle adjustment.

Similar to previous studies, feature extraction and matching are performed using a sub-Harris operator coupled with the scale-invariant feature transform algorithm, which can find evenly distributed corresponding points even in the overlapping areas of remote sensing images with illumination change and weak texture [30]. In traditional aerial photogrammetry, the poses of the airborne camera, that is, positions and orientations, must be known to provide the parameters of initial exterior orientation for performing aerial triangulation. However, low-altitude platforms, such as low-cost consumer-grade UAVs, are usually not mounted on high-precision equipment when obtaining the information of positions and orientations of cameras. Hence, traditional aerial triangulation relying on the parameters of initial exterior orientation may be unavailable for UAV-based aerial triangulation. UAV-based SfM algorithms have been applied to bank retreats at streams, and this study can generate DSM with smaller errors compared with the use of terrestrial laser scanning [31]. Therefore, SfM is used to estimate the poses of the airborne camera and reconstruct a sparse 3D geometry for the overlapping images without the help of initial exterior orientation parameters [18]. Notably, incremental SfM [32,33] is employed in this study to reconstruct the sparse 3D model increasingly and iteratively because it allows 3D reconstruction in an incremental process for repeated self-calibrating bundle adjustments (i.e., sparse bundle adjustment software [34]) to optimize the 3D model and interior and exterior orientation parameters.

Unlike DSM generation via interpolation of point-based elevation data [35], a novel method of region growing coupled with SGM for dense matching illustrated in Figure 4 is exploited to generate the pixel-by-pixel dense point clouds and reconstruct the finely detailed surface of the revetment. The SGM algorithm is a popular technique to minimize the image matching cost along several one-dimensional path directions through the images for image-based 3D reconstruction, and it may significantly increase the computational expense for most mapping applications which mainly deal with sets of overlapping images. Damage recognition of revetments along urban rivers requires high implementation efficiency, which is also highly valuable in enabling maintenance and improvements in advance before further deterioration occurs. SGM-based matching is one of the most time-consuming steps in photogrammetric point cloud generation, and thus it is of great significance to improve the efficiency of this step.

In this paper, to reduce the computational expense of redundant point clouds, an object space-based dense matching approach is exploited to satisfy the need of rapid 3D reconstruction for revetments. Unlike in photogrammetry software, such as Agisoft Metashape and Pix4Dmapper, the height of each grid in the revetment is utilized to calculate the pixel-by-pixel dense matching while considering the impact of adjacent 3D object points and obtaining the finely detailed surface of revetments. Different from many computer vision applications, our study on the surface of the revetment is focused on 3D construction (i.e., height value) of the top surface of the ground. This approach is not to generate normalized image stereo pairs, but to perform dense matching in the voxel object space. On this basis, there is no need to derive the matching cost of image space. Instead, all images can be used to represent the matching cost directly on each voxel, which is more suitable for UAV-based mapping applications. That is, semi-global optimization can be performed in voxel space, and image-based dense point clouds can be obtained directly. Unlike most SGM-based image matching [36,37], the sparse point clouds obtained from bundle adjustment can be used to simplify the image-based point cloud generation procedure. To be specific, by using prior knowledge of reconstructed objects, the search scope of the corresponding points can be narrowed in the voxel space, which contributes to improving the accuracy and efficiency of reconstruction. In this regard, the height values on the vertical sides are not essential. The object space-based approach can only compute the height values on the top surface and helps to reduce the computational expense of redundant point clouds. That is, the object space-based dense matching approach used in this paper is suitable for accurate, rapid, and cost-effective revetment damage recognition. The proposed region growing coupled with SGM mainly includes: (a) a triangulated irregular network (TIN) in 3D space is generated to initialize the 3D object surface; (b) inverse distance weighted interpolation is used to obtain initial height values; and (c) a region growing strategy is explored to gradually generate the pixel-by-pixel dense point clouds for surface reconstruction of revetment considering accuracy and efficiency.

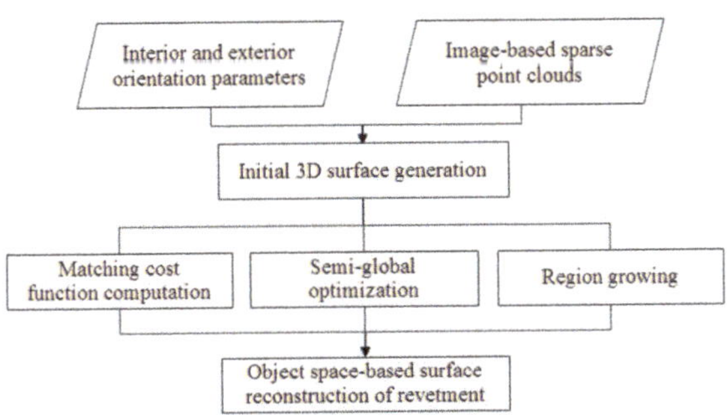

Figure 4. Object space-based surface reconstruction of revetment.

The innovation of our method is that we assume that the set of 3D sparse points obj_{set}^{sparse} is derived from SfM, and then we denote it by using the $P_{obj}^{i}(X,Y,Z)$ cell in the set obj_{set}^{sparse} in the ith 3D point at the object position (X,Y,Z) with $i \in \{1,\ldots,N\}$. As shown in Figure 5a, an object point obtained from the sparse 3D model is relevant to n 2D UAV remote sensing images, where n could be ≥2. We then denote $p_{img}^{i\to j}(x,y)$ position in the jth UAV remote sensing images at the image position (x,y). In this study, the corresponding points in the UAV remote sensing images reprojected from the 3D sparse points are considered salient correspondence and seeds, which are extended through region growing in the four neighborhoods illustrated in Figure 5b. Then, the pixel-by-pixel dense point clouds are iteratively determined by using the cost function and SGM algorithm with a known epipolar geometry, shown in Figure 5c.

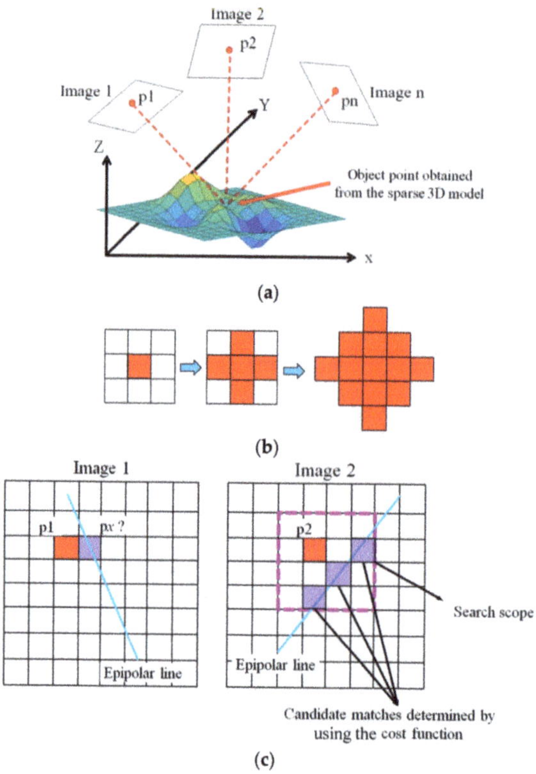

Figure 5. Region growing coupled with semi-global matching (SGM). (**a**) Object point backprojected onto multiple views. (**b**) Region growing in the four neighborhoods. (**c**) Process of candidate matches determined by using the cost function under the constraint of epipolar lines in two views.

Specifically, a set obj_{set}^{dense} of dense point clouds is initially assigned by using the set obj_{set}^{sparse}. Assuming that all 3D points $P_{obj}^{i}(X,Y,Z)\big| i \in \{1,\ldots,N\}$ in the obj_{set}^{dense} are seeds and backprojected onto the relevant images, the $p_{img}^{i\to j}(x,y)$ position in the first relevant image is considered the seed and extended with region growing in the four neighborhoods $\mathbb{R}^4$. For example, the query point p_x is fixed in Image1, and the correspondences in other relevant images are determined by using the SGM algorithm with a known epipolar geometry. On the basis of the SGM algorithm [36,38], the 3D points in the direction of region growing are determined and saved in set obj_{set}^{dense}. We repeat these dense matching steps until no 3D point can be added into the set obj_{set}^{dense}. Although the SGM algorithm can appropriately generate dense point clouds, some local areas with weak texture are likely reconstructed poorly.

In this study, we introduce a novel approach of 3D scene patching to generate the 3D points in these local areas. A triangulated irregular network (TIN) is established by using the set obj_{set}^{dense} of dense point clouds. Subsequently, the coordinates of 3D points within the TIN can be calculated by using the weighted interpolation of inverse distance, which is expressed as

$$Z = \frac{1}{m}\sum_{i=1}^{m} w_i Z_i, \left(w_i = \frac{1}{d_i}\right), \quad (1)$$

where Z is the height value of an unknown 3D point, m is the number of surrounding 3D points of the unknown 3D point, Z_i is the height value of the ith surrounding 3D point, w_i is the weight corresponding to Z_i, and d_i is the distance between the unknown 3D point and the ith known surrounding 3D point. The algorithm of the proposed dense matching method is expressed below.

Algorithm 1: Region growing coupled with SGM

Input: 3D sparse points obj_{set}^{sparse}, exterior orientation parameters EOP, and UAV remote sensing images.
Parameters: 3D dense points obj_{set}^{dense}, four neighborhoods $\mathbb{R}^4$, query point p_x, relevant images $p_{img}^{i\to j}(x,y)$, disparity d, minimum cost path $L_r(p_x, d)$, new 3D point $P_{obj}^{new}(X,Y,Z)$, and unknown 3D points $obj_{set}^{dense,unknown}$.
Initialize $obj_{set}^{dense} \leftarrow obj_{set}^{sparse}$.
repeat
  for each 3D point $P_{obj}^{i}(X,Y,Z) \big| i \in \{1,\ldots,N\}$ in set obj_{set}^{dense} **do**
    Assign $P_{obj}^{i}(X,Y,Z)$ as a seed.
    Reproject $P_{obj}^{i}(X,Y,Z)$ onto $p_{img}^{i\to j}(x,y)$.
    Compute epipolar geometry based on EOP.
    for k = 1 to $\mathbb{R}^4$ **do**
      Calculate $L_r(p_x, d)$ corresponding to p_x in $p_{img}^{i\to j}(x,y)$ with known epipolar geometry.
    end for
    Compute the coordinate of $P_{obj}^{new}(X,Y,Z)$ using SGM and aerial triangulation.
    Update obj_{set}^{dense} by adding the new 3D point.
  end for
until no 3D point need to be added.
Find $obj_{set}^{dense,unknown}$ in the local areas that have not been reconstructed well
Establish TIN.
for each 3D point in $obj_{set}^{dense,unknown}$ **do**
  Compute the coordinate of the 3D point by inverse distance weighted interpolation.
  Update obj_{set}^{dense} by adding the 3D point.
end for

3.2. Damage Signature Generation

Dense point clouds derived from UAV photogrammetry can generate a finely detailed geometry structure of the revetment and be regarded as an alternative to the visual inspection method. The categories of revetment damage are mainly collapse and crack. On flat ground, the place of collapse is usually characterized by an uneven region below the surface height of the ground with an irregular boundary, and a crack is typically shown as a linear object.

Unlike flat ground, revetments along urban rivers are built in a sloping pattern. Therefore, we attempt to transform the dense point clouds into a slope intensity image for damage recognition in this study on the basis of the assumption that the revetment is constructed with a fixed slope angle. Ideally, the values of the slope intensity image located in the revetment regions are approximately equal in this case. Then, a slope intensity image is generated via slope calculation, which is performed to identify the slope in each cell of the rasterized surface of the dense point clouds using the slope module of the ArcGIS software. A portion of the revetment surface may likely be covered with vegetation (e.g., grass), which appears as the 3D points of the fluctuating height values within the dense point clouds. UAV photogrammetry clearly has limitations in surveying the surface of the revetment in the presence of vegetation, thus possibly affecting the accuracy of revetment damage recognition. To eliminate the influence of vegetation in the slope intensity image, vegetation removal is preliminarily conducted with a gamma-transform green leaf index [39]. Subsequently, damage recognition is performed using a proposed operator called SMGO, which is designed to extract the abnormal regions with different sizes in the slope intensity image. Specifically, omnidirectional (horizontal, vertical, and diagonal) gradient operation is conducted using a self-adaptive operator with degraded weights. Hence, a variable gradient operator is used in each cell to determine whether it belongs to the damaged or nondamaged region. A multiscale architecture is introduced into this operator for the recognition of damaged regions with different sizes.

The main goal of this study is to identify the damaged areas of the revetment. Thus, automatic revetment recognition is an essential task in determining the dense point clouds. In this study, we extract the area of interest (AOI) or the area covered by the revetment from the slope intensity image. On the basis of the assumption that the AOIs of the revetment have approximately equal slope angles, previous studies [39] using SLIC and superpixel merging are jointly used to extract the revetment regions from the intensity map, as shown in Figure 6. First, the slope intensity image is segmented into a set of superpixels in terms of similar slope values. Second, the superpixels are merged into a series of regions on the basis of the approximately equal slope values. Third, the AOI of the revetment is determined by using the average slope value of the slope intensity image. The following are the main steps of revetment region extraction.

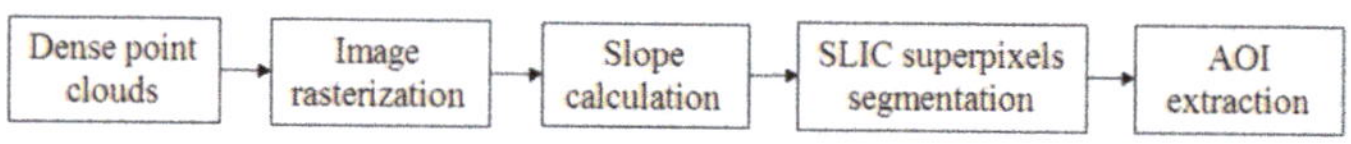

Figure 6. Area of interest (AOI) extraction.

Step 1: The dense point clouds derived from low-cost UAV photogrammetry are rasterized using the grid size $\Delta d \times \Delta d$, where Δd is the resolution of the UAV remote sensing images.

Step 2: The slope of the rasterized image is computed, and the intensity image $I_{\text{slope}}(x, y)$ of the slope is generated using ArcGIS software.

Step 3: The intensity image $I_{\text{slope}}(x, y)$ is segmented into superpixels with the SLIC algorithm, and the superpixels are merged into a series of regions on the basis of the approximately equal slope values.

Step 4: The AOI of the revetment is determined by using the slope value of the region within [*slope_value*−10°, *slope_value*+10°], where *slope_value* is the average slope of multiple samples in this study.

Then, the feature image $I_{\text{damage}}(x, y)$ of damage is generated from the intensity $I_{\text{slope}}(x, y)$ via the proposed operator SMGO. Mathematically, the gradients $grad(x, y)$ in each cell (x, y) are computed as

$$grad(x, y) = \frac{\partial I_{\text{slope}}}{\partial x}\boldsymbol{i} + \frac{\partial I_{\text{slope}}}{\partial y}\boldsymbol{j} + \frac{\partial I_{\text{slope}}}{\partial diag_{\text{L}}}\boldsymbol{k} + \frac{\partial I_{\text{slope}}}{\partial diag_{\text{R}}}\boldsymbol{l}, \tag{2}$$

where $\frac{\partial I_{\text{slope}}}{\partial x}$, $\frac{\partial I_{\text{slope}}}{\partial y}$, and $\frac{\partial I_{\text{slope}}}{\partial diag}$ are the gradients in the horizontal, vertical, and diagonal directions, respectively. The multiscale architecture in SMGO is illustrated in Figure 7, and two scales are shown. The adjacent areas surrounding the cell $p \in I_{\text{slope}}(x, y)$ are defined on the basis of the following equation:

$$r = \text{INT}(k \cdot \sigma + 0.5), \tag{3}$$

where r is the radius of the area surrounding the cell p, INT(.) is the integer operation, and σ is the initial scale factor of the SMGO, set to 1.6 in this study. $k \in \{1, 2, 3, \ldots s | s \geq 2\}$ is the set of multiple factors, which are key values in determining the scope of the area surrounding the cell p. Gradient calculation is performed on each scale on the basis of suboperators, which are illustrated in Figure 7c–f (Scale 1) and Figure 7h–k (Scale 2). The gradient of each suboperator is mathematically calculated using the following convolutional operation:

$$grad(x, y) = G(x, y, k\sigma) * I_{\text{slope}}(x, y), \tag{4}$$

where $G(.)$ denotes the matrices of weights in the gradient operator and is defined by the nonlinear inverse distance as

$$G(x, y, k\sigma) \Leftarrow \frac{8}{\sqrt{\pi k^2 \sigma^2}} e^{-\frac{(\Delta x^2 + \Delta y^2)}{2k^2\sigma^2}}, \tag{5}$$

where $(\Delta x, \Delta y)$ is the shift between the adjacent cell and the center (x, y). Then, the matrices of the suboperators can be determined. For example, Figure 7c–f are represented by the matrices $\begin{bmatrix} -1.91 & -2.32 & -1.91 \\ 0 & 0 & 0 \\ 1.91 & 2.32 & 1.91 \end{bmatrix}$, $\begin{bmatrix} -1.91 & 0 & 1.91 \\ -2.32 & 0 & 2.32 \\ -1.91 & 0 & 1.91 \end{bmatrix}$, $\begin{bmatrix} 0 & 0 & -1.29 & 0 & 0 \\ 0 & -1.91 & 0 & 0 & 0 \\ -1.29 & 0 & 0 & 0 & 1.29 \\ 0 & 0 & 0 & 1.91 & 0 \\ 0 & 0 & 1.29 & 0 & 0 \end{bmatrix}$, and $\begin{bmatrix} 0 & 0 & -1.29 & 0 & 0 \\ 0 & 0 & 0 & -1.91 & 0 \\ 1.29 & 0 & 0 & 0 & 1.29 \\ 0 & 1.91 & 0 & 0 & 0 \\ 0 & 0 & 1.29 & 0 & 0 \end{bmatrix}$, respectively. Similar to the output of the neural network, a max activation function is utilized to determine the gradient of cell p by using the maximum value of all the suboperators. The mathematical expression of the max activation function is

$$\max grad(x, y) = \max\left\{\left|\frac{\partial I_{\text{slope}}}{\partial x}\right|, \left|\frac{\partial I_{\text{slope}}}{\partial y}\right|, \left|\frac{\partial I_{\text{slope}}}{\partial diag_{\text{L}}}\right|, \left|\frac{\partial I_{\text{slope}}}{\partial diag_{\text{R}}}\right|\right\}. \tag{6}$$

Notably, the number of scales is not fixed but adaptive. If the gradient G is less than the given value t_{gradient}, then the value k is not increased. In this study, at least two scales of SMGO are needed to establish the multiscale architecture. The algorithm of the proposed SMGO is given as Algorithm 2.

Algorithm 2: Gradient calculation using SMGO

Input: intensity image $I_{\text{slope}}(x, y)$ with width W and height H, constant value σ, and gradient threshold t_{gradient}.
Parameters: multiple factor k and radius r of the area surrounding the cell p.

for $col - 1$ to W **do**
  for $row = 1$ to H **do**
    repeat
      $r \leftarrow \text{INT}(k \cdot \sigma + 0.5)$
      suboperators $\Leftarrow \frac{8}{\sqrt{\pi k^2 \sigma^2}} e^{-\frac{(\Delta x^2 + \Delta y^2)}{2k^2\sigma^2}}$
      Compute the gradients $grad(x, y)$ using suboperators.
      Gradient G located in $(x, y) \leftarrow \max\left\{\left|\frac{\partial I_{\text{slope}}}{\partial x}\right|, \left|\frac{\partial I_{\text{slope}}}{\partial y}\right|, \left|\frac{\partial I_{\text{slope}}}{\partial diag_{\text{L}}}\right|, \left|\frac{\partial I_{\text{slope}}}{\partial diag_{\text{R}}}\right|\right\}$.
      $k \leftarrow k + 1$.
    until gradient G $< t_{\text{gradient}}$ and $k > 2$.
  end for
end for

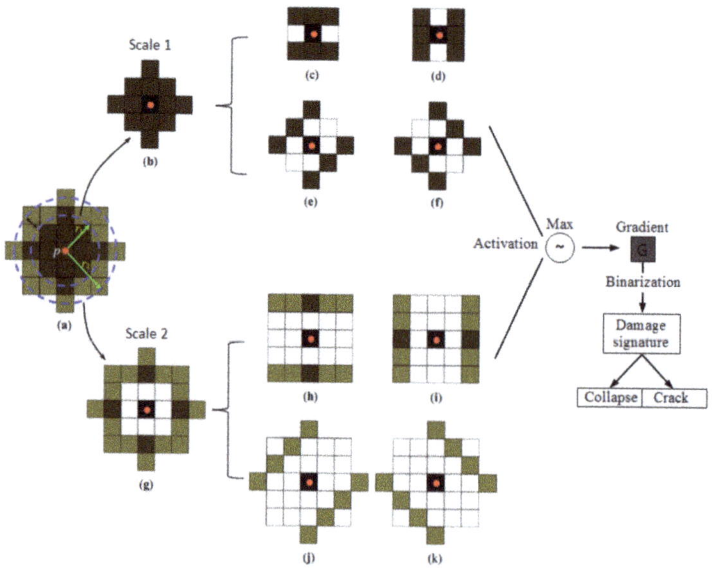

Figure 7. Multiscale architecture of the proposed self-adaptive and multiscale gradient operator (SMGO) and damage signature generation. Scope of the area surrounding a cell in (**a**) Scales 1 and 2 for gradient computation and (**b**) Scale 1. (**c–f**) Kernels of the gradient computation in Scale 1. (**g**) Scope of the area surrounding a cell in Scale 2. (**h–k**) kernels of the gradient computation in Scale 2. The white and nonwhite grids denote the null and nonzero values, respectively.

After the damage signature generation, the damaged regions $\mathbb{R}^{\text{damage}}$ are determined by a binary operation based on a given condition, which can be defined as

$$\left[grad_{\mathbb{R}}(x,y) - mean\left(grad_{\text{img}}\right)\right] > 3.0 \cdot std\left(grad_{\text{img}}\right),\ (x,y) \in \mathbb{R}^{\text{damage}}, \quad (7)$$

where $mean(\cdot)$ and $std(\cdot)$ denote the calculation of mean and standard deviation in the damage signature map. If the $grad_{\mathbb{R}}(x,y)$ satisfies this condition, it is considered to be within a damage region. Then, the collapse and crack are separated from the damaged regions via two criteria, i.e., if $Area\left(\mathbb{R}^{\text{damage}}\right) > 0.25\ \text{m}^2$ and $Perimeter\left(\mathbb{R}^{\text{damage}}\right)/Area\left(\mathbb{R}^{\text{damage}}\right) < 1.5$, then the damaged region $\mathbb{R}^{\text{damage}}$ is defined as a collapse; otherwise, the $\mathbb{R}^{\text{damage}}$ is considered to be a crack, where $Area(\cdot)$ and $Perimeter(\cdot)$ denote the calculation of area and perimeter that can be conducted using ArcGIS software.

4. Results

In the experiments, dense point clouds are generated by using the proposed method and implemented through C++ programming and an open-source library (i.e., OpenCV). Our software mainly includes distortion correction, sparse matching, dense matching, absolute orientation, image stitching, DSM generation, and orthophoto generation. The sparse matching module includes two ways, i.e., match all images without any supporting information and GPS/IMU supported trajectory matching, and the dense matching module is run based on the sparse matching results. The performance of low-cost UAV-based (i.e., DJI Mavic Air) mapping is critical in the accurate reconstruction of the revetment surface for damage recognition. Take Part 1 as an example, the distribution of the GCPs and CPs (i.e., check points) are laid out widely and evenly in the survey areas, as shown in Figure 8. The residual error and root mean square error (RMSE) were calculated on the basis of 13 and 15 CPs for Parts 1 and 2, respectively, and measured on RTK GPS. Their corresponding 3D points were determined from the dense point clouds. The X, Y, and Y RMSE values are calculated using Equation (8), and the error statistics of the CPs are summarized in Figure 9 and Table 1. Additionally, the re-projection errors $RMSE_{\text{img}}$ of the check points (CPs) are calculated using Equation (9), and the error statistics are summarized in Table 2. Figure 10 (Part 1) illustrates the pixel-by-pixel dense point clouds textured with colors from the UAV remote sensing images. The revetment consists of 1.96×10^7 points, which correspond to the density of approximately 963 points/m^2 and the grid size of 3.2 cm × 3.2 cm. The use of the proposed dense matching method reconstructs the fine details of the revetment surface. The results show that the X and Y RMSE values obtained via the proposed dense matching method were less than 4 cm, which is a relatively small horizontal error. Moreover, the vertical RMSE value or the Z RMSE value was less than 6 cm and the re-projection errors are less than one pixel. Therefore, these RMSE values seemed fairly satisfactory for high-precision reconstruction of the revetment surface. The accuracy was deemed sufficient for recognizing damage signatures on the surface of the revetments along urban rivers.

$$\begin{cases} RMSE_X = \sqrt{\frac{\sum(X_{\text{dense}}-X_{\text{GCP}})^2}{n}} \\ RMSE_Y = \sqrt{\frac{\sum(Y_{\text{dense}}-Y_{\text{GCP}})^2}{n}} \\ RMSE_Z = \sqrt{\frac{\sum(Z_{\text{dense}}-Z_{\text{GCP}})^2}{n}} \end{cases}, \tag{8}$$

$$RMSE_{\text{img}} = \sqrt{\sum_{i=1}^{n}\sum_{j=1}^{m}\rho_{ij}\left\|P(X_i, C_j) - x_{ij}\right\|^2 / \sum_{i=1}^{n}\sum_{j=1}^{m}\rho_{ij}}, \tag{9}$$

where X_i and C_j denote a 3D point and a camera, correspondingly; $P(X_i, C_j)$ is the predicted projection of point X_i on camera C_j; x_{ij} is the observed image point; $\|\cdot\|$ denotes the operation of L2-norm; ρ_{ij} is an indicator function with $\rho_{ij} = 1$ if point X_i is visible in camera C_j; otherwise, $\rho_{ij} = 0$.

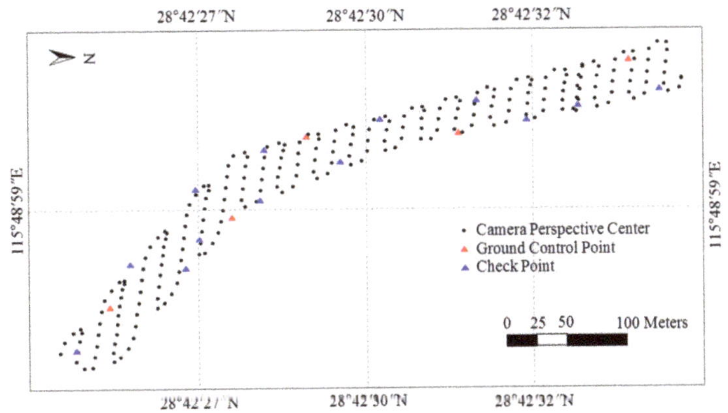

Figure 8. Placements of ground control points and check points.

Table 1. Comparison of the obtained RMSE values of CPs via Pix4Dmapper 4.4, Agisoft Metashape Professional 1.5.3, and the object space-based approach.

Area	Method	RMSE X (cm)	RMSE Y (cm)	RMSE Z (cm)	Total RMSE (cm)
Part 1	Pix4Dmapper	3.85	3.84	5.70	4.56
	Agisoft Metashape	5.08	4.53	6.59	5.47
	object space-based	3.76	3.72	5.67	4.48
Part 2	Pix4Dmapper	3.83	4.27	5.07	4.42
	Agisoft Metashape	4.89	4.63	6.43	5.38
	object space-based	3.49	3.30	5.31	4.13

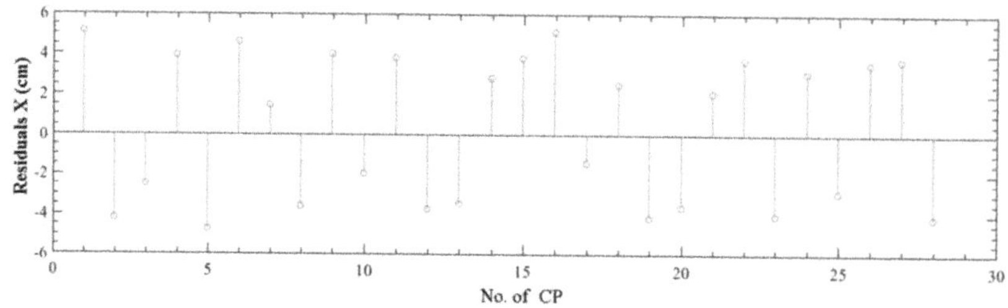

(**a**)

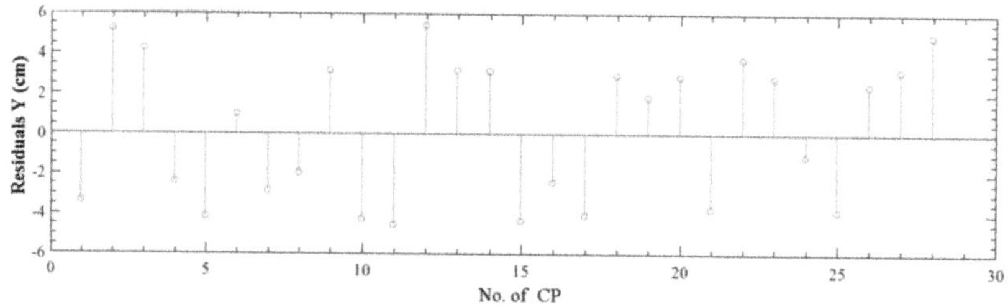

(**b**)

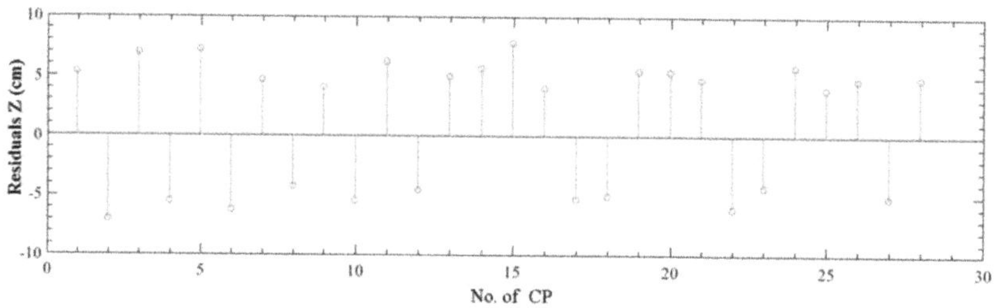

(**c**)

Figure 9. Residuals of 28 check points (CPs) for Part 1 and Part 2 measured on real-time kinematic (RTK) GPS and their corresponding 3D points determined from the dense point clouds. Residuals X, Y, and Z are shown in (**a**–**c**) respectively.

Table 2. Comparison of the re-projection errors $RMSE_{img}$ of CPs via Pix4Dmapper 4.4, Agisoft Metashape Professional 1.5.3, and the object space-based approach.

Area	Method	RMSE (Pixel)
Part 1	Pix4Dmapper	0.611
	Agisoft Metashape	0.679
	object space-based	0.597
Part 2	Pix4Dmapper	0.752
	Agisoft Metashape	0.783
	object space-based	0.730

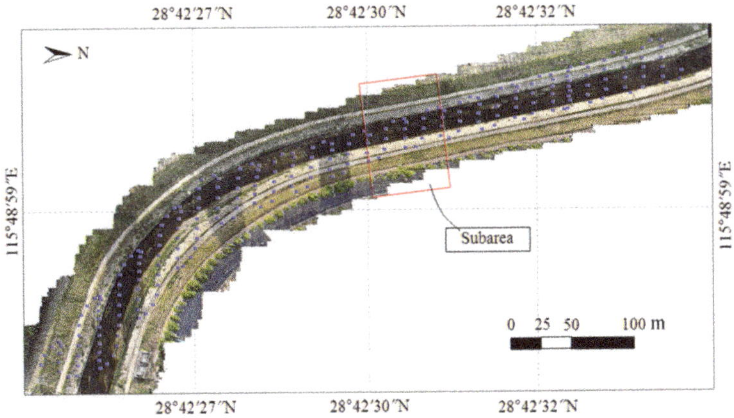

(a)

(b)

Figure 10. Dense point clouds derived from UAV photogrammetry. (**a**) Top view of the dense point clouds. (**b**) Oblique view of the dense point clouds. The viewpoints of the camera are marked by blue dots.

We also compared the performance of surface reconstruction with that of the commercial software such as Agisoft Metashape Professional 1.5.3 (www.agisoft.com) and Pix4Dmapper 4.4 (www.pix4d.com), which are widely used photogrammetric software for 3D surface reconstruction and revetment monitoring [19,23]. In order to balance accuracy and efficiency, medium precision is set to perform sparse and dense matching in Agisoft Metashape Professional 1.5.3, and the default settings are used in Pix4Dmapper 4.4. To evaluate the effects of the multiscale architecture in the proposed SMGO, the non-multiscale gradient operator (NMGO) is compared with our method. The gradient intensity images with the values normalized from 0 to 1 shown in Figure 11h,i are correspondingly generated by the non- and multiscale gradient operators, respectively.

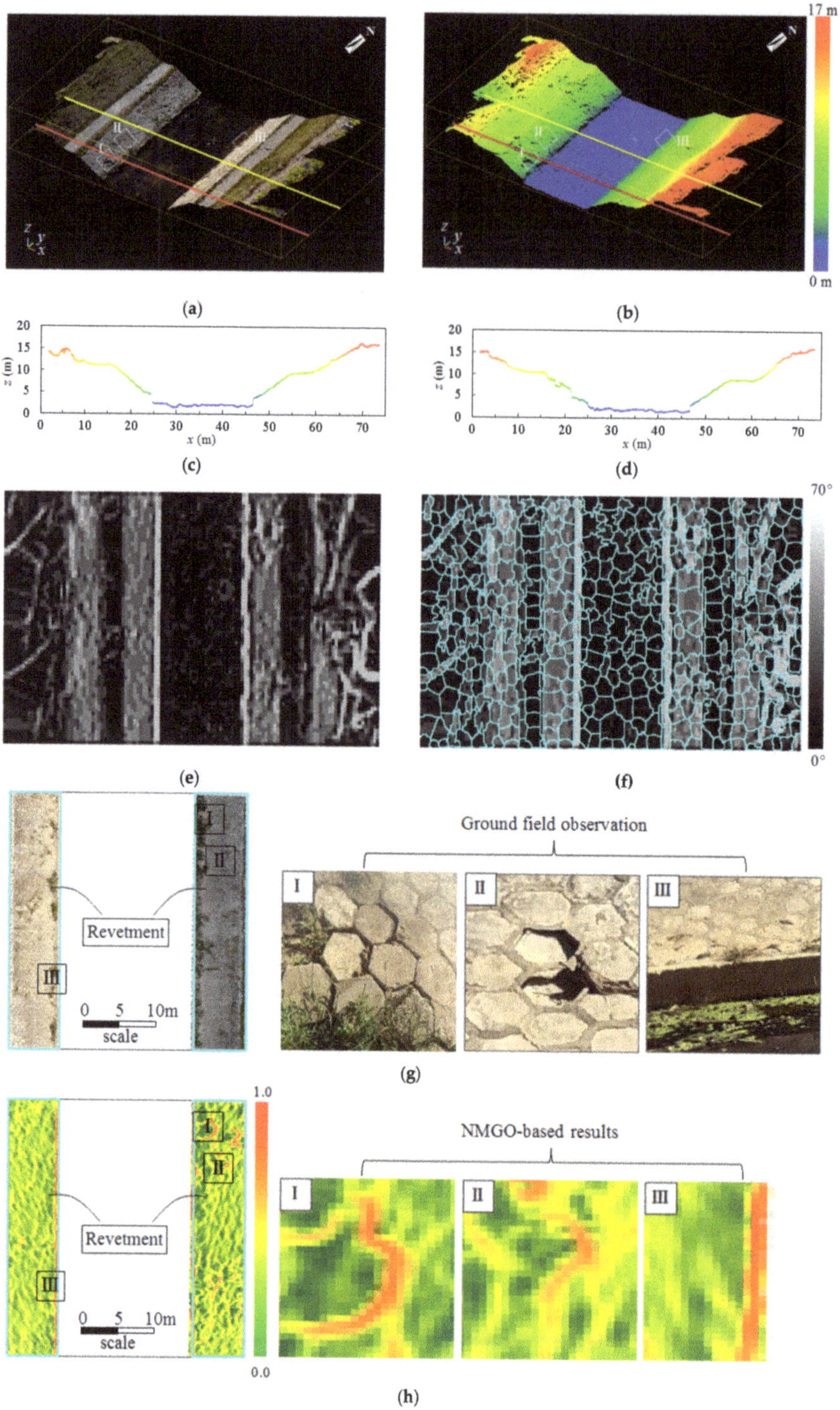

Figure 11. *Cont.*

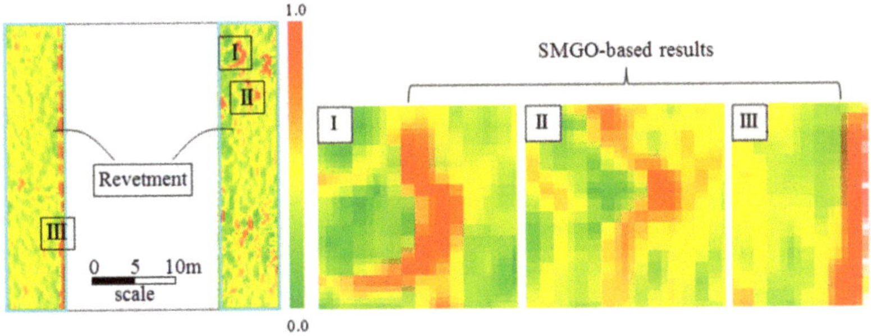

(i)

Figure 11. Revetment damage recognition along an urban river. The dense point clouds, depth map, and slope intensity image of the subarea in Figure 10a are shown in (**a**,**b**,**e**), respectively. (**c**) Cross-section without damage. (**d**) Cross-section with a collapse. (**f**) Superpixels marked by the cyan boundaries that are generated via the simple linear iterative clustering (SLIC)-based algorithm. The true color (RGB) point cloud of the revetment is exhibited in (**g**), and the enlarged three damaged regions of I, II, and III from the ground field observation are also shown in (**g**). In addition, the gradient intensity images generated using the non-multiscale gradient operator (NMGO) and SMGO are shown in (**h**,**i**), respectively.

The indicators *Precision*, *Recall* and *F1_score* are used to evaluate the proposed method in our experiments as follows:

$$Precision = \frac{TP}{TP + FP}, \tag{10}$$

$$Recall = \frac{TP}{TP + FN}, \tag{11}$$

$$F1\_score = 2 \cdot \frac{Precision \cdot recall}{Precision + recall}, \tag{12}$$

where *TP* is the number of damaged regions that are correctly identified, *FP* is the number of damage regions that are incorrectly identified, and *FN* is the number of unrecognized damaged regions. Table 3 lists the statistical results of *Precision*, *Recall* and *F1_score* for collapse and crack recognition. Furthermore, the field visual inspection, NMGO-based damage recognition, and the proposed method are compared in Table 3. It should be noted that the true value of the collapse and crack is obtained through manual inspection. To be specific, the experimental areas are divided into grids on the map, and then the professionals check in detail whether there is any collapse or crack in each grid. If there is, the coordinates are marked using a GPS measuring instrument. For a fair comparison, field visual inspection is conducted according to the commonly used process by three different surveyors, and the average values of *TP*, *FP*, and *FN* are calculated.

Table 3. Comparison of the three indicators obtained through field visual inspection, NMGO-based method, and our method.

Site	Category	Number	Indicator (%)	Method		
				Field Visual Inspection	NMGO-Based	Our Method
Part 1	Collapse	14	*Precision*	86.67	73.33	92.85
			Recall	92.85	78.57	92.85
			F1_score	89.66	75.86	92.85
	Crack	36	*Precision*	91.18	79.41	89.18
			Recall	86.11	75.00	91.67
			F1_score	88.57	77.14	90.41

Table 3. *Cont.*

Site	Category	Number	Indicator (%)	Method		
				Field Visual Inspection	NMGO-Based	Our Method
Part 2	Collapse	18	*Precision*	84.21	73.68	89.47
			Recall	88.89	77.78	94.44
			F1_score	86.49	75.67	91.89
	Crack	54	*Precision*	88.46	82.97	90.91
			Recall	85.18	72.22	92.59
			F1_score	86.79	77.23	91.74

5. Discussion

In this study, the proposed dense matching method performs better at surface reconstruction than Pix4Dmapper and Agisoft Metashape in terms of the RMSE values shown in Tables 1 and 2. Notably, the time consumption of the proposed method is only 87% of Pix4Dmapper and Agisoft Metashape in the same operating environment. These results can be attributed to the dense matching achieved by the object space-based approach, which only computes the height values on the top surface of the ground and reduces the computational expense of redundant point clouds. To offer a detailed description, as shown in Figure 10a, the extracted subarea illustrates the details of the geometry structure, and Figure 11a,b present the corresponding results of the subarea. Two examples of cross-sections of dense point clouds derived from UAV mapping are demonstrated in Figure 11c,d with (marked by a red line) and without damage (marked by a yellow line), respectively. Subsequently, Figure 11e,f show the slope intensity image generated via the slope calculation and the superpixels segmented with the SLIC-based algorithm [39], respectively. Figure 11g exhibits the revetment regions obtained through superpixel merging on the basis of similar gradients and adjacency, and the enlarged three damaged regions of I, II, and III from the ground field observation (i.e., RGB ground photos) are also shown.

In terms of visual assessment, the profile in Figure 11d is the geometry structure corresponding to region I. In Figure 11h,i, it can be seen that the region detected by the SMGO algorithm is more consistent with the geometric structure than that detected by the NMGO. The NMGO has difficulty identifying all the damaged regions and ignores the spatial continuity of cracks or can even cause edges in over-recognition. By comparison, the proposed SMGO can extract the damaged regions within accurate boundaries and improve the accuracy of the revetment damage recognition by highlighting the gap between the damaged and nondamaged areas. Therefore, the proposed SMGO enables collapse and crack with a height drop relative to the surrounding areas to be detected. This finding is attributed to the revetment damage recognition using the proposed SMGO, which can achieve feature extraction in all the orientations with the multiscale operator in the horizontal, vertical, and diagonal directions. Notably, the strip regions with vertical drop (e.g., region III shown in Figure 11h) close to the river are also detected but not considered damaged regions in this study. In addition, the proposed method achieves better performance than the two other methods in terms of *Precision*, *Recall* and *F1_score*, especially in crack recognition. For field visual inspection, the inconspicuous cracks may easily be ignored and manual recognition is easily affected in this case, that is, crack damage often exists in other types of damage (e.g., collapse). As mentioned above, the NMGO-based method ignores the feature of damage and performs poorly.

6. Conclusions

This study aims to achieve revetment damage recognition along urban rivers through dense point clouds derived from low-cost UAV photogrammetry. In this study, two improvements of the proposed approach confirm that our method can be used as an effective alternative to field visual inspection for revetment (with slope protection) damage recognition along urban rivers. (1) Region growing coupled with SGM is proposed to generate the pixel-by-pixel dense point clouds from UAV remote sensing images and reconstruct the fine details of the high-precision revetment surface. This reconstruction is considered satisfactory in terms of the horizontal error <4 cm and vertical error <6 cm relative to GCPs. (2) On the basis of the in situ visual assessment and quantitative analysis (e.g., at least 90% of the *Precision*, *Recall*, and *F1_score* values), the accuracy of revetment damage recognition is confirmed after comparing the results of the field visual inspection and the NMGO-based method. Notably, UAV-based mapping can offer a new possibility in fully measuring, monitoring, and understanding revetment damage with low-cost operation. UAV-based mapping presents a technology that has the potential to transform how revetment damage recognition is observed and investigated. Furthermore, it could help the government and local authorities develop revetment management plans and provide evidence for maintenance or improvements.

This study is suitable for recognizing the damage signatures in revetments designed with slope protection. The use of the proposed method on revetments with steep slopes still needs further investigation because the nadir orientation of a camera for photogrammetry has difficulty achieving high-precision surface reconstruction of steep revetments. In future studies, we will optimize the proposed approach by using oblique photogrammetry and deep learning to achieve satisfactory damage recognition of steep revetments.

Author Contributions: Ting Chen proposed the framework of revetment damage recognition and the paper. Haiqing He wrote the source code of dense matching. Dajun Li designed the experiments and revised the paper. Puyang An and Zhenyang Hui generated the datasets and performed the experiments. All authors have read and agreed to the published version of the manuscript.

Funding: This study was financially supported by the National Natural Science Foundation of China (41861062 and 41401526) and the Natural Science Foundation of Jiangxi Province of China (20171BAB213025 and 20181BAB203022).

Acknowledgments: The authors thank Jing Yu for providing datasets. The authors also want to thank the anonymous reviewers for their constructive comments that significantly improved our manuscript.

Conflicts of Interest: The authors declare no conflicts of interest.

References

1. Chibana, T. Urban river management: Harmonizing river ecosystem conservation. In *Urban Environmental Management Technology*; Springer: Tokyo, Japan, 2008; pp. 47–66.
2. Osman, A.M.; Thorne, C.R. Riverbank stability analysis. I: Theory. *J. Hydraul. Eng.* **1988**, *114*, 134–150. [CrossRef]
3. Hesp, P. Foredunes and blowouts: Initiation, geomorphology and dynamics. *Geomorphology* **2002**, *48*, 245–268. [CrossRef]
4. Tarrant, O.; Hambidge, C.; Hollingsworth, C.; Normandale, D.; Burdett, S. Identifying the signs of weakness, deterioration, and damage to flood defense infrastructure from remotely sensed data and mapped information. *J. Flood Risk Manag.* **2017**, *11*, 317–330. [CrossRef]
5. Baghdadi, N.; Gratiot, N.; Lefebvre, J.-P.; Oliveros, C.; Bourguignon, A. Coastline and mudbank monitoring in French Guiana: Contributions of radar and optical satellite imagery. *Can. J. Remote Sens.* **2004**, *30*, 109–122. [CrossRef]
6. Royet, P. Rapid and Cost-Effective Dike Condition Assessment Methods: Geophysics and Remote Sensing. FloodProbe. 2012. Available online: http://www.floodprobe.eu/partner/assets/documents/Floodprobe-D3.2_V1_3Dec2012.pdf (accessed on 1 September 2019).
7. Hagenaars, G.; Luijendijk, A.; de Vries, S.; de Boer, W. Long term coastline monitoring derived from satellite imagery. *Coast. Dyn.* **2017**, *122*, 1551–1562.
8. Choi, C.E.; Cui, Y.; Au, K.Y.K.; Liu, H.; Wang, J.; Liu, D.; Wang, H. Case study: Effects of a partial-debris dam on riverbank erosion in the Parlung Tsangpo river, China. *Water* **2018**, *10*, 250. [CrossRef]
9. Rosser, N.J.; Petley, D.N.; Lim, M.; Dunning, S.A.; Allison, R.J. Terrestrial laser scanning for monitoring the process of hard rock coastal cliff erosion. *Q. J. Eng. Geol. Hydrogeol.* **2005**, *38*, 363–375. [CrossRef]

10. Longoni, L.; Papini, M.; Brambilla, D.; Barazzetti, L.; Roncoroni, F.; Scaioni, M.; Ivanov, V.I. Monitoring riverbank erosion in mountain catchments using terrestrial laser scanning. *Remote Sens.* **2016**, *8*, 241. [CrossRef]
11. Cheng, Y.-J.; Qiu, W.; Lei, J. Automatic extraction of tunnel lining cross-sections from terrestrial laser scanning point clouds. *Sensors* **2016**, *16*, 1648. [CrossRef]
12. Thoma, D.P.; Gupta, S.C.; Bauer, M.E.; Kirchoff, C.E. Airborne laser scanning for riverbank erosion assessment. *Remote Sens. Environ.* **2005**, *95*, 493–501. [CrossRef]
13. Yang, B.; Hwang, C.; Cordell, H.K. Use of LiDAR shoreline extraction for analyzing revetment rock beach protection: A case study of Jekyll island state park, USA. *Ocean Coast. Manag.* **2012**, *69*, 1–15. [CrossRef]
14. Pye, K.; Blott, S.J. Assessment of beach and dune erosion and accretion using LiDAR: Impact of the stormy 2013-14 winter and longer term trends on the Sefton coast, UK. *Geomorphology* **2016**, *266*, 146–167. [CrossRef]
15. Ternate, J.R.; Celeste, M.I.; Pineda, E.F.; Tan, F.J.; Uy, F.A.A. Floodplain modelling of Malaking-ilog river in southern Luzon, Philippines using LiDAR digital elevation model for the design of water-related structures. In Proceedings of the 2nd International Conference on Civil Engineering and Materials Science, Seoul, Korea, 26–28 May 2017; pp. 1–9.
16. Drummond, H.; Weiner, H.M.; Kaminsky, G.M.; McCandless, D.; Hacking, A. Assessing bulkhead removal and shoreline restoration using boat-based lidar. In Proceedings of the Salish Sea Ecosystem Conference, Seattle, WA, USA, 5 April 2018.
17. Cook, K.L. An evaluation of the effectiveness of low-cost UAVs and structure from motion for geomorphic change detection. *Geomorphology* **2017**, *278*, 195–208. [CrossRef]
18. He, H.; Chen, T.; Zeng, H.; Huang, S. Ground control point-free unmanned aerial vehicle-based photogrammetry for volume estimation of stockpile carried on barges. *Sensors* **2019**, *19*, 3534. [CrossRef] [PubMed]
19. Pitman, S.J.; Hart, D.E.; Katurji, M.H. Application of UAV techniques to expand beach research possibilities: A case study of coarse clastic beach cusps. *Cont. Shelf Res.* **2019**, *184*, 44–53. [CrossRef]
20. Hemmelder, S.; Marra, W.; Markies, H.; De Jong, S.M. Monitoring river morphology & bank erosion using UAV imagery-a case study of the river Buëch, Hautes-Alpes, France. *Int. J. Appl. Earth Obs. Geoinf.* **2018**, *73*, 428–437.
21. Hallermann, N.; Morgenthal, G.; Rodehorst, V. Vision-based deformation monitoring of large scale structures using unmanned aerial systems. *IABSE Symp. Rep.* **2014**, *102*, 2852–2859. [CrossRef]
22. Nakagawa, M.; Yamamoto, T.; Tanaka, S.; Noda, Y.; Kashimoto, K.; Ito, M.; Miyo, M. Location-based infrastructure inspection for sabo facilities. In Proceedings of the International Archives of the Photogrammetry, Remote Sensing and Spatial Information Sciences, Volume XL-3/W3, ISPRS Geospatial Week, La Grande Motte, France, 28 September–3 October 2015; pp. 257–262.
23. Kubota, S.; Kawai, Y.; Kadotani, R. Accuracy validation of point clouds of UAV photogrammetry and its application for river management. In Proceedings of the International Archives of the Photogrammetry, Remote Sensing and Spatial Information Sciences, Volume XLII-2/W6, International Conference on Unmanned Aerial Vehicles in Geomatics, Bonn, Germany, 4–7 September 2017; pp. 195–199.
24. Starek, M.J.; Giessel, J. Fusion of UAS-based structure-from-motion and optical inversion for seamless topo-bathymetric mapping. In Proceedings of the IEEE International Geoscience and Remote Sensing Symposium (IGARSS), Fort Worth, TX, USA, 23–28 July 2017; pp. 2999–3002.
25. Jayson, P.-N.; Appeaning Addo, K.; Amisigo, B.; Wiafe, G. Assessment of short-term beach sediment change in the Volta Delta coast in Ghana using data from Unmanned Aerial Vehicles (Drone). *Ocean Coast. Manag.* **2019**, *182*, 104952. [CrossRef]
26. Pires, A.; Chaminé, H.I.; Piqueiro, F.; Pérez-Alberti, A.; Rocha, F. Combing coastal geoscience mapping and photogrammetric surveying in maritime environments (Northwestern Iberian Peninsula): Focus on methodology. *Environ. Earth Sci.* **2016**, *75*, 196. [CrossRef]
27. DJI. Mavic Air User Manual. 2018. Available online: https://dl.djicdn.com/downloads/phantom_4_pro/Phantom+4+Pro+Pro+Plus+User+Manual+v1.0.pdf (accessed on 15 December 2018).
28. He, H.; Yan, Y.; Chen, T.; Cheng, P. Tree height estimation of forest plantation in mountainous terrain from bare-earth points using a DoG-coupled radial basis function neural network. *Remote Sens.* **2019**, *11*, 1271. [CrossRef]

29. Puliti, S.; Ørka, H.O.; Gobakken, T.; Næsset, E. Inventory of small forest areas using an unmanned aerial system. *Remote Sens.* **2015**, 7, 9632–9654. [CrossRef]
30. He, H.; Chen, X.; Liu, B.; Lv, Z. A sub-Harris operator coupled with SIFT for fast images matching in low-altitude photogrammetry. *Int. J. Signal Process. Image Process. Pattern Recognit.* **2014**, 7, 395–406. [CrossRef]
31. Hamshaw, S.D.; Bryce, T.; Rizzo, D.M.; O'Neil-Dunne, J.; Frolik, J.; Dewoolkar, M.M. Quantifying streambank movement and topography using unmanned aircraft system photogrammetry with comparison to terrestrial laser scanning. *River Res. Appl.* **2017**, 33, 1354–1367. [CrossRef]
32. Wu, C. Towards linear-time incremental structure from motion. In Proceedings of the 3DV-Conference, International Conference on IEEE Computer Society, Seattle, WA, USA, 29 June–1 July 2013; pp. 127–134.
33. Schönberger, J.L.; Frahm, J.M. Structure-from-motion revisited. In Proceedings of the IEEE Conference on Computer Vision and Pattern Recognition (CVPR), Las Vegas, NV, USA, 27–30 June 2016; pp. 4104–4113.
34. Sba: A Generic Sparse Bundle Adjustment C/C++ Package. 2018. Available online: http://users.ics.forth.gr/~{}\{\}lourakis/sba/ (accessed on 5 August 2019).
35. Bhattacharya, A.; Arora, M.; Sharma, M. Usefulness of adaptive filtering for improved digital elevation model generation. *J. Geol. Soc. India* **2013**, 82, 153–161. [CrossRef]
36. Hirschmuller, H. Accurate and efficient stereo processing by semi-global matching and mutual information. In Proceedings of the IEEE Computer Society Conference on Computer Vision and Pattern Recognition (CVPR'05), San Diego, CA, USA, 20–25 June 2005; pp. 807–814.
37. Humenberger, M.; Engelke, T.; Kubinger, W. A census-based stereo vision algorithm using modified semi-global matching and plane fitting to improve matching quality. In Proceedings of the IEEE Computer Society Conference on Computer Vision and Pattern Recognition (CVPR), San Francisco, CA, USA, 13–18 June 2010; pp. 77–84.
38. Viola, P.; Wells, W.M. Alignment by maximization of mutual information. *Int. J. Comput. Vis.* **1997**, 24, 137–154. [CrossRef]
39. He, H.; Zhou, J.; Chen, M.; Chen, T.; Li, D.; Cheng, P. Building extraction from UAV images jointly using 6D-SLIC and multiscale Siamese convolutional networks. *Remote Sens.* **2019**, 11, 1040. [CrossRef]

International Journal of
Geo-Information

MDPI

Article

A Vector Line Simplification Algorithm Based on the Douglas–Peucker Algorithm, Monotonic Chains and Dichotomy

Bo Liu [1], Xuechao Liu [1], Dajun Li [1,*], Yu Shi [1], Gabriela Fernandez [2] and Yandong Wang [3,*]

1 Faculty of Geomatics, East China University of Technology, 418# Guanglan Road, Nanchang 330013, China; liubo@ecut.edu.cn (B.L.); liuxuechao1991@163.com (X.L.); yushi19930807@outlook.com (Y.S.)

2 Department of Geography, Center for Human Dynamics in the Mobile Age (HDMA), San Diego State University, 5500 Campanile Drive, San Diego, CA 92182-4493, USA; gfernandez2@sdsu.edu

3 State Key Laboratory of Information Engineering in Surveying, Mapping and Remote Sensing, Wuhan University, 129# Luoyu Road, Wuhan 430079, China

* Correspondence: djli@ecut.edu.cn (D.L.); ydwang@whu.edu.cn (Y.W.)

Received: 11 March 2020; Accepted: 16 April 2020; Published: 17 April 2020

Abstract: When using the traditional Douglas–Peucker (D–P) algorithm to simplify linear objects, it is easy to generate results containing self-intersecting errors, thus affecting the application of the D–P algorithm. To solve the problem of self-intersection, a new vector line simplification algorithm based on the D–P algorithm, monotonic chains and dichotomy, is proposed in this paper. First, the traditional D–P algorithm is used to simplify the original lines, and then the simplified lines are divided into several monotonic chains. Second, the dichotomy is used to search the intersection positions of monotonic chains effectively, and intersecting monotonic chains are processed, thus solving the self-intersection problems. Two groups of experimental data are selected based on large data sets. Results demonstrate that the proposed experimental method has advantages in algorithmic efficiency and accuracy when compared to the D–P algorithm and the Star-shaped algorithm.

Keywords: Line Simplification; Douglas-Peucker Algorithm; Monotonic Chain; Dichotomy

---

1. Introduction

With the development of remote-sensing technology, sensor technology, and Web 2.0, the large amounts of obtained spatial vector data produce great challenges in data storage, processing, and transmission. To enhance the processing capability for massive spatial vector data, new vector data simplification algorithms with high efficiency and robustness are urgently needed.

There are many classical methods used to simplify vector data, including the Douglas–Peucker algorithm (D–P algorithm) [1], Ramer algorithm [2], and other algorithms [3–9]. The D–P algorithm [1] and Ramer algorithm [2] use a given distance tolerance to determine which vertices on a line are to be eliminated or retained. Lang [3] used a perpendicular distance tolerance to filter data, but the method was too time consuming [1]. Based on a sequential set of five procedures, McMaster [4] presented a conceptual model to process linear digital data. This employed method used the perpendicular distance tolerance proposed by Lang [4] to simplify the lines and used smoothing techniques to produce the most aesthetically acceptable results. Based on selecting local minima and maxima, an algorithm for compressing digital contour data has been developed by Li [5]. The new algorithm was more efficient than the D–P algorithm, but the result remained the same as the D–P algorithm. Visvalingam and Whyatt [6] used the "effective area" to simplify the line features and discussed the influence of rounding errors on a version of the Ramer–Douglas–Peucker algorithm [1,2] for line simplification. To show how to make robust, precise, and reproducible geographic information systems

(GIS) algorithms, Ratschek et al. [7] proposed a robust version of the R-D–P algorithm. Based on recognizing line shapes and filtering them against cartographic rules, Wang and Muller [8] proposed a Bend Simplify algorithm. The bend simplify algorithm attempts to simulate manual line simplification by using cartographic rules, and it is typically used to simplify naturally occurring features such as lakes and stream channels [10]. Based on the Li–Openshaw algorithm [11], the D–P algorithm, and the orthogonal simplification method, Samsonov and Yakimova [12] proposed a methodology and generalization model for the geometric simplification of heterogeneous line datasets.

The line simplification results processed by the above algorithms consisted of a set of original polyline vertices with no "Steiner" points. Other researchers have applied the "Steiner points" [13] to simplify the linear features [14,15]. The concept of Steiner points originates from the discipline of computational geometry and is referred to as a point or a set of points that are introduced when solving a geometric optimization problem to improve upon solutions based only on the original set of points [13,16]. On the basis of the traditional D–P algorithm [1], Cromley [14] used "Steiner points" to simplify a line; the experimental results showed that the proposed method is computationally faster than the traditional D–P algorithm [1]. Based on the method proposed by Li and Openshaw [11], Raposo [15] presented a scale-specific cartographic line simplification algorithm by using a hexagonal tessellation instead of a square grid. The hexagonal quantization algorithm draws from sampling and map-resolution theory as well as the concept of vertex clustering from computer graphics to yield a method which is simple and effective.

The experimental results addressed in the line simplification algorithms above show that good results have been achieved for each method and have been successfully applied to the corresponding fields. This has to be due to all the advantages concerning the D–P algorithm—it is highly effective at preserving the shape of the line, unique in vector curve compression at the presence of the threshold values and, above all, precise in a higher position, which is thereby often used to simplify lines [16,17]. However, the D–P algorithm is found to be flawed in that a large area deviation might be caused [18]. In addition, the method only addresses the curves themselves rather than the topological relations of the curves, thus leading to self-intersection problems [19]. Therefore, many scholars have improved the D–P algorithm in order to solve these self-intersection problems. A hierarchical representation scheme for planar curves was proposed by Ho and Kim [20], which used natural approximation and efficient localization. It was effective in removing self-intersections in all possible approximations for a curve using the cross-link technique, reducing computation time remarkably. Mantler and Snoeyink [21] introduced a new algorithm. They defined a notion of safe sets, which are fragments of linear features that can be simplified without introducing intersections or topology changes. This algorithm can also help to identify a collection of safe sets using the Voronoi diagram of points, but it is required to produce a Voronoi diagram, the efficiency of the algorithm is limited. To solve the self-intersection problems, Avelar and Müller [22] proposed an algorithm to compute the topological relations when compressing the polyline features. In this algorithm, simple geometric operations are used and tested step-by-step to check whether the topological relations have changed after compression. If the topological relations are not changed, the algorithm will terminate; otherwise, the topological relations will be maintained. Wu and Marquez [23] proposed a star-shaped algorithm (ST algorithm) to simplify the curves. The original curves are first scanned first and then divided into "Star" areas. Finally, the D–P algorithm is applied to compress the "Star" areas. The star-shaped algorithm solves the self-intersection problems, but it has the worst case $O(nm)$ time complexity, where n is the number of input vertices and m depends on the number of star-shaped regions, the time consumption and efficiency of this method is relatively high.

Most of the above improved D–P algorithms solved the self-intersection problems to simplify the linear objects; however, the algorithms used have the disadvantages of low efficiency and complex steps. To solve the self-intersection problems when using the D–P algorithm to simplify the linear objects and improve the efficiency of the algorithm, a new vector line simplification algorithm that combines the D–P algorithm, monotonic chains and dichotomy is proposed in this paper. There are four main stages: first, the D–P algorithm is used to process the original lines; second, the monotonic chain method is used to divide the simplified lines into monotonic chains if the simplified lines have self-intersection problems; third, the dichotomy is used to quickly and accurately locate the self-intersection position of the simplified lines, process the self-intersection problems, and obtain the final result; finally, the experimental results are presented in this part, and the results of the experiments show that our proposed method demonstrates a more effective and higher performance.

The remainder of this paper is organized as follows. The basic theories, methods and steps of the new algorithm are introduced in Section 2. Experimental results and analysis are reported in Section 3. Conclusions are drawn in Section 4.

2. Methodology

In this section, we will first introduce the basic theories of the D–P algorithm, monotonic chains and the dichotomy method; then, the basic steps of the improved algorithm are introduced in further detail. A flow chart of the proposed research method is shown in Figure 1.

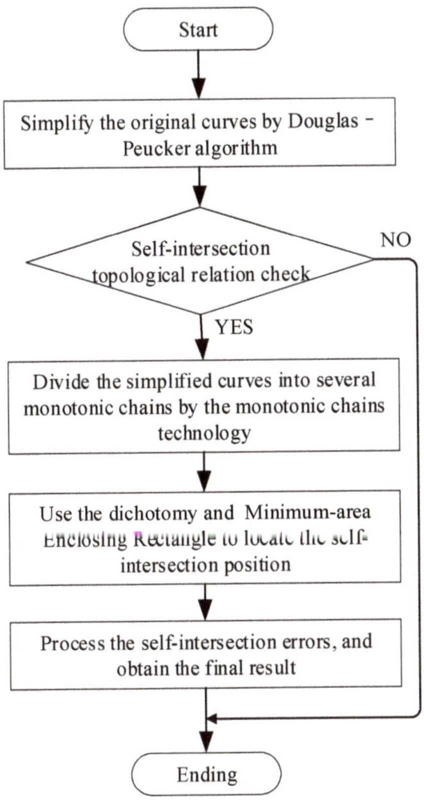

Figure 1. The flowchart of the proposed method.

2.1. *Basic Theory of the Douglas–Peucker (D–P) Algorithm*

The D–P algorithm is a classic algorithm used for curve compression. The algorithm is used to simplify polylines by deleting non-feature vertices and retaining the feature vertices. The basic theory and computational steps of the D–P algorithm are as follows [1,24]:

Step 1: For a curve L, which is composed of N coordinate vertices, the coordinate vertices set V is written as $V = \{v_1, v_2, \ldots, v_i, \ldots, v_N\}, (i = 1, 2, \ldots, N)$. First, connect the first vertex v_1 and the last vertex v_N, to obtain a new straight line $L_{v_1v_N}$. Second, calculate the shortest distances between the remaining vertices $\{v_2, \ldots, v_{N-1}\}$ and the new straight line $L_{v_1v_N}$ and obtain the shortest distance sets $D = \{D_2, \ldots D_k, \ldots, D_{N-1}\}$ (D_k is the shortest distance between vertex v_k and the new straight line $L_{v_1v_N}$);

Step 2: Select the maximum distance ($D_{\max}$) with shortest distance D, $D_{\max} = D_k$ (D_k is the shortest distance between vertex v_k and the new straight line $L_{v_1v_N}$). Given a distance ε as the distance threshold, if $D_{\max} < \varepsilon$, then the remaining vertices $\{v_2, \ldots, v_{N-1}\}$ from vertices set $V = \{v_1, v_2, \ldots, v_N\}$ are deleted, the given curve L is compressed into a straight line $L_{v_1v_N}$ and the D–P algorithm is finished. If $D_{\max} \geq \varepsilon$, then the vertices set $V = \{v_1, v_2, \ldots, v_N\}$ is divided into two subsets V_t and V_s, that is, $V = V_t + V_s$ ($V_t = \{v_1, v_2, \ldots, v_k\}$, $V_s = \{v_k, v_{k+1}, \ldots, v_N\}$);

Step 3: For the subsets V_t and V_s, repeat step 1 and 2, respectively. If all of the calculated shortest distances are less than the giving distance threshold (ε), then end the D–P algorithm.

2.2. Monotonic Chains and Dichotomy

The theory of the monotonic chain is mainly derived from computational geometry [12,25]. For the curve L, the monotonic chain is defined as follows:

Monotonic chain: For a curve L, which is composed of M coordinate vertices, the coordinate vertices set V is expressed as $V = \{v_1, v_2, \ldots, v_i, \ldots, v_M\}, (i = 1, 2, \ldots, M)$; x_i is the X-axis coordinate of vertex v_i, and y_i is the Y-axis coordinate of vertex v_i. In the direction of the X-axis, for the coordinate vertices set $X = \{x_1, x_2, \ldots, x_i, \ldots, x_M\}, (i = 1, 2, \ldots, M)$, if $x_i \leq x_{i+1}$ $(i = 1, 2, \ldots, M)$ or $x_i > x_{i+1}$ $(i = 1, 2, \ldots, M)$, the curve L will be called a monotonic increasing (or decreasing) chain of the X-axis. Similarly, in the direction of the Y-axis, for the coordinate vertices set $Y = \{y_1, y_2, \ldots, y_i, \ldots, y_M\}, (i = 1, 2, \ldots, M)$ $(i = 1, 2, \ldots, M)$, if $y_i \leq y_{i+1}$ $(i = 1, 2, \ldots, M)$ or $y_i > y_{i+1}$ $(i = 1, 2, \ldots, M)$, the curve L will be called a monotonic increasing (or decreasing) chain of the Y-axis.

Dichotomy: Dichotomy is one of the most commonly used search algorithms for ordinal sequences and has a high search efficiency [12,13]. Given the target element t and the ordered sequence $K = \{k_1, k_2, \ldots, k_i, \ldots, k_U\}, (i = 1, 2, \ldots, U)$, $(t \in K)$, to search for the target element t from K, the basic theory of dichotomy is as follows:

Step 1: For the target element t, compare t with the intermediate element $k_{\frac{U}{2}}$ from the sequence K. If $t \neq k_{\frac{U}{2}}$, then K will be divided into two parts: K_1 and K_2, $K = K_1 \cup K_2$, $K_1 = \left\{k_1, k_2, \ldots, k_i, \ldots, k_{\frac{U}{2}}\right\}$ $(i = 1, 2, \ldots, U)$, $K_2 = \left\{k_{\frac{U}{2}}, k_{\frac{U}{2}+1}, \ldots, k_j, \ldots, k_U\right\}$ $\left(j = \frac{U}{2}, \frac{U}{2} + 1, \ldots, U\right)$.

Step 2: For the target element t, if $t \geq k_{\frac{U}{2}}$, then execute step 1 in the K_2 until the target element t is found from the ordered K_2; if $t < k_{\frac{U}{2}}$, then execute step 1 in the K_1 until the target element t is found from the ordered K_1.

In vector spatial data structure, it is well known that a simple curve is composed of a number of line segments. For a curve L, there are N coordinate vertices: $P = \{p_1, p_2, \ldots, p_i, \ldots, p_N\}, (i = 1, 2, \ldots, N)$, and the curve L is composed of some line segments, such as: $L = L'_{1,2} + L'_{2,3} + \ldots + L'_{i,j} + \ldots + L'_{N-1,N}$ (N is the number of the coordinate vertices). Figure 2, shows that curve L is composed of 26 coordinate vertices $(0, 1, 2, \ldots, 25)$. In the Gauss-Krueger plane rectangular coordinate system, the horizontal axis was the Y-axis, and the vertical axis was the X-axis. Along the Y-axis, L could be divided into two monotonic chains L'_i $(i = 0, 1, 2, \ldots, 13)$ and L'_{ji} $(j = 13, 14, \ldots, 25)$. For L'_i, along the Y-axis, vertex p_0 is the smallest, and the vertex p_{13} is the biggest, and L'_i is a monotonic increasing chain; For L'_j, along the Y-axis, vertex p_{13} is the biggest, and the vertex p_{25} is the smallest, and L'_j is a monotonic decreasing chain. When using the D–P algorithm to process the curve L, it should be noted that if the final result has self-intersection problems, it has been caused by the corresponding monotonic chains L'_i and L'_j.

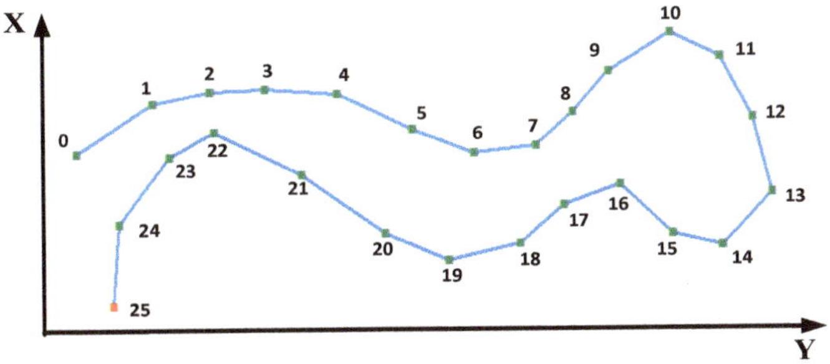

Figure 2. A schematic chart of the monotonic chain.

2.3. *The New Vector Line Simplification Algorithm based on the D–P Algorithm, Monotonic Chains and Dichotomy*

This paper used the monotonic chains and dichotomy to solve the self-intersection problem in spatial line simplification when processed by the D–P algorithm. In our proposed method, we firstly use the D–P algorithm to simplify the original polyline M, and obtain the simplified polyline T; Secondly, we check the self-intersection problems of T. If T does not have self-intersection problems, then we end this proposed method, otherwise, we use monotonic chain technology to quickly divide the T into several sequential monotonic chains; Thirdly, the dichotomy, MER (minimum-area enclosing rectangle, which refers to the rectangle with the smallest area that encloses the polyline) and geometric calculation method are used to process the sequential monotonic chains, in order to quickly locate the positions of the self-intersection problems of the sequential monotonic chains and solve the self-intersection problems, to obtain the final results.

This strategy of the proposed method does not only take the curve characteristics of a polyline into account, but also improves the time consumption of the proposed method. The main steps of the proposed method are described below.

Step 1: Use the D–P algorithm to process one curve M (There aren't self-intersection errors of M) and obtain a new curve T.

Step 2: Check the self-intersection problems of the T; if there are self-intersection errors, then perform step 3; otherwise, T is the final result of line simplification.

Step 3: For T, after step 2 of processing, if there are self-intersection errors, according to the sequence of the coordinate vertices, use the monotonic chain technology (as described in Section 2.2) to divide T into several sequential monotonic chains $T_1{}', T_2{}', \ldots, T_i{}', \ldots, T_j{}', \ldots, T_n{}'$ $(i, j \in [1, n])$.

Step 4: For monotonic chains $T_i{}'$ and $T_j{}'$, which include and coordinate vertices, respectively, if $n \geq m$, then use the dichotomy to quickly divide $T_i{}'$ into two monotonic chains: $L'_{1,t}$ and $L'_{t,n}$ ($t = \frac{n}{2}$, when n was even; or $t = \frac{n}{2} + 1$, when n was odd, n is an integer, and $n > 1$), $L'_{1,t}$ and $L'_{t,n}$ are also two monotonic chains. Similarly, if $n < m$, then divide $T_j{}'$ into two monotonic chains $S'_{1,t}$ and $S'_{t,m}$ ($t = \frac{m}{2}$, when m was even; or $t = \frac{m}{2} + 1$, when m was odd, m is an integer, and $m > 1$), $S'_{1,t}$ and $S'_{t,m}$ are also two monotonic chains.

Step 5: If $n \geq m$, calculate the MER of $L'_{1,t}$, $L'_{t,n}$ and T'_j, respectively, as $R_{L_{1,t}}$, $R_{L_{t,n}}$ and R_{T_j}. Similarly, if $n < m$, calculate the MER of $S'_{1,t}$, $S'_{t,m}$ and T'_i, respectively, as $R_{S_{1,t}}$, $R_{S_{t,n}}$ and R_{T_i}.

Step 6: For $R_{L_{1,t}}$, $R_{L_{t,n}}$ and R_{T_j}, if $R_{L_{1,t}} \cap R_{T_j} = R_{L_{t,n}} \cap R_{T_j} = \phi$, then there is a non-intersection between $T_i{}'$ and $T_j{}'$; If $R_{L_{1,t}} \cap R_{T_j} \neq \phi$, and $R_{L_{t,n}} \cap R_{T_j} = \phi$, then there may be an intersection problem between the monotonic chain $L'_{1,t}$ and $T_j{}'$, and there is a non-intersection between $L'_{t,n}$ and $T_j{}'$; If $R_{L_{1,t}} \cap R_{T_j} = \phi$, and $R_{L_{t,n}} \cap R_{T_j} \neq \phi$, there may be an intersection problem between the monotonic chain $L'_{t,n}$ and $T_j{}'$, and there is a non-intersection between $L'_{1,t}$ and $T_j{}'$; If $R_{L_{1,t}} \cap R_{T_j} \neq \phi$, and $R_{L_{t,n}} \cap R_{T_j}$ $\neq \phi$, then there may be an intersection problem between the monotonic chain $L'_{1,t}$ and $T_j{}'$, and there may be an intersection problem between the monotonic chain $L'_{t,n}$ and $T_j{}'$. Using the same method, we can calculate whether there are intersection problems between $R_{S_{1,t}}$, $R_{S_{t,n}}$, and R_{T_i}.

Step 7: Process all of the sequential monotonic chains $T_1', T_2', \ldots, T_i', \ldots, T_j', \ldots, T_n'$ $(i, j \in [1, n])$ using step 4, step 5, and step 6, until all the intersection problems of the monotonic chains are found.

Step 8: After processing by step 1 to step 7, all the intersection problems of the monotonic chains are found. In this step, we take an example to show how the proposed method deals with these intersection problems.

For one curve T, which is processed by the D–P algorithm as shown in Figure 3a, T includes 25 coordinate vertices. Using step 2 and step 3, we can obtain three monotonic chains T_1', T_2' and T_3' (as shown in Figure 3b); T_1' contains six coordinate vertices $(P_1, \ldots, P_6)$, and P_1 and P_6 are the end vertices of T_1'; T_2' contains nine coordinate vertices $(P_6, \ldots, P_{14})$, and P_6 and P_{14} are the end vertices of T_2'; T_3' contains 12 coordinate vertices $(P_{14}, \ldots, P_{25})$, and P_{14} and P_{25} are the end vertices of T_3'. After using step 4, step 5, step 6 and step 7, there is one intersection problem between T_1' and T_3', and there is another intersection problem between T_2' and T_3'.

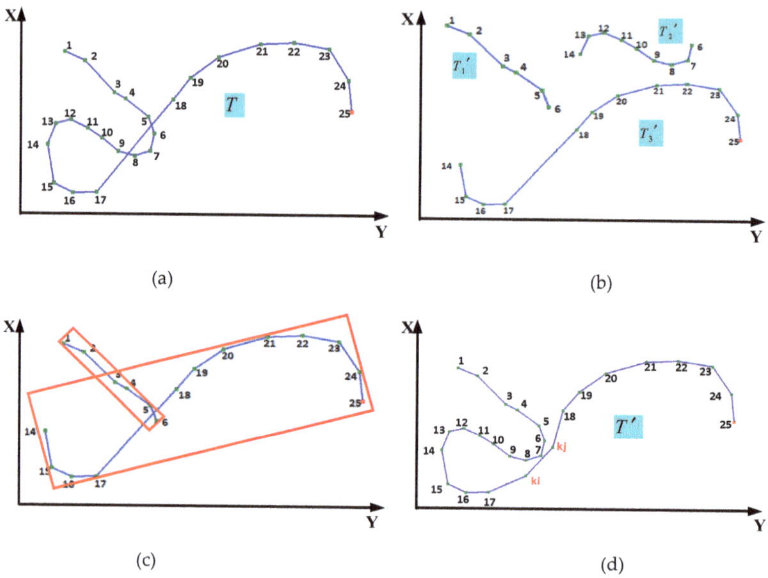

Figure 3. (**a**) The curve T which processed by D–P algorithm; (**b**) three monotonic chains T_1', T_2' and T_3' processed by the monotonic chain technology; (**c**) minimum-area enclosing rectangle (MER) of T_1' and T_3'; (**d**) the final result T'.

Using Figure 3c as an example, after processing by step 6, assuming that there is one intersection problem between T_1' and T_3', to obtain the intersection line segment $K_{5,6}$ and $K_{17,18}$ by the geometric calculation method [12,25] and obtain the coordinate vertices P_5, P_6, and P_{17}, P_{18}. If there are coordinate vertices $\left(v_p, v_{p+1}, \ldots v_i, \ldots, v_q, i \in [p, q]\right)$ (p, q are two integers) between P_{17} and P_{18} that belong to the original curve M, then calculate the shortest distance between the vertices $\left(v_p, v_{p+1}, \ldots v_i, \ldots, v_q, i \in [p, q]\right)$ and the line segment $K_{17,18}$ and find the maximum value ($D_{\max}$) of the shortest distance and the corresponding coordinated point P_i.

Connect $P_{17}P_i$, and P_iP_{18} and obtain two new monotonic chains T_{17i}' and T_{i18}'. Calculate whether there are intersection problems between the two new monotonic chains T_{17i}', T_{i18}' and the monotonic chain T_1'. If there are no intersection problems, then conclude this algorithm; the monotonic chain T_3' will be divided into two new monotonic chains T_{17i}' and T_{i18}'. If there are other intersection errors, then re-execute step 8 and step 9 until there is no intersection error between T_1' and T_3'.

Similarly, if there are coordinate vertices between P_5 and P_6 that belong to the original curve M, re-execute steps 8 and 9 until there is no intersection error between T_1' and T_3'.

Execute steps 4 to 8 until there are no intersection errors between all the monotonic chains, and then obtain the final result T'. Figure 3d shows the final result, k_i and k_j are two coordinate vertices from the original curve M.

Step 9: After processed by step 1 to step 8, all of the intersection problems have been processed, then end the proposed method, and obtain the final results.

3. Experiments and Analysis

We select two groups of experimental data to verify the validity of the proposed algorithm. The first group of data is the road line of Jiangxi Province in China. Its total length is approximately 1.56×10^5 km, and the data volume is approximately 92,000 bytes, including approximately 5.13×10^6 vertices. The second group of data is the land use line of Dingnan County in Jiangxi Province in China. Its total length is approximately 1.41×10^4 km, and the data volume is approximately 26,000 bytes, including approximately 1.24×10^6 vertices.

3.1. Assessment

In this study, we adopted a number of different methods to simplify the two groups of data and compare the performance of the proposed method. This is due to the ST algorithm [23], which is also based on the D–P algorithm, which could solve the self-intersection problems, in this paper, we compared the performance of the proposed method with the ST algorithm and the D–P algorithm.The scale of the experimental data is 1:10,000, and the results in target proportions of the original vertices are 60% and 70%, respectively. As a result, the two groups of the data are displayed as large volumes. Thus it is difficult to show case further details, in the same experimental environment. Moreover, we chose six self-intersection problems from the two groups of data instead. The simplified results of the six self-intersection problems are shown in Figure 4.

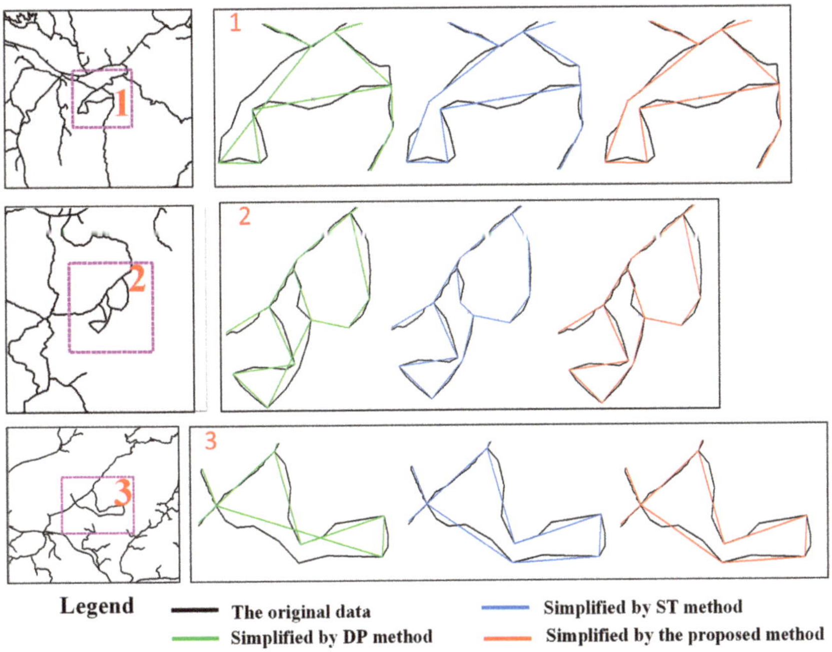

(a)

Figure 4. *Cont.*

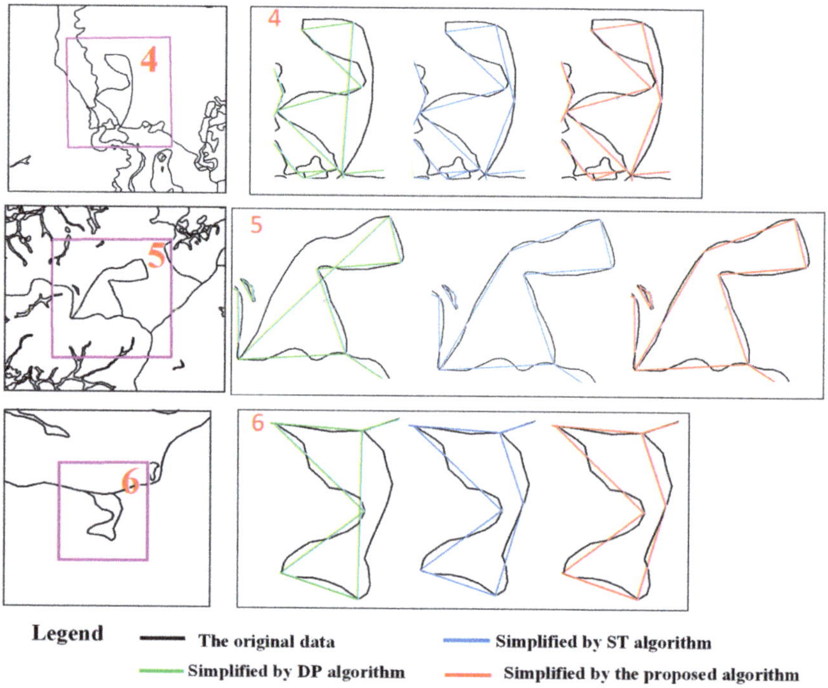

(b)

Figure 4. The six simplified results of the three applied methods from the two groups of data. (**a**) The three self-intersection problems identified from the first group of data; (**b**) the three self-intersection problems identified from the second group of data. Notes: DP algorithm is the Douglas–Peucker algorithm proposed by Douglas and Peucker [1]; ST algorithm is the star-shaped algorithm proposed by Wu and Marquez [23].

As is shown in Figure 4, the simplified results brought up from each group of data, the D–P algorithm produced self-intersection problems, but the proposed method could process self-intersection problems as well as the ST algorithm. To compare the performance of the different methods, four metrics are selected, including time consumption, mean vector displacement [3,26], Hausdorff distance (HD) [27], and standardized measure of displacement (SMD) [28].

Time consumption indicates how much time the algorithm takes.

Mean vector displacement is computed as the average displacement of the vector between the original vertices and the simplified version of the same vertices.

The Hausdorff distance (HD) between the two geometric objects is the largest minimum distance between points on one object to the other [27].

Standardized measure of displacement (SMD) is defined by Joao [28], and the calculation formula is demonstrated as follows:

$$SMD(\%) = (1 - (W - O)/W) \times 100 \quad (1)$$

W is the distance from the coordinate vertices which has the maximum displacement between the original polyline and the simplified polyline to the straight line. This is obtained by connecting the first and last nodes of the polyline, and O is the actual displacement of the coordinate vertices between the original polyline and the simplified polyline.

3.2. Results

The simplification assessment metrics of the data processed data by method are shown as follows: the resulting statistics are computed using the two groups of experimental datasets.

(1) Time consumption: the time consumption results of the three line simplification methods are shown in Figure 5, and the time consumption is measured in milliseconds (ms).

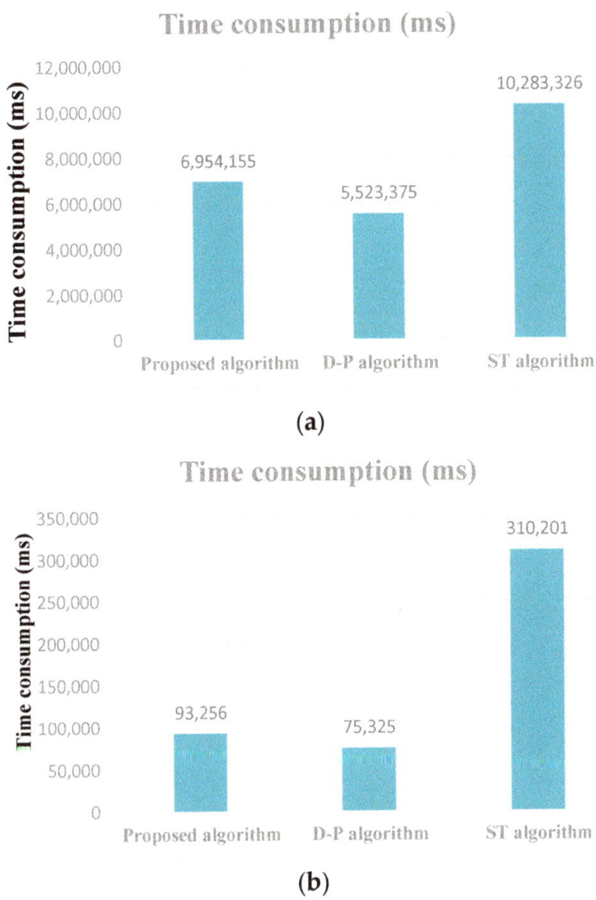

Figure 5. Time-consumption results. (**a**) Time consumption of the first group of data; (**b**) time consumption of the second group of data.

(2) Mean vector displacement: the mean vector displacement results of the three line simplification methods are shown in Figure 6. The mean vector displacement is measured in meters (m).

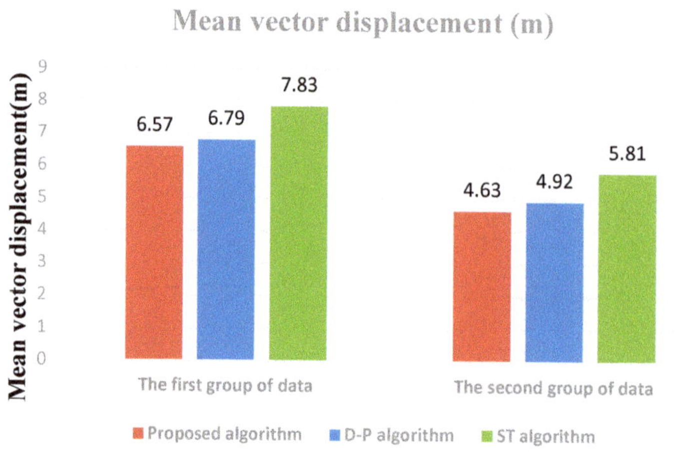

Figure 6. The mean vector displacement results (m).

(3) Hausdorff distance (HD): the Hausdorff distance (HD) results of the three line simplification methods are shown in Figure 7. The HD is measured in meters (m).

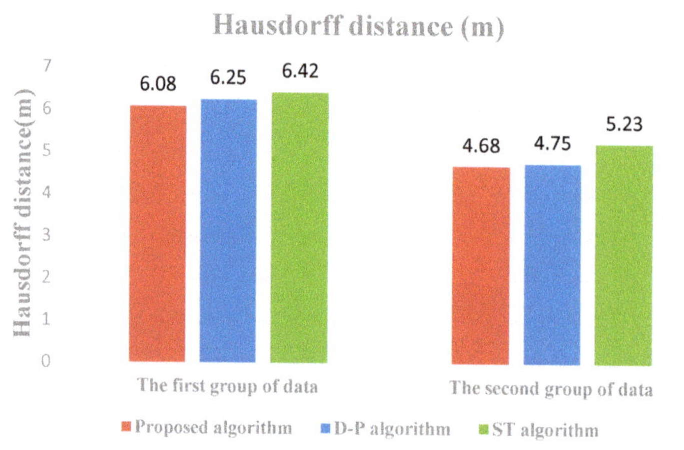

Figure 7. The Hausdorff distance results (m).

(4) Standardized measure of displacement (SMD): the standardized measure of displacement (SMD) results of the three line simplification methods are shown in Figure 8.

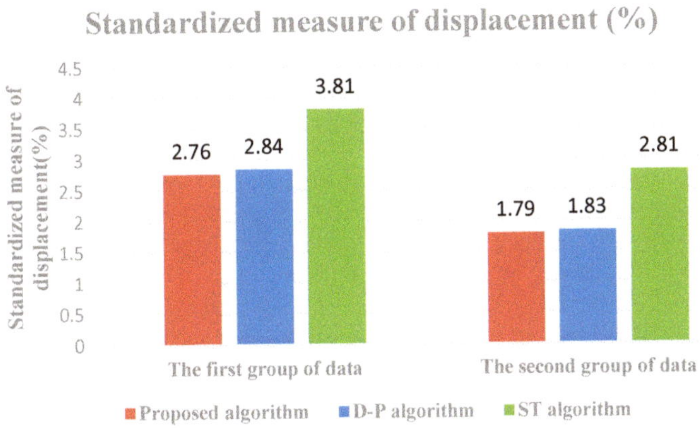

Figure 8. The standardized measure of displacement results (%).

3.3. Analysis

From Figure 5 to Figure 8, we observe the following:

(1) The proposed method can be used effectively for vector line simplification. We used two groups of data to verify the proposed method. It can be shown from the experimental results that the proposed method can not only solve the problem of self-intersection caused by the D–P algorithm but also has a high execution efficiency. Figure 4 shows the six results of simplifying the two groups of data using the three methods. In the six self-intersection problems as shown in the Figure 4, the polylines of six regions demonstrate complex curves with a number of hierarchical bends. As shown in Figure 4, the D–P algorithm produces self-intersection problems, but the proposed method and ST algorithm avoid these problems. This is due to the ST algorithm also being based on the D–P algorithm. As a result, the three methods identified the same experimental results in some cases.

(2) The D–P algorithm was found to have $O(nm)$ the worse case time and $O(n \log n)$ expected time, where n was the number of input vertices and m was the number of simplified polyline segments [23]; The ST algorithm was composed of mainly three steps; the worst case $O(nm)$ time complexity, where n was the number of input vertices and m depended on the number of star-shaped regions [23]. The proposed method in this paper involved two key steps: first, we first used the D–P algorithm to simplify the original curves and checked the self-intersection problems; then, we used the monotonic chain and dichotomy methods to address the self-intersection values. In the first step, our algorithm had the same execution time as the D–P algorithm. In the second step, our algorithm was carried out in $O(m \log m)$ time, while m was the number of self-intersection segments of simplified polylines.

The time consumption results of the three methods for processing the two groups of data are shown in Figure 5. For the first group of data, the time consumption results of the D–P algorithm, the proposed method and ST algorithm are 5,523,375 ms, 6,954,155 ms, and 10,283,326 ms, respectively. For the second group of data, the time consumption results of the D–P algorithm, the proposed method and ST algorithm are 75,325 ms, 93,256 ms, and 310,201 ms, respectively. It can be seen from the experimental results of each group of data that the time consumption of the proposed method is slightly higher than the D–P algorithm with the proposed method and, after the procession of the D–P algorithm, we use monotonic chains and dichotomy to modify the self-intersection problems. It is obvious that the time consumption of the proposed method is much lower than the ST algorithm. This is because the monotonic chains and dichotomy have high search efficiency and can quickly find and solve the self-intersection problems that are processed by the D–P algorithm.

(3) We use mean vector displacement to measure the location accuracy. As shown in Figure 6, the first group of data, the mean vector displacement results of the D–P algorithm, the proposed

method and ST algorithm are 6.79 m, 6.57 m, and 7.83 m, respectively, The second group of data showed that the mean vector displacement results of the D–P algorithm, the proposed method and ST algorithm are 4.92 m, 4.63 m, and 5.81 m, respectively. For each group of data, the mean vector displacement of the proposed method is similar to the D–P algorithm but much lower than the ST algorithm.

(4) Figure 7 shows the Hausdorff distance of the three methods for processing the two groups of data. For the first group of data, the Hausdorff distances of the D–P algorithm, the proposed method and ST algorithm are 6.25 m, 6.08 m, and 6.85, respectively. The second group of data showed, the Hausdorff distances of the D–P algorithm, the proposed method and ST algorithm are 4.75 m, 4.68 m, and 5.23 m, respectively. For each group of the data, the Hausdorff distance of the proposed method is similar to the D–P algorithm and the ST algorithm.

(5) We also used a standardized measure of displacement (SMD) to measure the location accuracy. As shown in Figure 8, the first group of data, the SMDs of the D–P algorithm, the proposed method, and ST algorithm are 3.46%, 3.58%, and 4.25%, respectively, The second group of data showed that the SMDs of the D–P algorithm, the proposed method and ST algorithm are 1.83%, 1.79%, and 2.81%, respectively. For each group of data, the mean vector displacement of the proposed method is similar to the D–P algorithm but much lower than the ST algorithm.

4. Conclusions

Vector line simplification is widely used in computer graphics, GIS, and others. The D–P algorithm is one of the most widely used methods for vector line simplification. When professionals use the D–P algorithm to address complex curves, we find it is easy to produce self-intersection problems. To further expand the application of the D–P algorithm, in this paper a new line simplification algorithm that combines the D–P algorithm, monotonic chains, and dichotomy is proposed. In the end, two experiments are designed to compare the results of our proposed method with the D–P algorithm and ST algorithm. From the result analysis, it is clear that the proposed algorithm has several advantages: (1) compared with the D–P algorithm, the proposed algorithm has the same execution efficiency but without self-intersection problems; (2) compared with the ST algorithm, the proposed method has the same ability to solve self-intersection problems but has better execution efficiency. At the same time, the proposed algorithm also has shortcomings to be further studied: (1) the proposed method focuses on the removal of self-intersection problems, however, the area preservation problems after the polyline simplification are not considered; (2) similar to the D–P algorithm, this proposed method does not consider the bending characteristics of the curves. In conclusion, these two thematic shortcomings will be the focus of our future research.

Author Contributions: Methodology, Bo Liu, Xuechao Liu, and Dajun Li; Software, Yu Shi; Visualization, Xuechao Liu and Bo Liu; Writing–original draft, Bo Liu, Xuechao Liu and Dajun Li; Writing–review and editing, Gabriela Fernandez and Yandong Wang. All authors have read and agreed to the published version of the manuscript.

Funding: This work was funded by the National Natural Science Foundation of China (Nos. 41201395, 416014160), the National Key Research and Development Program of China (No. 2016YFB0501403), the Key Laboratory of Earth Observation and Geospatial Information Science of NASG (201811), and the China Scholarship Council Foundation of China (No. 201808360267).

Conflicts of Interest: The authors declare there is no conflicts of interest regarding the publication of this paper.

References

1. Douglas, D.H.; Peucker, T.K. Algorithms for the reduction of the number of points required to represent a digitized line or its caricature. *Can. Cartogr.* **1973**, *10*, 112–122. [CrossRef]
2. Ramer, U. An iterative procedure for the polygonal approximation of plane curves. *Comput. Graph. Image Process.* **1972**, *1*, 244–256. [CrossRef]
3. Lang, T. Rules for robot draughtsmen. *Geogr. Mag.* **1969**, *42*, 50–51.
4. McMaster, R.B. The integration of simplification and smoothing algorithms in line generalization. *Can. Cartogr.* **1989**, *26*, 101–121. [CrossRef]

5. Li, Z.L. An Algorithm for Compressing Digital Contour Data. *Cartogr. J.* **1988**, *25*, 143–146. [CrossRef]
6. Visvalingam, M.; Whyatt, J. *Line generalisation by repeated elimination of the smallest area. Technical Report, Discussion Paper 10, Cartographic Information Systems Research Group (CISRG)*; The University of Hull: Hull, UK, 1992.
7. Ratschek, H.; Rokne, J.; Leriger, M. Robustness in GIS algorithm implementation with application to line simplification. *Int. J. Geogr. Inf. Sci.* **2001**, *15*, 707–720. [CrossRef]
8. Wang, Z.S.; Muller, J.-C. Line Generalization Based on Analysis of Shape Characteristics. *Cartogr. Geogr. Inf. Syst.* **1998**, *25*, 3–15. [CrossRef]
9. Zhao, Z.; Saalfeld, A. Linear-Time Sleeve-Fitting Polyline simplification algorithms. In Proceedings of the AutoCarto 13, Seattle, WA, USA, 7–10 April 1997; Published by American Congress on Surveying and Mapping & American Society for Photogrammetry and Remote Sensing, Maryland. pp. 214–223, ISBN -1-57083-043-6.
10. Gary, R.H.; Wilson, A.D.; Archuleta, C.M.; Thompson, F.E.; Vrabel, J. *Production of a National 1:1000000-Scale Hydrography Dataset for the United States: Feature selection, Simplification, and Refinement*; U.S. Geological Survey Scientific Investigations Report 2009–5202. Revised May 2010; U.S. Geological Survey: Reston, VA, USA, 2010; 22p. [CrossRef]
11. Li, Z.L.; Openshaw, S. Algorithms for automated line generalization based on a natural principle of objective generalization. *Int. J. Geogr. Inf. Sci.* **1992**, *6*, 373–389. [CrossRef]
12. Samsonov Timofey, E.; Yakimova, O.P. Shape adaptive geometric simplification of heterogeneous line datasets. *Int. J. Geogr. Inf. Sci.* **2017**, *31*, 1485–1520. [CrossRef]
13. de Berg, M.; van Kreveld, M.; Overmars, M.; Overmars, M.; Schwarzkopf, O. *Computational Geometry: Algorithms and Applications*, 2nd ed.; Springer: Berlin, Germany, 2000.
14. Cromley, R.G. Principal axis line simplification. *Comput. Geosci.* **1992**, *18*, 1003–1011. [CrossRef]
15. Raposo, P. Scale-specific automated line simplification by vertex clustering on a hexagonal tessellation. *Cartogr. Geogr. Inf. Syst.* **2013**, *40*, 427–443. [CrossRef]
16. Kronenfeld, B.J.; Stanislawski, L.V.; Buttenfield, B.P.; Tyler, B. Simplification of polylines by segment collapse: Minimizing areal displacement while preserving area. *Int. J. Cartogr.* **2020**, *6*, 22–46. [CrossRef]
17. Shi, W.Z.; Cheung, C.K. Performance Evaluation of Line Simplification Algorithms for Vector Generalization. *Cartogr. J.* **2006**, *43*, 27–44. [CrossRef]
18. Mi, X.J.; Sheng, G.M.; Zhang, J.; Bai, H.X.; Hou, W. A new algorithm of vector date compression based on the tolerance of area error in GIS. *Sci. Geogr. Sin.* **2012**, *32*, 1236–1240.
19. Saalfeld, A. Topologically consistent line simplification with the Douglas-Peucker algorithm. *Cartogr. Geogr. Inf. Sci.* **1999**, *26*, 7–18. [CrossRef]
20. Ho, P.S.; Kim, M.H. A hierarchical scheme for representing curves without self-intersections. In Proceedings of the 2001 IEEE Computer Society Conference (CVPR 2001), Kauai, HI, USA, 8–14 December 2001. [CrossRef]
21. Mantler, A.; Snoeyink, J. Safe sets for line simplification. In *10th Annual Fall workshop on Computational Geometry*; Stony Brok University: New York, NY, USA, 2000; Available online: http://citeseerx.ist.psu.edu/viewdoc/summary?doi.10.1.1.32.402 (accessed on 29 March 2020).
22. Avelar, S.; Müller, M. Generating topologically correct schematic maps. In *Proceedings of the 9th International Symposium on Spatial Data Handling*; Technical Report; Swiss Federal Institute of Technolog Zurich: Zurich, Switzerland, 2000; pp. 4–28. [CrossRef]
23. Wu, S.T.; Marquez, M.R.G. A non-self-intersection Douglas-Peucker algorithm. In Proceedings of the 16th Brazilian Symosium on Computer Graphics and Image Processing (SIBGRAPI), Sao Carlos, Brazil, 12–15 October 2003. [CrossRef]
24. Ebisch, K. Short note: A correction to the Douglas-Peucker line generalization. *Comput. Geosci.* **2002**, *28*, 995–997. [CrossRef]
25. Yan, H.W.; Wang, M.X.; Wang, Z.H. *Computational Geometry: Spatial Data Processing Algorithm*; Science Press: Beijing, China, 2012.
26. White, E.R. Assessment of line-generalization algorithms using characteristic points. *Cartogr. Geogr. Inf. Sci.* **1985**, *12*, 17–28. [CrossRef]

27. Hangouët, J.F. Computation of the Hausdorff distance between plane vector polylines. In *Auto-Carto XII: Proceedings of the International Symposium on Computer-Assisted Cartography, Charlotte, North Carolina*; American Congress on Surveying and Mapping & American Society for Photogrammetry and Remote Sensing: Gaithersburg, MD, USA, 1995; Volume 4, pp. 1–10. ISBN-1-57083-019-3.
28. Joao, E.M. *Gauses and Consequences of Map Generalization*; Taylor and Francis: London, UK, 1998.

International Journal of *Geo-Information*

MDPI

Article

A Multi-Scale Water Extraction Convolutional Neural Network (MWEN) Method for GaoFen-1 Remote Sensing Images

Hongxiang Guo [1,2], Guojin He [1,3,4,*,†], Wei Jiang [5,6,†], Ranyu Yin [1,2], Lei Yan [1,2] and Wanchun Leng [1,2]

1 Aerospace Information Research Institute, Chinese Academy of Sciences, Beijing 100094, China; guohx@radi.ac.cn (H.G.); yinry@radi.ac.cn (R.Y.); yanlei@aircas.ac.cn (L.Y.); lengwch@radi.ac.cn (W.L.)
2 College of Resource and Environment, University of Chinese Academy of Sciences, Beijing 100049, China
3 Satellite Remote Sensing Technology Department, Key Laboratory of Earth Observation Hainan Province, Sanya 572029, Hainan, China
4 Satellite Remote Sensing Technology Department, Sanya Institute of Remote Sensing, Sanya 572029, Hainan, China
5 State Key Laboratory of Simulation and Regulation of Water Cycle in River Basin, China Institute of Water Resources and Hydropower Research, Beijing,100038, China; jiangwei@iwhr.com
6 Remote Sensing Technology Application Center, Research Center of Flood and Drought Disaster Reduction of the Ministry of Water Resources, Beijing,100038, China
* Correspondence: hegj@radi.ac.cn
† These authors contributed equally to this work.

Received: 12 January 2020; Accepted: 22 March 2020; Published: 25 March 2020

Abstract: Automatic water body extraction method is important for monitoring floods, droughts, and water resources. In this study, a new semantic segmentation convolutional neural network named the multi-scale water extraction convolutional neural network (MWEN) is proposed to automatically extract water bodies from GaoFen-1 (GF-1) remote sensing images. Three convolutional neural networks for semantic segmentation (fully convolutional network (FCN), Unet, and Deeplab V3+) are employed to compare with the water bodies extraction performance of MWEN. Visual comparison and five evaluation metrics are used to evaluate the performance of these convolutional neural networks (CNNs). The results show the following. (1) The results of water body extraction in multiple scenes using the MWEN are better than those of the other comparison methods based on the indicators. (2) The MWEN method has the capability to accurately extract various types of water bodies, such as urban water bodies, open ponds, and plateau lakes. (3) By fusing features extracted at different scales, the MWEN has the capability to extract water bodies with different sizes and suppress noise, such as building shadows and highways. Therefore, MWEN is a robust water extraction algorithm for GaoFen-1 satellite images and has the potential to conduct water body mapping with multisource high-resolution satellite remote sensing data.

Keywords: convolutional neural network; water body extraction; GaoFen-1; multiple scales; deep learning

1. Introduction

Water is the basic substance for human society's production and development [1]. Surface water bodies play important roles in Earth's material and energy cycles [2,3]. Since satellite remote sensing data can capture large-scale surface information in little time and with low costs, the data have been used in water body surveys [4]. Multiple remote sensing data, including optical data [5] and radar data [6], have been used for water body information extraction. The current water information extraction methods include the threshold method [7], machine learning [8,9], and deep learning [10,11], etc. The threshold method is a conventional method for water body extraction. The threshold method selects an appropriate threshold to distinguish water bodies and other objects in one or more bands [7]. Because the spectral characteristics of water in the near-infrared (NIR) band are significantly different from those of other objects, the NIR band is very popular in threshold segmentation [12]. To further highlight the difference between water bodies and surrounding features, water indexes have been developed [13]. However, the water index method has some problems. One is that objects with similar spectral characteristics, such as mountain shadows, cloud shadows, and highways, can be easily confused with water bodies, which makes it difficult to select thresholds. In addition, the threshold selected in large-scale water extraction may not be applicable to local areas [14]. With the development of machine learning, traditional machine learning algorithms, such as decision tree (DT) [15], support vector machine (SVM) [6], and random forest (RF) [9], have been widely used in water body extraction. These algorithms perform classification by using artificially designed features, including spectral and textural features. However, artificially designed features require considerable professional domain knowledge and artificially designed features are usually based on a specific scale of images. A standard way to extract artificially designed features from images at multiple scales is resampling the images to different scales and extract features based on the images with different scales. Thus, the process requires intensive computation with time consuming. In addition, different feature vectors are needed for different images and the feature vectors have great impacts on the final classification results. These issues make applying machine learning to water extraction challenging.

Deep learning is a popular method in image processing during the past several years [16,17]. Convolutional neural networks (CNNs) have been used in scene classification [18], semantic segmentation [19], and object detection [20,21]. The advantage of CNNs is to capture the features from raw images directly by multiple convolutional layers [22], which can avoid the complex feature processing. CNNs for semantic segmentation are capable of performing image classification at pixel level, which is important for information extraction from remote sensing images. In CNNs, the shallow convolutional layers are able to capture the pixel position information and the deep convolutional layers are used to label the pixels [22]. The fully convolutional network (FCN) is the first end-to-end CNN designed for semantic segmentation [19]. FCN extracts abstract features from the input image and labels each pixel in the feature maps extracted by the last convolutional layer. However, FCN loses information contained in low-level features extracted by shallow convolutional layers. In recent years, many models, such as Unet [23] and Deeplab V3+ [24], have been developed to improve the performance of CNNs for semantic segmentation in the field of computer vision. CNNs are gradually being applied to water information extraction with remote sensing images. In [10], CNN was firstly used for water body extraction in Landsat ETM+ images. The structure of the CNN contained only two convolutional layers and a fully connected layer. The shallow structure allows it to capture only low-level features which results in poor robustness in complex scenes. In addition, the input tile (19 $\times$ 19) is small in the CNN model. Thus, it cannot be used to extract features at large scales. With the improvement of the spatial resolution of satellite images [25], various methods based on deep learning have been proposed for water body extraction in high-resolution images. A CNN method that combines the super pixel was proposed by Chen, Y, et al. [11]. The core idea is to combine artificial designed features and CNN extraction features. However, the process reduces the fluidity of the water extraction and misses some useful information during forward propagation. In recent years, end-to-end CNNs, such as fully convolutional network (FCN) [26] and DeepWaterMap [27] have been applied to water body extraction. These end-to-end CNNs greatly improved the accuracy and efficiency of water body extraction. There are still challenges in the application of CNNs in water body extraction: (1) In the process of forward propagation, the resolution of feature maps is reduced due to the repeated max-pooling layers, which leads to the loss of detailed water body information. (2) The receptive fields of pixels are different in the feature maps extracted by the convolutional layers at different depths, which allows these feature maps to contain feature information at different scales [22].

The combination of the features extracted at multiple scales in water body extraction still needs to be explored.

This paper aims to propose an improved convolutional neural network (CNN), named multi-scale water extraction convolutional neural network (MWEN), for water body extraction for GaoFen-1 images. For the first challenge, the encoder-decoder structure is used in the MWEN inspired by the Unet [23]. The encoder extracts the features from the input images and obtains feature maps with low resolution. The role of the decoder is to map the feature maps to the input resolution feature maps. For the second challenge, a structure, named the multi-scale feature extractor (MTFE), is proposed to capture features at multiple scales. Objects exist at various scales in remote sensing images and geological correlations may exist between adjacent objects. Features extracted by CNNs at different scales contain various information [28]. In the MTFE, four dilated convolutional layers with different dilation rates are used to learn features from images with different receptive fields.

The structure of the remainder of this article is as follows. First, GaoFen-1 high-resolution remote sensing satellite images in Beijing-Tianjin-Hebei region, Zhejiang province, and Tibet province in China are collected for the dataset and preprocessed. Then, four CNNs are employed to extract water body information. Finally, the accuracies of these algorithms are compared based on five accuracy metrics and a visual comparison.

2. Materials and Methods

2.1. Data

In this study, 24 GaoFen-1 images (17 for training and 7 for testing) located in Beijing-Tianjin-Hebei region, Zhejiang province, and Tibet province in China were collected as the experiment dataset and these images are showed in Figure 1. Four multispectral bands with a spatial resolution of 8 m and panchromatic band with a spatial resolution of 2 m are included in GaoFen-1 images. The radiation resolution of both the panchromatic band and multispectral bands is 16 bits. The spectral and textural characteristics of the water bodies in different regions are quite different, and the environments surrounding the water bodies are complex. To test the universality of these CNNs for water body extraction, environment characteristics, such as spectral, textural, season, water environment characteristics and confusing areas, such as shadows, highways, and ice are considered in the dataset. The detail information of the dataset is shown in Table 1.

Table 1. Detailed information of dataset.

Images	Location	Acquisition Times	Water Types	Major Confusing Objects
a1-a8	Tibet province	July, 2014 and August, 2016	Plateau lake, Plateau river, Saline lake	Cloud shadows, Saline land
b1-b7	Beijing-Tianjin-Hebei region	January, September and October, 2019	Agricultural water, town water, city water	Building shadows, sports field, highways.
c1-c9	Zhejiang province	April, 2017 and October, 2019	Agricultural water, town water, woodland water, city water	Mountain shadows, wetland, roads

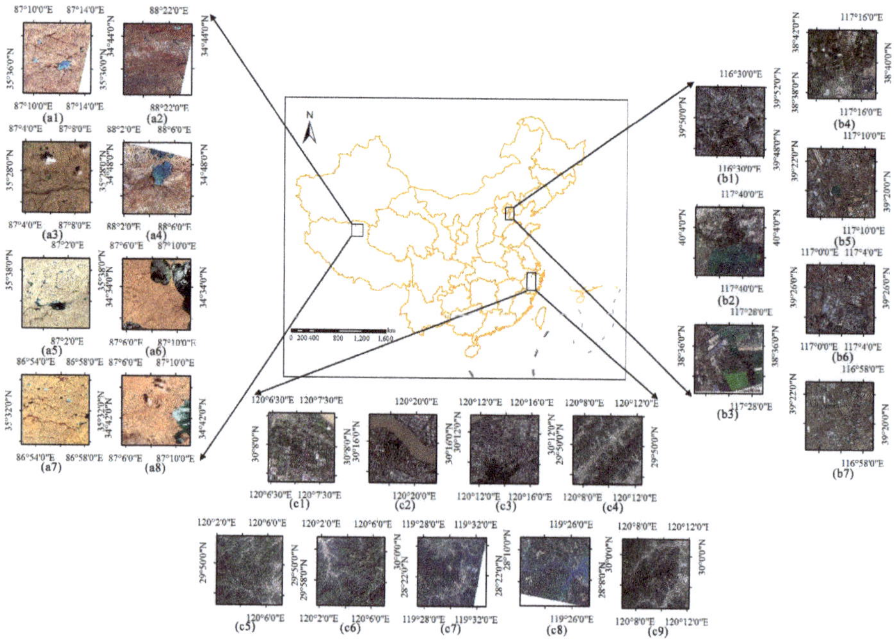

Figure 1. The GaoFen-1 (GF-1) dataset (a1, a3, a5, a6, a7, a8, b1, b2, b5, b6, b7, c1, c2, c3, c4, c5, and c6 are used for training images. a2, a4, b1, b3, b4, c7, and c8 are used for test images.).

2.2. Methods

The methods can be divided into four parts: image preprocessing, sample generation, water information extraction, and accuracy assessment. In the image preprocessing part, the Rational Polynomial Coefficient (RPC) model is used to geometrically correct these images [29]. Then, the multispectral and panchromatic images fusion was conducted using PANSHARP method [30]. The image preprocessing part was conducted based on the PCI Geo Imaging Accelerator software. The geometric errors of the images after preprocessing were within 1 pixel. In the second part, the water bodies in the fused images are labeled. These images and labels are clipped to 512 × 512 pixels and divided into a training dataset and a validation dataset. In the third step, MWEN (multi-scale water extraction convolutional neural network), MWEN "without MTFE", FCN, Unet, and Deeplab V3+ are employed to extract the water bodies. Finally, the accuracy comparison for different methods are conducted using visual comparison and quantitative evaluation metrics. The flowchart is shown in Figure 2.

2.2.1. Sample Generation

The labels in the dataset are from the fusion images and cover all water types mentioned in Section 2.1. The labels consist of water areas and background areas. All the labels in the dataset are binary images, where 1 represents water body and 0 represents background. All of the images were labeled via visual interpretation. These images were divided into training images and test images (17 for training and 7 for test). Both the training images and test images contain all water types mentioned in Table 1. These training images and training labels were clipped to samples with 512 × 512 pixels. A training sample library containing 13,509 samples from training images was obtained. The samples in the training sample library contains all water pixels in training images. Some areas without surface water bodies are also contained in these samples. The training sample library was divided into two parts. Ninety percent of the training samples were used as the training dataset and the remaining small part was used for the validation dataset. The role of the validation dataset is to reflect the generalization ability of the model parameters and indicate whether the model is overfitting during training process. Both the validation dataset and training dataset were from the training images, which reduced the generalized representation of the validation dataset. To get a more generalized training model, the samples from the images other than the training image are needed for the validation dataset. In this study, a random part of each image in the test images was selected and clipped to 512 × 512 pixels to enrich the validation dataset. The final validation dataset consisted of 1651 samples from test images and 1350 samples from the training images.

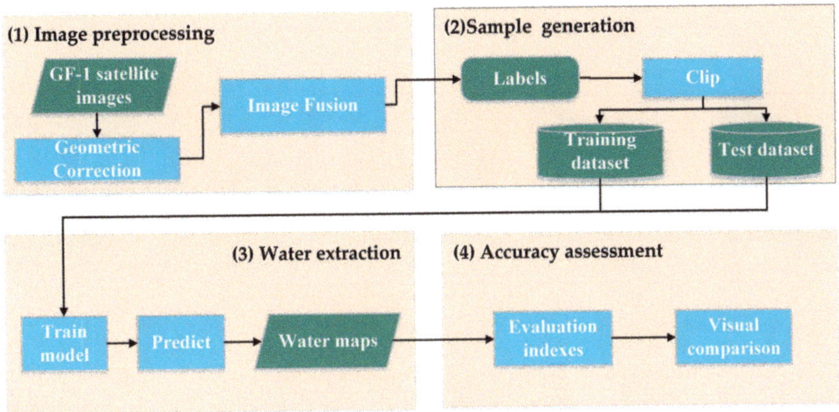

Figure 2. Flowchart of this study.

2.2.2. Multi-Scale Feature Extractor

Dilated convolution was originally used for the wavelet transform [31] and has been used in convolutional neural networks for semantic segmentation [32]. The convolution kernel with holes (or gaps) is used in the dilated convolution. The number of gaps inserted in the kernel depends on the dilation rate r. The dilation rate is prerequisite when a convolution kernel is defined. The dilated convolution with filter dilation rates of 0, 1, and 2 are shown in Figure 3. The kernel with a dilation rate of 0 is the same as the standard convolution kernel. The convolution kernels with different dilation rates have different receptive fields. The combination of dilated convolutions with different dilation rate kernels can capture the features at different scales.

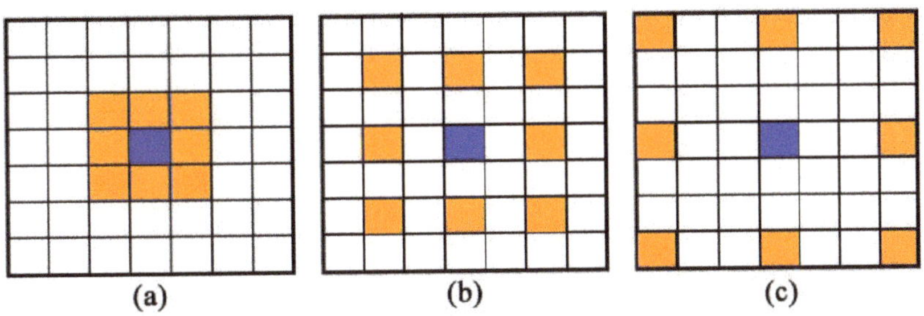

Figure 3. Dilated convolution kernels with different rates. (**a**), (**b**), (**c**) are dilated convolution kernels with dilation rate 0, 1, 2, respectively.

In remote sensing images, the sizes of water bodies are diverse and there are many confusing objects in high-resolution images, such as building shadows, mountain shadows, and sports fields, whose spectral characteristics are similar to those of water body. The combination of features extracted at multiple scales is important in dealing with these issues. In this study, a structure, named multi-scale feature extractor (MTFE) is proposed. Dilated convolutions with various rates are used in the MTFE to extract the features at multiple scales. The structure of the MTFE is given in Figure 5. An example of feature extraction at multiple scales by dilated convolution with different rates is shown in Figure 4. As we can see in Figure 4b, the standard convolution (dilated convolution with a rate of 0) can only get the information of the surrounding 9 pixels, all of which lie in building shadows. It is difficult to identify the pixel at the center of the convolution kernel because shadows and water bodies have similar spectral characteristics. In the dilated convolutions with rates of 2, 4, and 8, the features are extracted at different scales and the information of the buildings and woods is captured. The combination of extracted features at these different scales is important for the distinction of building shadows.

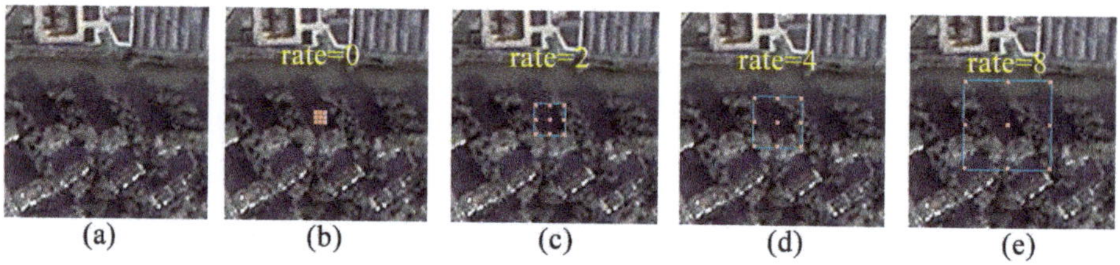

Figure 4. Examples of dilated convolution with different rates. (**a**) is the sample image; (**b–e**) are examples of dilated convolution with dilation rate 0, 2, 4, 8, respectively.

2.2.3. Convolutional Neural Networks (CNNs) for Water Extraction

A multi-scale water extraction convolutional neural network (MWEN) for surface water information extraction is proposed. The structure of the MWEN is shown in Figure 5. The MWEN can be divided into three parts: encoder, multi-scale feature extractor (MTFE), and decoder. In the first part, the input data are encoded by the encoder and feature maps with an output stride of 16 are obtained. In the multi-scale feature extractor (MTFE) part, the feature maps from the encoder are fed to four dilated convolutions with different rates. These dilated convolutions with different rates can learn features at different scales. Then, the feature maps generated by these dilated convolutions are concatenated and integrated by three convolutional layers. In the decoding part, the feature maps are decoded by the decoder to obtain the water segmented images.

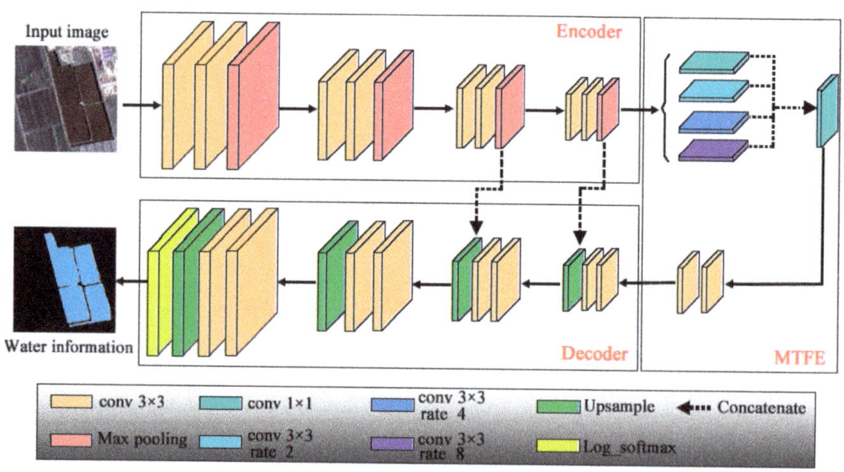

Figure 5. The structure of the multi-scale water extraction convolutional neural network (MWEN).

To examine the importance of MTFE to the segmentation results, both of the MWEN structure "with MTFE" and "without MTFE" were trained for water body extraction. The other three kinds of convolutional neural networks (CNNs) used for semantic segmentation, the FCN [33], Unet [23], and DeepLab V3+ [24], were also selected in this study for comparison. The water body extraction process using CNNs contains three steps: data augmentation, forward propagation, and model training.

- Data augmentation: Date augmentation is performed before training. In this step, the input samples are randomly processed in three ways, including flipping, zooming, and panning. All samples in the training dataset are randomly processed before every training epoch, and the number of training samples for every training epoch does not change. The data augmentation results for the three samples are shown in Figure 6.

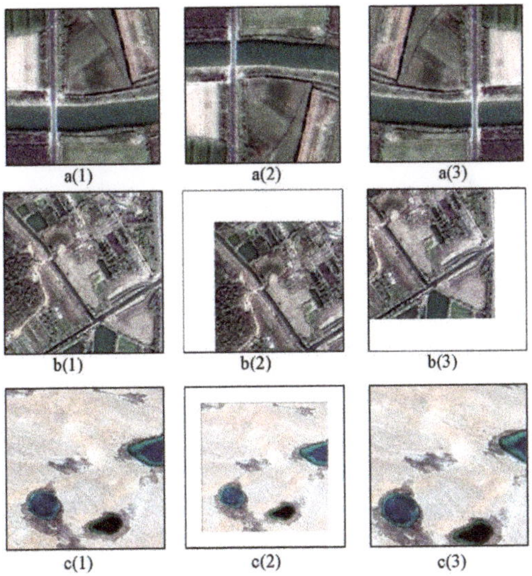

Figure 6. Data augmentation of the three samples. a(2) and a(3) are the results of flipping a(1), b(2) and b(3) are the results of panning b(1), and c(2) and c(3) are the results of zooming c(1).

Then, the data are normalized. The fused GF-1 data have a radiation resolution of 16 bits, with DN values ranging from 0 to 65535. To improve the accuracy and training efficiency of convolutional neural networks (CNNs), the input images are normalized. The normalization converses each input image into a feature map with a mean of 0 and a variance of 1. The formulas are as follows:

$$\mu = \frac{1}{w \times h \times c} \sum_{i=1}^{w} \sum_{j=1}^{h} \sum_{z=1}^{c} DN_{i,j,z} \tag{1}$$

$$\sigma^2 = \frac{1}{w \times h \times c} \sum_{m=1}^{w} \sum_{n=1}^{h} \sum_{z=1}^{c} (DN_{m,n,z} - \mu)^2 \tag{2}$$

$$\overline{DN_{m,n,z}} = \frac{DN_{m,n,z} - \mu}{\sqrt{\sigma^2}} \tag{3}$$

where μ is the average of the input image array, and w, h, and c are the width, height, and the number of channels of the input image, respectively. $DN_{m,n,z}$ is the DN value of the pixel in row n, column m, and channel z. σ^2 is the variance of the input image. $\overline{DN_{m,n,z}}$ is the DN value of the pixel in row n, column m, and channel z after normalization.

- Forward propagation: The normalized sample is fed into the CNN and a feature map is obtained after forward propagation. The output of the CNN is a feature map with a size of 512 × 512 × channels (where the channels are the number of classes). In this study, the number of channels is 2 (water bodies and backgrounds). Then, the feature map is activated by an activation function. The log softmax function is used as the activation function and the argmax function [34] is used to get the final water maps in this study. The formula of the activation function for each pixel in the feature maps is as follows:

$$P_{(m)} = \log(\frac{e^m}{\sum_{n=1}^{c} e^n}) \quad (4)$$

where $P_{(m)}$ is the data value of the pixel in channel m. c is the number of classes (2 in this study to reflect the water and background).

- Model training: The cross-entropy loss function [35] and the back propagation algorithm [36] are used when training the CNNs. The mean cross-entropy and the sparse categorical accuracy [37] are calculated between the labels and the predicted maps by the CNN forward propagation. To minimize the cross entropy, the Adam optimizer [38] is applied to identify the weights and biases in the back-propagation process. In this study, the weights of the CNNs model are trained on training dataset and weights with the highest parse categorical accuracies on the validation dataset are selected as the training results.

2.2.4. Accuracy Assessment

The performances of these convolutional neural networks (CNNs) are thoroughly evaluated via visual comparison and five evaluation metrics. The visual comparisons contain the comparison between MWEN "with MTFE" and "without MTFE" and the comparison between MWEN, FCN, Unet, and Deeplab V3+ on regions with different types of surface water bodies and confusing objects. Regarding the evaluation metrics, five evaluation metrics are used to evaluate the accuracy in this study, including the Overall Accuracy (OA) [30], the True Water Rate (TWR), the False Water Rate(FWR), the Water Intersection over Union (WIoU) [30], and the Mean Intersection over Union (MIoU) [39]. The definitions and formulas of these indicators are listed in Table 2.

Table 2. Five evaluation metrics for the accuracy assessment.

Evaluation Index	Definition	Formula
OA	The ratio of the correctly classified number of pixels and the total number of pixels	$OA = \frac{TP+TN}{TP+TN+FP+FN} \times 100\%$
TWR	The ratio of the number of properly classified water pixels and the number of labeled water pixels	$TWR = \frac{TP}{TP+FP} \times 100\%$
FWR	The ratio of the number of misclassified water pixels and the number of labeled water pixels	$FWR = \frac{FP}{FP+TP} \times 100\%$
WIoU	The ratio of the intersection and the union of the ground truth water and the predicted water area.	$WIoU = \frac{TP}{FN+TP+FP}$
MIoU	The average IoU for all classes (water and background)	$MIoU = \frac{1}{k+1} \sum_{i=0}^{k} \frac{TP}{FN+TP+FP}$

where TP, TN, FN, and FP represent the numbers of pixels of true water, true background, false background, and false water, respectively.

3. Results

3.1. Model Training

The training processes were conducted using Python3.6, Keras, and TensorFlow on a NVIDIA Titan GPU with cuDNN 10.0 acceleration. The categorical accuracies on the training dataset and validation dataset are calculated at the end of each training epoch. The weights with the highest categorical accuracies are used for water extraction in next steps. The highest validation accuracies of these models are shown in Table 3. The training accuracy and validation accuracy curves are shown in Figure 7. The training and validation accuracy curves of these models grow slowly after the 15th epoch and some even show downward trends after the 25th epoch. There is a large gap between the training accuracy curve and the validation accuracy curve of the Deeplab V3+. The Deeplab V3+ appeared to overfit when it is directly used in water body extraction from remote sensing images. The efficiency of training models is affected by many factors. The efficiency of the CNNs are simply compared via the number of trainable parameters and training time in this study. The efficiency comparison of these CNNs are shown in Table 4. The FCN has the most parameters but less training time. The Deeplab V3+ has the longest train time due to its complex and deep model structure. The MWEN and Unet have fewer parameters and less training time.

Table 3. The highest validation accuracy of CNN models in training process.

CNN	MWEN	MWEN without MTFE	FCN	Unet	Deeplab V3+
Highest validation accuracy	0.987	0.981	0.978	0.983	0.957

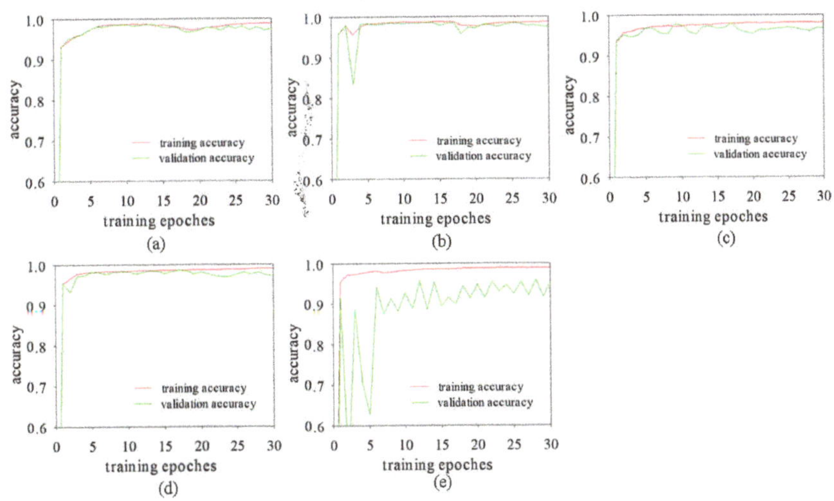

Figure 7. The training and validation accuracy curves of the convolutional neural networks (CNN) models. (**a**), (**b**), (**c**), (**d**), and (**e**) represent training and validation accuracy curves of the MWEN, MWEN "without multi-scale feature extractor (MTFE)", fully convolutional network (FCN), Unet, and Deeplab V3+, respectively.

Table 4. The efficiency comparison of the various methods.

CNN	Number of Trainable Parameters (Million)	Training Time (s/epoch)
MWEN	3.72	1343
MWEN without MTFE	1.57	1161
FCN	5.71	1345
Unet	3.11	1366
Deeplab V3+	4.11	2161

3.2. Water Extraction Results on the Test Dataset

The results of the water body extraction using these CNNs on the test images are shown in Figure 8. As can be seen from the figure, the water body prediction results of these CNNs are different. For Regions a and g, more confusing objects are contained in these two regions than the others, which makes the CNNs more prone to make mistakes. The roads and the building shadows are misclassified using Unet and Deeplab V3+ in these two regions. For Regions e and f, there are some detailed water bodies that are missed by the FCN and MWEN "without MTFE". Although performances of these CNNs are similar in Regions b, c, and d across these images, there are still differences in details. Some details are derived from these results and shown in Section 3.3. Figure 8 shows that MWEN has the capability to capture detailed water and suppresses noise better than the others.

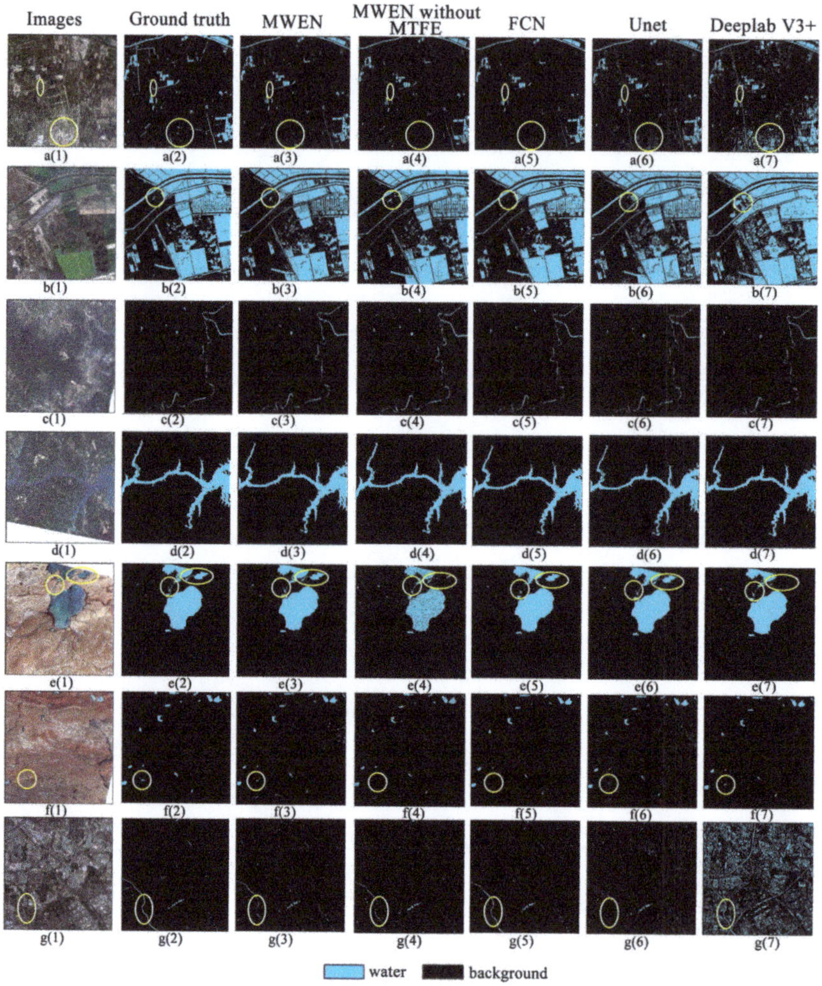

Figure 8. The results classified by the four CNNs on the test dataset. (a1–g1) are the original images, (a2-g2), (a3-g3), (a4–g4), (a5–g5), (a6–g6), (a7–g7) are the water body information extracted by artificial interpretation, MWEN, MWEN "without MTFE", FCN, Unet, Deeplab V3+, respectively. The areas in yellow circles are the areas water bodies greatly differ. Blue parts of the pictures represent the extracted water bodies and black parts of the pictures represent the background.

3.3. Accuracy Analysis

To analyze the universality of the MWEN method, different water types are analyzed. The accuracy comparisons via the evaluation metrics are shown in Section 3.3.1, the comparisons between MWEN "with MTFE" and "without MTFE" are shown in Section 3.3.2, and the accuracy comparisons via the visual comparison between MWEN, FCN, Unet, and Deeplab V3+ are shown in Sections 3.3.3 and 3.3.4.

3.3.1. Accuracy Comparisons via the Evaluation Metrics

To quantitatively analyze the water body extraction accuracy, the metrics mentioned in 2.2.3 were calculated based on the water maps predicted by the CNNs and the ground truth. Results are summarized in Table 5. As can be seen from the table, the MWEN outperforms the others in the OA, FWR, WIoU, and MIoU [30]. Deeplab V3+ is one of the best CNNs for semantic segmentation. In this study, Deeplab V3+ performs poorly in the OA, FWR, WIoU, and MIoU, but it performs the best in the TWR. Deeplab V3+ may be suitable for datasets with complex scenes, but it appears to be overfitting when training for water extraction.

Table 5. Water body extraction accuracies of the various methods.

CNN	OA (%)	TWR (%)	FWR (%)	WIoU	MIoU
MWEN	98.62	92.34	0.61	0.880	0.932
MWEN without MTFE	98.35	91.58	0.86	0.863	0.916
FCN	98.52	91.40	0.62	0.870	0.927
Unet	98.18	92.82	1.16	0.849	0.914
Deeplab V3+	91.82	96.92	8.81	0.566	0.737

3.3.2. Performance Comparison for MWEN and MWEN "Without Multi-Scale Feature Extractor (MTFE)"

Feature maps extracted by CNN at different scales contain various information. In this study, the multi-scale feature extractor (MTFE) is proposed to capture the features at multiple scales. In order to examine the importance of features extracted by MTFE for water extraction, results containing ponds and rivers with different sizes, and building shadows are derived from the result water maps mentioned in Section 3.2. The comparisons between the MWEN "with MTFE" and "without MTFE" are shown in Figure 9.

For the pools with different sizes in Figure 9a, both of the MWEN "with MTFE" and "without MTFE" can identify larger ponds, but the latter has obvious disadvantages for addressing the smaller pool information in Figure 9(a4). Moreover, tiny rivers cannot be identified by the MWEN "without MTFE" in Figure 9(b4,c4). Regarding confusing objects, the highway and some building shadows are mixed by the MWEN "without MTFE" in Figure 9(d4,e4). This may result from the relevance information between objects, such as the relationship between buildings and shadows, being ignored by MWEN "without MTFE". The relevance information may be contained in the features extracted by the convolution kernel with a large expansion rate. Figure 9 shows that MTFE plays an important role in extracting water bodies with various sizes and suppressing noise.

3.3.3. Performance Comparison for Different Water Types

Different surface water bodies, including open ponds, plateau rivers and lakes, city waters and agricultural water bodies, are taken from the results to assess the universality of the MWEN algorithm. The performances of the MWEN are compared with those of the FCN, Unet, and Deeplab V3+ based on the visual inspection. The performance comparison is shown in Figure 10.

For the open pools in Figure 10a, the comparison shows that all four CNNs are able to extract the large open pools. The smaller open pools are missed when using the FCN in Figure 10(a4). The results for agricultural waters show that detailed boundary information is missing by the FCN and Deeplab

V3+ in Figure 10(b4,c4,c6). Rough boundaries and mixing between water and wetlands appear when using the Unet in Figure 10(c5). Regarding plateau rivers and lakes, it can clearly be seen that the parts of rivers and lakes are missing by the FCN and Deeplab V3+ in Figure 10(d4,d6,e4,e6). The results for small puddle and tiny rivers in city demonstrate that the small puddle and tiny rivers are missed by the FCN and Unet in Figure 10(f4,g4,g5). Affected by urban buildings and other objects, the results extracted by the Unet and Deeplab V3+ contain more noises in Figure 10(f5,f6,g6).

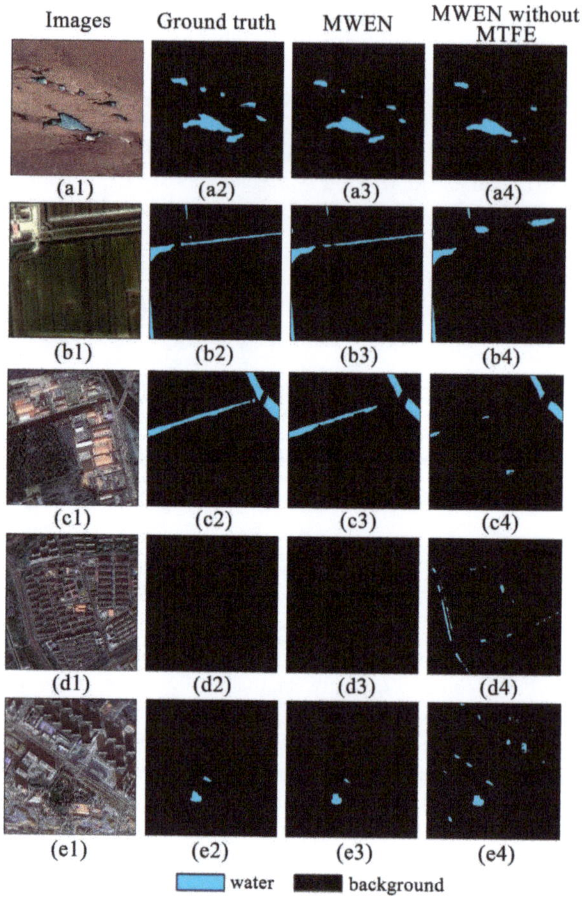

Figure 9. Results comparison between MWEN "with MTFE" and "without MTFE". (a1–e1) are the images, (a2–e2) are the ground truth, (a3–e3) are the water maps extracted by the MWEN "with MTFE", (a4–e4) are the water maps extracted by the MWEN "without MTFE".

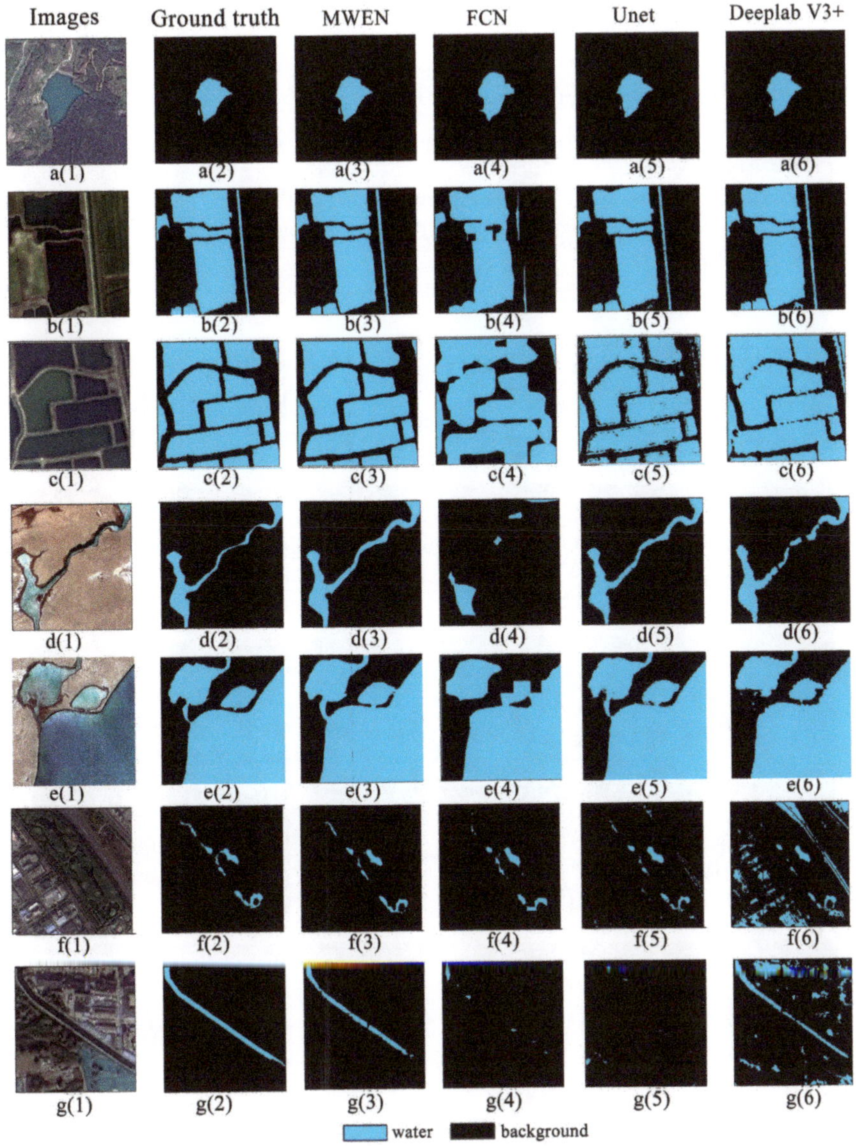

Figure 10. Typical surface water classification results. a(1) is the original image with open pools, and a(2–6) represent the water body information extracted from a(1) by artificial interpretation, MWEN, FCN, Unet, DeepLab V3+, respectively. Additionally, b, c, d, e, f, g give the experimental results of water body extraction from different images with agricultural water, plateau river, plateau lakes, small water bodies, and tiny rivers, respectively. Blue parts of the pictures are the extracted water bodies and black parts of the pictures are the backgrounds.

From Figure 10, it can be seen that MWEN performs better than the other algorithms. The FCN loses much detailed information for surface water body, which leads to blurred boundaries and the absence of small water bodies. Unet and Deeplab V3+ can better extract detail information of the water body compared with FCN but may be confused with objects with spectral characteristics to similar water. Figure 10 shows that the MWEN has the ability to extract different types of water bodies and the universal performance is better than other.

3.3.4. Performance Comparison for Confusing Areas

In high-resolution remote sensing images, some objects have spectral features or texture features similar to those of water bodies. It is a challenge to distinguish water bodies from these objects. To examine the reliability of these CNNs in distinguishing water bodies from confusing areas, the water body extraction results for confusing areas, such as building shadows, sports fields, and highways, are shown in Figure 11.

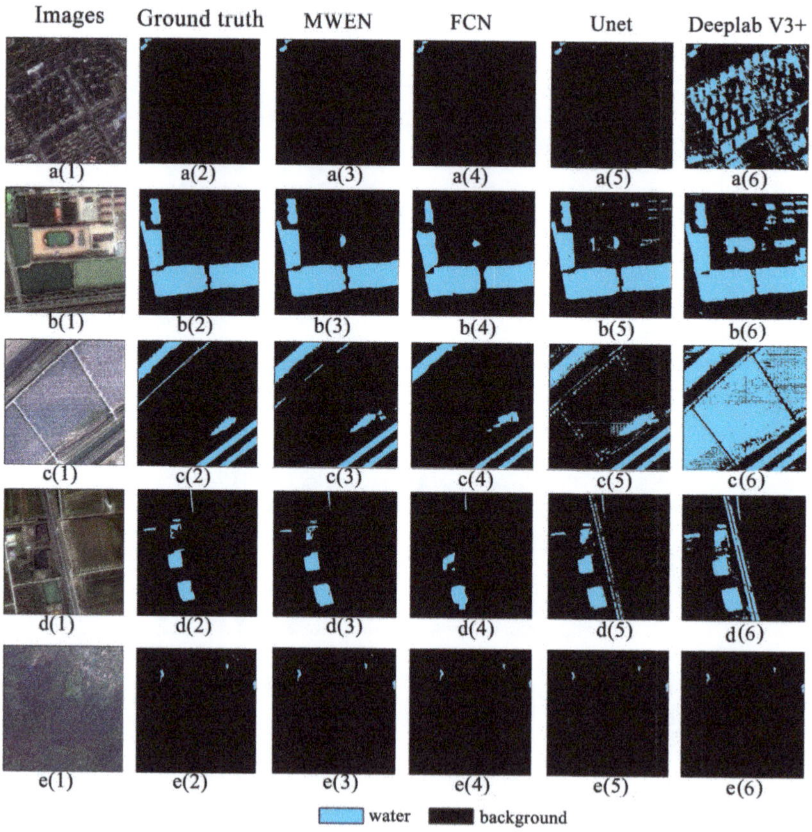

Figure 11. The results of the four methods for confusing regions. a(1) is the original image with building shadows, and a(2–6) represent the water body information extracted from a(1) by artificial interpretation, MWEN, FCN, Unet, DeepLab V3+, respectively. Additionally, b, c, d, e give the experimental results of water body extraction from different images with playgrounds, shade net, highways and mountain shadows, respectively. Blue parts of the pictures are the water bodies and black parts of the pictures are the backgrounds.

For the building shadows shown in Figure 11a, the MWEN, FCN, and Unet can better suppress noise, while Deeplab V3+ does not remove the building shadows, which may be caused by overfitting during training. Figure 11b demonstrates that all of these CNNs cannot clearly remove the noises from the sports field, but the MWEN and FCN perform better than the others. For the areas in Figure 11c,d, the Unet and Deeplab V3+ obviously mix the surface water body and other objects. For the mountain shadow area in Figure 11e, all four CNNs can clearly remove the noise. The performance comparison in confusing areas shows that the noises from the sports field, shade net and highway still exist in the results based on Unet and Deeplab V3+. The MWEN and FCN achieve better performances in suppressing the noise than the others.

4. Discussion

With the improvement in the temporal and spatial resolution of remote sensing data [25], many meaningful works have been conducted on water body information extraction with high-resolution remote sensing data [40,41]. Deep learning has been a hot topic in recent years [42], and it shows great promise in water body extraction with high-resolution remote sensing data. In this study, a new CNN named MWEN is proposed for water body extraction for GaoFen-1 images. The extraction accuracy of water bodies on the test dataset is evaluated by five evaluation metrics and visual comparison. The results show that MWEN has the ability to extract water bodies with different sizes and can accurately capture the boundaries of water bodies. In addition, MWEN can suppress noise better than Unet and Deeplab V3+.

The different performance in water body extraction may relate to the structures of these CNNs. FCN has been applied to water body extraction in previous research [26]. The FCN based methods extract features by several convolutional layers from the image and then perform water body segmentation based only on the low-resolution feature maps extracted by the last convolutional layer. The water maps are mapped to the original image resolution by upsampling. However, the upsampling process is not sensitive to the details in the image, which leads to small water bodies to be ignored and the boundaries of water bodies are smoothed. The Unet combines the structure of the encoder and decoder, and features at multiple scales are fused through skip connection between the encoder and decoder [23]. This is good for extracting the accurate boundaries of water bodies and capturing detailed information in the image. However, the Unet fuses too many low-level features extracted by the shallow convolutional layers. These low-level feature maps may be related to mistakes for noises that have similar spectral characteristics with water bodies. Deeplab V3+ is one of the state-of-the-art CNNs in the field of computer vision [24]. Deeplab V3+ uses ASPP pyramids to extract features at multiple scales and uses a decoder to restore the resolution of the feature maps. The Deeplab V3+ does not perform well in this study, which may be related to its complex structure. It may be suitable for pixel-level segmentation in complex scenes. It is prone to overfit in water body extraction. Motivated by the Unet [23] and Deeplab V3+ [24], the MWEN is proposed in this study. In the MWEN, the MEFT structure is proposed for capturing features at multiple scales and the encoder-decoder structure is used to restore the resolution. Compared with Deeplab V3+, the MWEN contains fewer convolutional layers and fewer trainable parameters, which effectively suppresses overfitting. The structure of MWEN makes it perform better in water body extraction for high-resolution images. Although MWEN obtains good accuracy on the test images, there are factors that affect the classification accuracy.

One is that new challenges appear in high-resolution image water extraction compared to mid-resolution images. The noise in water extraction based on medium resolution images, such as mountain shadows [42], can be easily distinguished in high-resolution images. Small water bodies may be difficult to extract in medium-resolution images, but they can be easily identified in high-resolution images. However, building shadows, highways, dark lawns, and dark roofs may result in new errors. In this study, the MWEN performs better in suppressing noise compared to the Unet and Deeplab V3+, but it does not completely remove the noise, such as noise from sports fields. In addition, very detailed water information is contained in high-resolution images, which brings new challenges for more accurate water body extraction.

The other is the dataset. The CNN with trained weights can perform well on images similar to the samples in the sample library. Its applicability to images that are quite different from the samples in the sample library needs further study. A dataset based on high-resolution remote sensing images containing multiple types of water bodies and easily confused areas, such as shadows, is needed.

Although the dataset proposed in this article contains common water bodies and easily confused areas, which can meet some data requirements in certain areas, the sample library needs to be enriched in the future.

5. Conclusions

Convolutional neural networks have been shown to have strong image classification and semantic segmentation abilities for remote sensing images. A new convolutional neural network named the MWEN for water body extraction for GF-1 high-resolution satellite images is proposed in this study. Three CNNs that conduct semantic segmentation in computer vision field are employed for comparison. The performances of the water body extraction results are evaluated based on five evaluation metrics and visual comparisons. The conclusions are as following:

(1) The performance of the MWEN is better than that of the FCN, Unet, and DeepLab V3+ when extracting surface water according to the visual comparison. The quantitative metrics show that results of the MWEN on the OA, TWR, FWR, WIoU, and MIoU are better than those of the others.

(2) The comparison between MWEN "with MTFE" and "without MTFE" demonstrates that the combination of features extracted at multiple scales is important to water extraction. The MTFE is helpful for dealing with confusing areas and water bodies with different sizes.

(3) Compared with the FCN and Unet, the results of the MWEN show that it can accurately extract water bodies in different scenes, such as the details of city water and plateau lakes. In addition, the MWEN has the ability to suppress noises, such as mountain shadows, highways, vegetation shadows, and dark lawns.

With the further enrichment of dataset, the MWEN has the application potential in large scale surface water mapping with high resolution satellite images, which can provide data support for surface water resource survey.

Author Contributions: Guojin He, Hongxiang Guo conceived of and designed the experiments. Hongxiang Guo. Wei Jiang and Ranyu Yin performed the experiments. Hongxiang Guo Wanchun Leng and Lei Yan made the dataset. Hongxiang Guo and Wei Jiang analyzed data. Hongxiang Guo wrote the whole paper, and all authors edited the paper. All authors have read and agreed to the published version of the manuscript.

Funding: This research was funded by The National Natural Science Foundation of China (61731022), the Second Tibetan Plateau Scientific Expedition and Research Program (STEP) (Grant No.2019QZKK0307) and the National Key Research and Development Program of China (2016YFA0600302).

Acknowledgments: The authors thank the anonymous reviewers and the editors for their valuable comments to improve our manuscript.

Conflicts of Interest: The authors declare no conflicts of interest.

References

1. Oki, T.; Kanae, S. Global hydrological cycles and world water resources. *Science* **2006**, *313*, 1068–1072. [CrossRef]
2. Van Oost, K.; Quine, T.A.; Govers, G.; De Gryze, S.; Six, J.; Harden, J.W.; Ritchie, J.C.; McCarty, G.W.; Heckrath, G.; Kosmas, C.; et al. The impact of agricultural soil erosion on the global carbon cycle. *Science* **2007**, *318*, 626–629. [CrossRef]
3. Wei, J.; Guojin, H.; Zhiguo, P.; Hongxiang, G.; Tengfei, L.; Yuan, N. Surface water map of china for 2015 (swmc-2015) derived from landsat 8 satellite imagery. *Remote Sens. Lett.* **2020**, *11*, 265–273.
4. Ji, L.Y.; Gong, P.; Wang, J.; Shi, J.C.; Zhu, Z.L. Construction of the 500-m resolution daily global surface water change database (2001-2016). *Water Resour. Res.* **2018**, *54*, 10270–10292. [CrossRef]
5. Fang, Y.; Ceola, S.; Paik, K.; McGrath, G.; Rao, P.S.C.; Montanari, A.; Jawitz, J.W. Globally universal fractal pattern of human settlements in river networks. *Earths Future* **2018**, *6*, 1134–1145. [CrossRef]
6. Lv, W.; Yu, Q.; Yu, W. Water extraction in sar images using glcm and support vector machine. In Proceedings of the 2010 IEEE 10th International Conference on Signal Processing Proceedings (Icsp2010), Beijing, China, 24–28 October 2010; pp. 740–743.

7. Xiao, Y.; Zhao, W.; Zhu, L. A study on information extraction of water body using band1 and band7 of tm imagery. *Sci. Surv. Mapp.* **2010**, *35*, 226–227.
8. Song, X.F.; Duan, Z.; Jiang, X.G. Comparison of artificial neural networks and support vector machine classifiers for land cover classification in northern china using a spot-5 hrg image. *Int. J. Remote Sens.* **2012**, *33*, 3301–3320. [CrossRef]
9. Ko, B.C.; Kim, H.H.; Nam, J.Y. Classification of potential water bodies using landsat 8 oli and a combination of two boosted random forest classifiers. *Sensors* **2015**, *15*, 13763–13777. [CrossRef] [PubMed]
10. Yu, L.; Wang, Z.; Tian, S.; Ye, F.; Ding, J.; Kong, J. Convolutional neural networks for water body extraction from landsat imagery. *Int. J. Comput. Intell. and Appl.* **2017**, *16*. [CrossRef]
11. Chen, Y.; Fan, R.S.; Yang, X.C.; Wang, J.X.; Latif, A. Extraction of urban water bodies from high-resolution remote-sensing imagery using deep learning. *Water* **2018**, *10*, 585. [CrossRef]
12. Frazier, P.S.; Page, K.J. Water body detection and delineation with landsat tm data. *Photogramm. Eng. Remote Sens.* **2000**, *66*, 1461–1467.
13. Gao, B.C. Ndwi—A normalized difference water index for remote sensing of vegetation liquid water from space. *Remote Sens. Environ.* **1996**, *58*, 257–266. [CrossRef]
14. Zhou, Y.A.; Luo, J.C.; Shen, Z.F.; Hu, X.D.; Yang, H.P. Multiscale water body extraction in urban environments from satellite images. *IEEE J. Sel. Topics Appl. Earth.Observ. Remote Sens.* **2014**, *7*, 4301–4312. [CrossRef]
15. Acharya, T.D.; Lee, D.H.; Yang, I.T.; Lee, J.K. Identification of water bodies in a landsat 8 oli image using a j48 decision tree. *Sensors* **2016**, *16*, 1075. [CrossRef]
16. Li, K.; Wan, G.; Cheng, G.; Meng, L.; Han, J. Object detection in optical remote sensing images: A survey and a new benchmark. *arXiv* **2019**, arXiv:1909.00133. [CrossRef]
17. Lu, D.; Weng, Q. A survey of image classification methods and techniques for improving classification performance. *Int. J. Remote Sens.* **2007**, *28*, 823–870. [CrossRef]
18. He, K.M.; Zhang, X.Y.; Ren, S.Q.; Sun, J. Deep residual learning for image recognition. In Proceedings of the 2016 IEEE Conference on Computer Vision and Pattern Recognition (CVPR), Las Vegas, NV, USA, 27–30 June 2016; pp. 770–778.
19. Long, J.; Shelhamer, E.; Darrell, T. Fully convolutional networks for semantic segmentation. *IEEE Trans. Pattern Anal. Mach. Intell.* **2014**, *39*, 640–651.
20. He, K.M.; Gkioxari, G.; Dollar, P.; Girshick, R. Mask r-cnn. In Proceedings of the 2017 IEEE International Conference on Computer Vision (Iccv), Venice, Italy, 22–29 October 2017; pp. 2980–2988.
21. Pan, H.D.; Chen, G.F.; Jiang, J. Adaptively dense feature pyramid network for object detection. *Ieee Access* **2019**, *7*, 81132–81144. [CrossRef]
22. Wu, Z.; Gao, Y.; Li, L.; Xue, J.; Li, Y. Semantic segmentation of high-resolution remote sensing images using fully convolutional network with adaptive threshold. *Connect. Sci.* **2019**, *31*, 169–184. [CrossRef]
23. Ronneberger, O.; Fischer, P.; Brox, T. U-Net: Convolutional Networks for Biomedical Image Segmentation. In Proceedings of the International Conference on Medical Image Computing and Computer-Assisted Intervention, Munich, Germany, 5–9 October 2015; Springer: Berlin/Heidelberg, Germany, 2015; pp. 234–241.
24. Chen, L.-C.; Zhu, Y.; Papandreou, G.; Schroff, F.; Adam, H. Encoder-decoder with atrous separable convolution for semantic image segmentation. In Proceedings of the European Conference on Computer vision (ECCV), Munich, Germany, 8–14 September 2018; pp. 801–818.
25. Ghamisi, P.; Rasti, B.; Yokoya, N.; Wang, Q.M.; Hofle, B.; Bruzzone, L.; Bovolo, F.; Chi, M.M.; Anders, K.; Gloaguen, R.; et al. Multisource and multitemporal data fusion in remote sensing a comprehensive review of the state of the art. *IEEE Geosci. Remote Sens. Mag.* **2019**, *7*, 6–39. [CrossRef]
26. Li, L.W.; Yan, Z.; Shen, Q.; Cheng, G.; Gao, L.R.; Zhang, B. Water body extraction from very high spatial resolution remote sensing data based on fully convolutional networks. *Remote Sens.* **2019**, *11*, 1162. [CrossRef]
27. Isikdogan, F.; Bovik, A.C.; Passalacqua, P. Surface water mapping by deep learning. *IEEE J. Sel. Topics Appl. Earth.Observ. Remote Sens.* **2017**, *10*. [CrossRef]
28. Yu, F.; Koltun, V. Multi-scale context aggregation by dilated convolutions. *arXiv* **2015**, arXiv:1511.07122.
29. Long, T.F.; Jiao, W.L.; He, G.J. Nested regression based optimal selection (nrbos) of rational polynomial coefficients. *Photogramm. Eng. Remote Sens.* **2014**, *80*, 261–269.
30. Peng, Y.; Zhang, Z.M.; He, G.J.; Wei, M.Y. An improved grabcut method based on a visual attention model for rare-earth ore mining area recognition with high-resolution remote sensing images. *Remote Sens.* **2019**, *11*, 987. [CrossRef]

31. Holschneider, M.; Kronland-Martinet, R.; Morlet, J.; Tchamitchian, P. A real-time algorithm for signal analysis with the help of the wavelet transform. In *Wavelets*; Springer: Berlin/Heidelberg, Germany, 1990; pp. 286–297.
32. Chen, L.C.; Papandreou, G.; Kokkinos, I.; Murphy, K.; Yuille, A.L. Deeplab: Semantic image segmentation with deep convolutional nets, atrous convolution, and fully connected crfs. *IEEE Trans. Pattern Anal. Mach. Intell.* **2018**, *40*, 834–848. [CrossRef]
33. Li, Y.; Qi, H.Z.; Dai, J.; Ji, X.Y.; Wei, Y.C. Fully convolutional instance-aware semantic segmentation. In Proceedings of the 30th IEEE Conference on Computer Vision and Pattern Recognition (Cvpr 2017), Honolulu, HI, USA, 21–26 July 2017; pp. 4438–4446.
34. Gould, S.; Fernando, B.; Cherian, A.; Anderson, P.; Cruz, R.S.; Guo, E. On differentiating parameterized argmin and argmax problems with application to bi-level optimization. *arXiv* **2016**, arXiv:1607.05447.
35. De Boer, P.-T.; Kroese, D.P.; Mannor, S.; Rubinstein, R.Y. A tutorial on the cross-entropy method. *Ann. Oper. Res.* **2005**, *134*, 19–67. [CrossRef]
36. Leung, H.; Haykin, S. The complex backpropagation algorithm. *IEEE Trans. Signal Process.* **1991**, *39*, 2101–2104. [CrossRef]
37. Von Davier, M. Bootstrapping goodness-of-fit statistics for sparse categorical data: Results of a monte carlo study. *Methods Psychol. Res. Online* **1997**, *2*, 29–48.
38. Bello, I.; Zoph, B.; Vasudevan, V.; Le, Q.V. Neural optimizer search with reinforcement learning. In Proceedings of the 34th International Conference on Machine Learning, Sydney, Australia, 6–11 August 2017; Volume 70, pp. 459–468.
39. Takikawa, T.; Acuna, D.; Jampani, V.; Fidler, S. Gated-scnn: Gated shape cnns for semantic segmentation. *arXiv* **2019**, arXiv:1907.05740.
40. Miao, Z.; Fu, K.; Sun, H.; Sun, X.; Yan, M. Automatic water-body segmentation from high-resolution satellite images via deep networks. *IEEE Geosci. Remote Sens. Lett.* **2018**. [CrossRef]
41. Yao, F.F.; Wang, C.; Dong, D.; Luo, J.C.; Shen, Z.F.; Yang, K.H. High-resolution mapping of urban surface water using zy-3 multi-spectral imagery. *Remote Sens.* **2015**, *7*, 12336–12355. [CrossRef]
42. Jiang, W.; He, G.; Long, T.; Ni, Y.; Liu, H.; Peng, Y.; Lv, K.; Wang, G. Multilayer perceptron neural network for surface water extraction in landsat 8 oli satellite images. *Remote Sens.* **2018**, *10*, 755. [CrossRef]

International Journal of
Geo-Information

MDPI

Article

Forecasting of Short-Term Daily Tourist Flow Based on Seasonal Clustering Method and PSO-LSSVM

Keqing Li *, Changyong Liang, Wenxing Lu, Chu Li, Shuping Zhao and Binyou Wang

School of Management, Hefei University of Technology, Hefei 230009, China; cyliang@hfut.edu.cn (C.L.); luwenxing@hfut.edu.cn (W.L.); 2019170692@mail.hfut.edu.cn (C.L.); zhaoshuping1753@hfut.edu.cn (S.Z.); 2013111070@mail.hfut.edu.cn (B.W.)

* Correspondence: lkqing1995@mail.hfut.edu.cn

Received: 12 October 2020; Accepted: 13 November 2020; Published: 13 November 2020

Abstract: The accurate prediction of tourist flow is essential to appropriately prepare tourist attractions and inform the decisions of tourism companies. However, tourist flow in scenic spots is a dynamic trend with daily changes, and specialized methods are necessary to measure it accurately. For this purpose, a tourist flow forecasting method is proposed in this research based on seasonal clustering. The experiment employs the K-means algorithm considering seasonal variations and the particle swarm optimization-least squares support vector machine (PSO-LSSVM) algorithm to forecast the tourist flow in scenic spots. The LSSVM is also used to compare the performance of the proposed model with that of the existing ones. Experiments based on a dataset comprising the daily tourist data for Mountain Huangshan during the period between 2014 and 2017 are conducted. Our results show that seasonal clustering is an effective method to improve tourist flow prediction, besides, the accuracy of daily tourist flow prediction is significantly improved by nearly 3 percent based on the hybrid optimized model combining seasonal clustering. Compared with other algorithms which provide predictions at monthly intervals, the method proposed in this research can provide more timely analysis and guide professionals in the tourism industry towards better daily management.

Keywords: seasonal clustering; short-term forecast; tourism flow forecast; optimization algorithm

1. Introduction

In recent years, owing to steady improvements in the standards of living, tourism has become an important part of leisure and lifestyle for people worldwide. According to data released by the World Travel Tourism Council, tourism was the third largest industry in the world in terms of the growth rate of Gross Domestic Product (GDP) in 2019. The growth rate of tourism was reportedly 3.5%, which was significantly greater than the global economic GDP growth rate of 2.5% [1]. In particular, the tourism industry created nearly 80 million jobs in China, accounting for 10.3% of the country's total labor force. At the same time, its output value was estimated to be 10.9 trillion Yuan, accounting for 11.3% of China's economy [1]. The rapid development of the world's tourism industry has promoted the vigorous development of China's own tourism industry. China's tourism industry has entered the stage of 'mass tourism', with people's willingness to travel constantly rising [2]. It is expected that the domestic tourism market will continue to thrive even in the post-epidemic era [3].

With the promotion of the economic improvement of the country and the region, the rapid development of tourism has also ushered in multiple problems pertaining to daily management services at tourist destinations, particularly at mountainous scenic spots, which play a pivotal role in Chinese tourism [4]. Their unique topography and landforms, extensive spatial range, poor natural conditions, and severe seasonal conditions make them inaccessible to personnel. In particular, the delivery of materials and resources, scheduling of arrangements for transportation, etc., pose particular challenges to management services in mountainous environments [5]. The effects of these challenges are primarily reflected in delays in passenger flow. All tourist destinations experience heavy tourist seasons and off-seasons, resulting in a serious seasonal imbalance in the tourist flow [6]. During the tourism season, spots are often overcrowded. This causes traffic congestion, overextends hotel, catering, and personnel supplies, leads to the overutilization of tourism resources and the environment, and degrades the quality of service for tourists, reducing overall tourist satisfaction. On the other hand, the oversaturation of tourists in specific spots also poses a threat to their own personal safety [6]. For example, on 4 October 2014, due to a surge in the number of tourists during the Golden Week, the passenger capacity at the Three Gorges scenic spot in Yichang, Hubei, was insufficient, resulting in hundreds of tourists being stranded at the terminal. On 2 October 2013, several tourists were stuck at the entrance of Jiuzhaigou Valley because of overcrowding. On 26 October 2014, the traffic was almost paralyzed at the Beijing Xiangshan area, leading to thousands of people being stranded at the bus station. Furthermore, the Golden Week of Tourism has been witness to a series of security incidents which have resulted in a poor travel experience for tourists [7]. However, during the off-season, the number of tourists at destinations are considerably low, resulting in idle hotels and wasted resources, materials, personnel, etc. These considerations corroborate the significance of the accurate forecast of tourist flow in the tourism industry.

Tourist flow forecasting can be divided into two categories: long-term forecasting and short-term forecasting. Both have important implications, and the determination of an accurate trend can aid professionals in the tourism industry [8,9], particularly with respect to problems such as optimal allocation of resources and managerial staff [10].

The forecasting of tourist flow in tourist destinations is affected by several factors, including weather [11], climate [12], and temperature [13]. Tourism is inherently seasonal [14] as the constraints of time and climate create inevitably unbalanced tourist flows [15]. Both natural seasons and artificial seasons defined by holidays and other institutional factors play a part in the determination of tourist flow [16]. Thus, both factors must be considered during prediction attempts. To the best of our knowledge, scant attention has been paid to seasonality in previous works on this topic. For instance, Huang and Min established a seasonal autoregressive average model combined with a difference method to eliminate seasonal effects on tourist flow forecasting, and experimentally verified its effectiveness [17]. However, these studies have focused solely on the elimination of seasonal influences on the prediction of tourist flow by proposing seasonal index adjustments or by establishing a seasonal model. Few studies have considered the influence of the alternatives of natural seasons in the forecast of tourist flow.

Tourist flows exhibit complicated non-linear variations. This makes it difficult to identify a relationship between the tourist flow later and the current influencing variables based on simple mathematical models. In recent years, with the development of machine learning, nonlinear models have been widely used in short-term time series forecasting. For instance, artificial neural network (ANN)-based methods and support vector machines (SVM) have already been used in the forecasting of tourist flow [18,19]. However, neural network-based models lack a systematic procedure for model construction because of their flexibility. This necessitates multiple trials to identify the optimal parameters required to obtain a reliable neural model [20]. Compared with ANN, SVM is more capable of avoiding problems such as data overfitting and local minima while maintaining positive features such as robustness. Moreover, SVM is less complicated than ANN in terms of parameter selection [21]. The LSSVM is an upgraded version of SVM that was developed to improve the accuracy of the standard SVM [22]. Compared to SVM, it is capable of using equality constraints instead inequalities, enabling it to solve sets of linear equations instead of being restricting to quadratic programming [23]. However, the prediction accuracy of the LSSVM algorithm is significantly dependent on the selection of two specific parameters [24]. To address this drawback, certain optimization algorithms, including the genetic algorithm (GA) and the fruit fly optimization algorithm, are used to identify the optimal values of the LSSVM parameters to enhance its prediction accuracy [25,26]. Among those intelligence-based optimization algorithms, PSO, proposed by Kennedy and Eberhart [27], has been widely used in optimization processes, model classification, machine learning, and neural network training [28] owing to its ease of implementation and its high coherence and coordination [29].

In addition to the development of such optimization algorithms, some studies have attempted to curate relevant information by analyzing comments on online forums. Certain researchers have used search engine data to forecast hotel demands [30,31] by designing a composite search index to forecast tourist flow [32]. Furthermore, Google Trends has been widely used to improve the performances of traditional models [10,33,34]. Related works have pointed out that combining different data sources and techniques can lead to higher accuracy [35]. Even price levels and web traffic have been as used as variables in certain studies [36]. User interactions on online forums have also been used to forecast tourist flows [37]. However, most of the methods are more suitable for long-term forecasting, rather than short-term forecasting.

As few research studies have been conducted to investigate short-term forecasting methods and substitutes to natural seasons in the forecasting process of tourist flows, we propose a seasonal clustering-based method, which can classify seasons based on their characteristics to address this shortcoming. We combine seasonal re-clustering and the PSO-LSSVM model and apply the combination for short-term daily tourist flow forecasting. The crucial hypothesis in this research is that seasonal clustering could improve tourist flow forecasting. Our results confirm the validity of the hybrid optimized model combining seasonal clustering and provide practically useful implications for management.

The remainder of this research is organized as follows. Section 2 presents the methods, including principles underlying the least squares support vector machine (LSSVM) and the particle swarm optimization (PSO) algorithms, and an illustration of the PSO-LSSVM procedure that considers seasonal clustering, and the experiments. Section 3 details their results. Section 4 is the discussion. Finally, Section 5 presents the conclusions, as well as the limitations and implications of this research.

2. Methods

2.1. Least Squares Support Vector Machine

The essential characteristic of LSSVM is that it is designed to utilize equality constraints and transform quadratic programming problems to problems of direct solution of quadratic equations. Consider a dataset (x_i, y_i), $x_i \in R^n$, $y \in R$, where x_i denotes the ith input item in an n-dimensional space and y_i denotes the output value corresponding to x_i, l is the total number of data points, $i = 1, 2, \cdots\cdots l$, and n is the number of dimensions of input variables. As a non-linear prediction model, the LSSVM model can be expressed as follows:

$$f(x) = w^T \phi(x) + b, \tag{1}$$

where w denotes the weight vector, b is the offset, and $\phi(x)$ represents a nonlinear transformation that maps the input data (x_i) into a high-dimensional feature space. According to the structure minimization principle, the optimization objective function of the LSSVM can be expressed as follows:

$$min \frac{1}{2} ||w||^2 + \frac{1}{2} C \sum_{i=1}^{l} e_i^2, \tag{2}$$

$$s.t. w^T \phi(X_i) + b + e_i = y_i, i = 1, 2, \cdots\cdots,$$

where e_i denotes the error and C represents a positive penalty coefficient. A Lagrange multiplier, λ_i, is introduced to solve the optimization problem. Hence, Equation (2) can be transformed into the following form:

$$L(w, \lambda_i, b, e_i) = \frac{1}{2}\|w\|^2 + \frac{1}{2}C\sum_{i=1}^{l} e_i^2 - \sum_{i=1}^{l} \lambda_i\left(w^T\phi(x_i) + b + e_i - y_i\right), \tag{3}$$

Next, the partial derivatives corresponding to each variable of Equation (3) are calculated:

$$\frac{\partial L}{\partial w} = w - \sum_{i=1}^{l} \lambda_i\phi(x_i) = 0 \Rightarrow w = \sum_{i=1}^{l} \lambda_i\phi(x_i), \tag{4}$$

$$\frac{\partial L}{\partial \lambda_i} = -\sum_{i=1}^{l}\left(w^T\phi(x_i) + b + e_i - y_i\right) = 0 \Rightarrow y_i = w^T\phi(x_i) + b + e_i, \tag{5}$$

$$\frac{\partial L}{\partial b} = \sum_{i=1}^{l} \lambda_i = 0, \tag{6}$$

$$\frac{\partial L}{\partial e_i} = \frac{1}{2} \times 2C\sum_{i=1}^{l} e_i - \sum_{i=1}^{l} \lambda_i = 0 \Rightarrow \lambda_i = Ce_i, \tag{7}$$

The variables, w and e_i, are then eliminated. This yields the following linear equation:

$$\begin{bmatrix} 0 \\ Y \end{bmatrix} = \begin{bmatrix} 0 & A^T \\ A^T & B + C^{-1}I \end{bmatrix}\begin{bmatrix} b \\ \lambda \end{bmatrix}, \tag{8}$$

where $Y = (y_1, y_2, \ldots\ldots, y_l)$, $A = (1, 1, \ldots\ldots, 1)^T$, $B_{ij} = \phi(x_i)^T\phi\left(x_j\right)$, $\lambda = (\lambda_1, \lambda_2, \ldots\ldots\lambda_l)$, and I denotes the unit matrix. Hence, the LSSVM can be expressed as follows:

$$y = \sum_{i=1}^{l} \lambda_i K(x, x_i) + b, \tag{9}$$

where $K(x, x_i)$ denotes the kernel function of a feature space.

2.2. Particle Swarm Optimization

A PSO algorithm begins by initializing a random group of particles and obtains the optimal solution after performing several iterative searches. During each iteration, the particles update their positions and velocities based on individual and global extrema. Let us assume that there is a total of N particles that are initialized and scattered in a D-dimensional space. Further, assume that the position of the ith particle is $X_i = (x_{i1}, x_{i2}, \cdots\cdots, x_{iD})$, and that the current best position for the ith particle is $local\_x_i = (local\_x_{i1}, local\_x_{i2}, \cdots\cdots, local\_x_{iD})$, whereas the best position found by the entire swarm is $global\_x_i = (global\_x_{i1}, global\_x_{i2}, \cdots\cdots, global\_x_{iD})$. In such a scenario, the new position of a particle after t time-instants is obtained by adding the velocity vector $V_i = (v_{i1}, v_{i2}, \cdots\cdots, v_{iD})$ to its current position. This can be expressed as follows:

$$x_{iD}^{(t+1)} = x_{iD}^t + wP \times v_{iD}^{t+1}, \tag{10}$$

The velocity of any particle is updated using the following formula:

$$V_{iD}^{t+1} = wV \times V_{iD}^t + c_1 \times rand \times \left(local_{x_{id}^t} - x_{id}^t\right) + c_2 \times rand \times \left(global_{x_{iD}^t} - x_{iD}^t\right), \tag{11}$$

where c_1, c_2 denote the acceleration coefficients, wV, wP represent the elasticity coefficients with initial values equal to 1, $rand$ denote two random numbers with uniform distributions in the range [0,1],

$local\_x^t_{id}$ is the best position identified by each individual particle, and $global\_x^t_{iD}$ is the best position identified by the global swarm.

2.3. Seasonal Clustering Approach

Several algorithms are used for clustering analysis, and they can be roughly divided into four categories [38]: (1) those based on cluster formation methodology, such as top-down, bottom-up, and analytical optimization techniques [39]; (2) those dependent on the cluster model obtained, such as stratification, centroids (e.g., K-means), distribution subspaces, and graph-based models; (3) those obtained via a membership function, which may be further subdivided into hard or soft clustering [40]; and (4) those that use groups to define the distinction between overlapping clusters and are less sensitive to noise because it becomes equally distributed among them [41].

The K-means clustering algorithm is a typical representative classification clustering algorithm due to its simplicity and effectiveness. It is particularly suitable for a simple clustering of big data. Considering that the primary characteristic of natural seasons is the change in weather [42], we attempt to analyze the correlation between climate-related factors and variations in daily tourist flow. The details of seasonal clustering are as follows.

Step 1: Analysis of the factors related to seasonal clustering.

Step 2: Input of the variables into the K-means algorithm to obtain the results of seasonal clustering.

2.4. Procedure of PSO-LSSVM Considering Seasonal Clustering

The present research primarily aims to prove that the use of seasonal clustering during the pre-processing of data is beneficial to the accurate prediction of daily tourist flow. Combined with historical tourist information, the PSO-LSSVM model is proposed to illustrate the positive impact of seasonal clustering on the prediction of tourist flow in tourist destinations. In the PSO-LSSVM model, the PSO algorithm is used as an optimization algorithm to optimize the regularization parameter (γ) and the kernel parameter (σ) of LSSVM. The considerations of seasonal clustering in PSO-LSSVM can summarized in the following steps.

Step 1. The natural seasons are clustered. The new natural season of the tourist destination combined with the spot's historical tourist data comprises a dataset. The original dataset is normalized and divided into training and test datasets.

Step 2. The parameters of the PSO algorithm, including population sizes, evolution times, and learning factors, are initialized.

Step 3. The swarm of particles is initialized with random individual velocities and positions.

Step 4. The various initialized parameters are fed into LSSVM, and then the fitness value of each particle is evaluated. In this research, the root mean squared error (RMSE) defined in the test dataset is used as the fitness function, as follows:

$$fitness = RMSE = \sqrt{\frac{1}{n}\sum_{i=1}^{n}(y_i - \hat{y}_i)^2}, \tag{12}$$

where n denotes the number data points in the dataset, and y_i and $\hat{y}_i$ represent the actual value and the estimated value, respectively. The local and global optima are then calculated following the fitness function.

Step 5. The velocity and position of each particle is updated using Equations (10) and (11).

Step 6. Steps 4 and 5 are repeated until the termination criterion is satisfied and the optimal values of the LSSVM parameters are obtained. The flow chart of the procedure of PSO-LSSVM is shown in Figure 1.

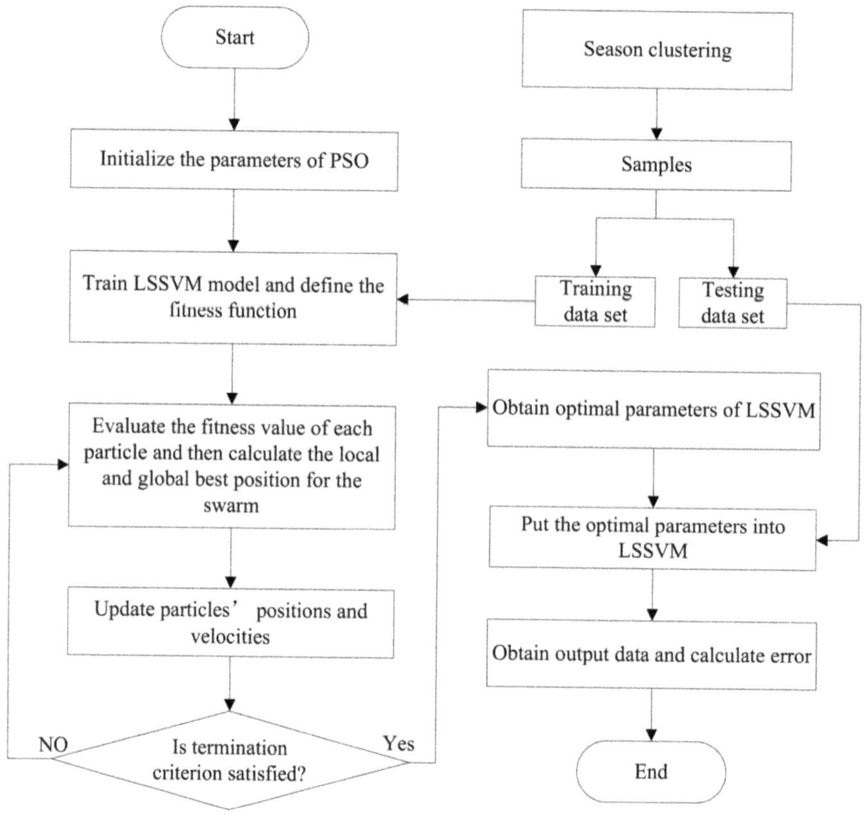

Figure 1. Flow chart of the procedure of the particle swarm optimization-least squares support vector machine (PSO-LSSVM).

In this research, to evaluate the forecasting accuracy, the mean absolute percentage error (MAPE) and RMSE are used as the evaluation criteria. It is evident that the values of MAPE or RMSE are inversely proportional to forecasting accuracy:

$$MAPE = \frac{1}{n}\sum_{i=1}^{n}\left|\frac{y_i - \hat{y}_i}{y_i}\right| \times 100\%, \quad (13)$$

$$RMSE = \sqrt{\frac{1}{n}\sum_{i=1}^{n}(y_i - \hat{y}_i)^2}, \quad (14)$$

where $y_i, \hat{y}_i$ denote the actual and evaluated data, respectively, and n denotes the total number of data points in the test dataset. It should be noted that the RMSE indicator only considers the annual average in the last row of the table as a supporting indicator. Consequently, the MAPE indicator is more suitable for prediction of daily trends.

2.5. Data Preprocessing

To improve the accuracy of prediction, it is necessary to normalize the original sequence of input variables. The following normalized formula is adopted in this research:

$$u = \frac{(u_{\max} - u_{min}) \times (x_i - x_{min})}{x_{\max} - x_{min}} + u_{min} \quad (15)$$

where u denotes the normalized value with uniform distribution in the range [0,1]; and u_{max} and u_{min} are the upper and lower limits, respectively. In this research, it is assumed that u_{max} and u_{min}. x_i denotes the tourist flow on the ith day in the original one-year data series, and x_{min} and x_{max} denote the minimum and maximum values of the original sequence, respectively.

2.6. Data Collection and Correlation Analysis

To verify the feasibility of the proposed algorithm, the dataset of the daily tourist flow at Mountain Huangshan during the period of 2014 to 2017 is accessed, the tourist flow data comes from our cooperation project with Huangshan Management Committee. Besides, we investigated the spot's historical temperature and weather for this research; the temperature is measured in degrees Celsius and the weather is measured in different categories such as sunny, cloudy, heavy snow, moderate snow, and so on. The tourist flow dataset contains both original regular daily tourist flow data and original tourist flow data on holidays. Four types of data are included in the data set: X_1, the daily tourist flow on a particular day; X_2, the tourist flow volume on the same day in the previous week; X_3, the tourist flow volume on the same day of the previous year; and Y, the daily tourist flow on the subsequent day. Each type contains 1461 data points. The relationship between the historical tourist flow, which includes X_1, X_2, X_3, and the daily tourist flow of the subsequent day is primarily determined by the respective correlation coefficients—the correlation coefficients between pairs of data items are proportional to the suitability of the selected factors as inputs to the model.

Table 1 presents the correlation coefficients between X_1, X_2, X_3, and Y. As expected, X_1 is observed to be superior to the other factors. Consequently, X_1 is selected as the input variable in the proposed model.

Table 1. Correlation coefficients between the daily tourist flow of tomorrow and each element of the historical tourist flow.

	X_1	X_2	X_3
Y	0.726	0.468	0.347

In addition, the severity of weather, weekday, and official holiday are also added to the model as dummy variables X_4, X_5 and X_6. $X_4 = \left\{ \begin{array}{c} 1 \\ 0 \end{array} \right.$, where 1 represents severe weather, such as blizzard, heavy snow, moderate snow, heavy rain, thunderstorms, and showers, which would significantly affect people's willingness to travel, and 0 represents non-severe weather, such as sunny, cloudy, and drizzle. X_5 represents a matrix which represents the day of the week. $X_6 = \left\{ \begin{array}{c} 1 \\ 0 \end{array} \right.$, where 1 represents an official holiday; 0 represents an ordinary day. The use of dummy variables is another difference between our research and previous ones. The incorporation of such factors allowed us to approach the problem of prediction from a more microscopic perspective.

2.7. Parameter Initialization and the Addition of Seasonal Factors

The initial parameters are set as follows, the size of the swarm is taken to be 30, maximum number of iterations is set as 300, and acceleration coefficients c_1 and c_2 are 2 and 2, respectively. To verify whether the ambient natural season affects the accuracy of prediction of the tourist flow on the subsequent day, a binary virtual variable-based approach is introduced to represent the different seasons; $s_i = \begin{cases} 1 \\ 0 \end{cases}$ (1 represents the ith natural season i =1, 2, 3, 4).

3. Results

The results of the experiments above are shown in this section.

3.1. Analysis of Influence of Original Natural Season

This research aims to investigate the effect of seasonal changes on tourist flow on the subsequent day. The daily tourist flow at scenic destinations varies dramatically over the different seasons, primarily because of the differences in temperature. In this part, the year is assumed to be divided into four seasons following the meteorological department's scheme: spring (March, April, and May), summer (June, July, and August), autumn (September, October, and November), and winter (December, January, and February) [15]. Figure 2 illustrates the distribution of the daily tourist flow on the subsequent day at Mountain Huangshan over the period of 2014 to 2017.

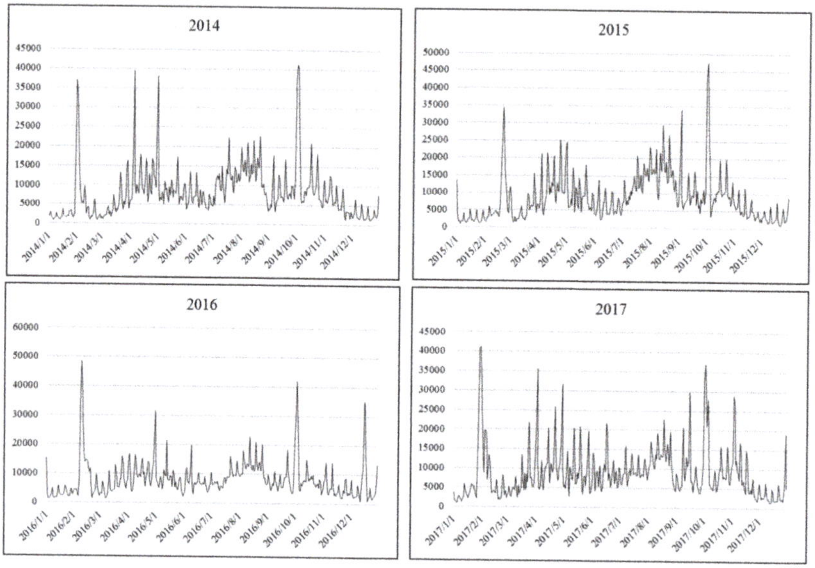

Figure 2. Daily tourist flow at Mountain Huangshan during 2014–2017.

It is clear from Figure 2 that due to the daily fluctuations in tourist flow, the distribution is complex and non-linear. Further, the daily tourist volume at Mountain Huangshan during the period from March to November is observed to remain high every year, whereas during December to January it appears to be consistently low. Further analysis of the data depicted in Figure 2 is presented in Tables 2 and 3.

Table 2. Total number of tourists during each season.

	Spring	Summer	Autumn	Winter
March 2014–February 2015	799,838	976,109	827,444	385,479
March 2015–February 2016	815,947	1,062,968	896,006	529,186
March 2016–February 2017	761,308	869,248	717,781	661,204

Table 3. Average number of tourists during each season.

	Spring	Summer	Autumn	Winter
March 2014–February 2015	8694	10,610	9093	4283
March 2015–February 2016	8869	11,554	9846	5815
March 2016–February 2017	8275	9448	7887	7346

Tables 2 and 3 reveal that the total tourist flow and the average tourist flow remain high during spring, summer, and autumn each year. It is further confirmed that the tourist flow is maximum during the summer and that it is the second highest during spring and autumn. The tourist volume in winter is significantly less than that during the other three seasons. Thus, it can be concluded that the tourist flows in different seasons are significantly different.

3.2. Predictions by the Models and Their Comparison before Seasonal Clustering

In this experiment, to satisfy the requirements of the model, the dataset is divided into a training dataset (2014–2016) and a test dataset (2017). To enhance the prediction accuracy, all the data are normalized using Equation (15) with a range of [0,1]:

$$y = \frac{x - x_{min}}{x_{max} - x_{min}}, \tag{16}$$

where y denotes the normalized data, x denotes the original input data, and x_{max}, x_{min} are the maximum and minimum values in the dataset, respectively.

Following that, the vectors $(X_1, X_4, X_5, X_6, S_1, S_2, S_3, S_4)$, including the natural seasons, are used as input variables in the predictive models, and the vectors (X_1, X_4, X_5, X_6), without considering seasonal factors, are used as input variables to the predictive models on a separate iteration for comparison purposes. Both the PSO-LSSVM algorithm and the LSSVM algorithm are adopted as predictive models for each of the two sets of input vectors. Table 4 presents the results of this experiment.

Table 4. Prediction results of PSO-LSSVM and LSSVM with different sets of parameters.

	(X_1, X_4, X_5, X_6)		$(X_1, X_4, X_5, X_6, S_1, S_2, S_3, S_4)$	
	LSSVM	**PSO-LSSVM**	**LSSVM**	**PSO-LSSVM**
January	42.52%	41.67%	38.65%	39.01%
February	40.00%	38.84%	34.82%	34.04%
March	29.88%	30.60%	36.81%	32.17%
April	34.04%	33.06%	31.38%	25.76%
May	42.74%	42.46%	43.06%	40.09%
June	28.44%	29.74%	30.14%	31.58%
July	10.07%	9.44%	9.10%	10.28%
August	13.44%	12.92%	13.93%	12.41%
September	26.73%	26.28%	27.05%	25.50%
October	20.72%	21.49%	22.07%	20.19%
November	24.85%	26.11%	28.18%	28.98%
December	34.28%	31.07%	24.47%	24.90%
Average MAPE	28.86%	28.53%	28.23%	27.08%
Average RMSE	4201	4214	4091	4030

(1) Table 4 reveals that the mean absolute percentage error corresponding to each month is not always better for the models that consider the seasonal factors than those of the models that do not. However, the average MAPE/RMSE scores of the two models are observed to be lower when they incorporate the seasonal factor within themselves. This establishes the fact that the ambient natural season is a factor that affects the accuracy of prediction.

(2) The annual mean absolute percentage error of the PSO-LSSVM model is observed to be better than that of the LSSVM model, which indicates that the PSO algorithm is an effective method to solve the optimization problem for the parameters in the LSSVM algorithm.

The prediction accuracies of PSO-LSSVM also demonstrate that the prediction errors corresponding to January, February, and May are relatively high when seasonal factors are not considered, and that the maximum prediction error is 42.46%. When the ambient natural season is considered, the high mean absolute percentage errors are, in particular, are observed to reduce by nearly 2.5%, even though the maximum prediction error remains high at 40.09%. This may be attributed to the fact that the daily tourist flow varies with the alternating seasons. Obtaining accurate forecasts simply based on the ambient natural seasonal factor is unrealistic. Hence, the pre-treatment of seasonal variation factor is necessary.

Therefore, PSO-LSSVM is verified to be an effective method for the accurate forecasting of daily tourist flow at tourist destinations. Further, the predictions verify that consideration of the ambient natural season reduces the prediction error by nearly 2%. However, given the differences in time and temperature, a simple incorporation of the seasonal factor cannot be expected to satisfactorily enhance the accuracy of forecasting. Hence, the pre-treatment of the seasonal variation factor is necessary.

3.3. Adjustment of Natural Seasons Based on K-Means

During the practical application of the predictive model, the climate changes from cold to warm or from warm to cold with the variation of seasons. In other words, the change of temperature within the same season might alter the trend of daily tourist flow at a destination, whereas the daily flow may be identical during successive months despite a season change between them if the difference in temperature is not palpable to tourists. Therefore, if the forecasting model considers the natural seasons directly, the accuracy of its predictions will be adversely affected. This leads to the necessity of pre-treating the seasonal variation factor.

Corresponding to each season, the daily tourist flow varies with the change of time and temperature. As is evident from the daily tourist data (from the cooperation project with Huangshan Management Committee) during the period from March, 2014 to February 2015 at Mountain Huangshan, the daily tourist volume varied in accordance with the maximum and minimum daily temperatures. Figure 3 illustrates the tourist flow over different seasons.

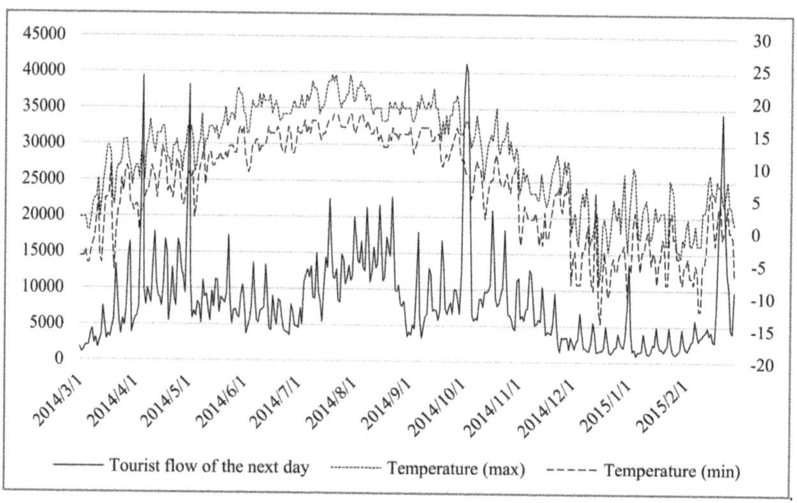

Figure 3. Daily tourist flow, along with the maximum and minimum daily temperatures, during March 2014 to February 2015.

As observed in the figure, the distribution of tourist flow over the four seasons exhibits an almost identical trend to that of daily temperatures, except for the sharp changes on four statutory holidays. Further conclusions can also be drawn from the data. During spring, the temperature in mountainous environments remains relatively low in early March, thereby lessening the daily tourist flow during that time. The data confirms that the daily number of tourists during this period is 2000 on average. With time, the temperature gradually rises as the climate becomes more comfortable. The climate becomes more suitable for travelling; thereby increasing the daily tourist flow at the mountain. Although summer is the hottest period of the year, the temperature at Mountain Huangshan stays consistent at 25 °C. Lu corroborated that Huangshan exhibits monsoon climate between June and August, which is quite conducive to travelling [14]. Moreover, the summer holidays are scheduled between July and August, during which people prefer to travel. Due to these factors, the daily tourist flow remains high during this period. In autumn, the overall temperature in mountainous destinations remains very comfortable during September and October, and the tourist flow remains high. However, the temperature starts to decrease in November, the number of people willing to visit the mountains lessens. Overall, in winter, the daily tourist flow at Mountain Huangshan remains low because of the low temperature. However, the tourist flow may exhibit increasing trends even in winter owing to the temporary rise in temperature, whereas during the majority of the season, the daily tourist flow exhibits the same distribution as the ambient temperature and humidity. Therefore, clustering the seasons at scenic tourist destinations according to the distribution of daily tourist flow is necessary.

Based on the analysis, the daily highest and lowest temperatures, the tourist flow of a particular day, and the time are selected as input variables. The K-means algorithm is adopted to adjust the natural season at the destination of Mountain Huangshan. Taking the data pertaining to 2014 as an example, the clustering results are shown in Figure 4.

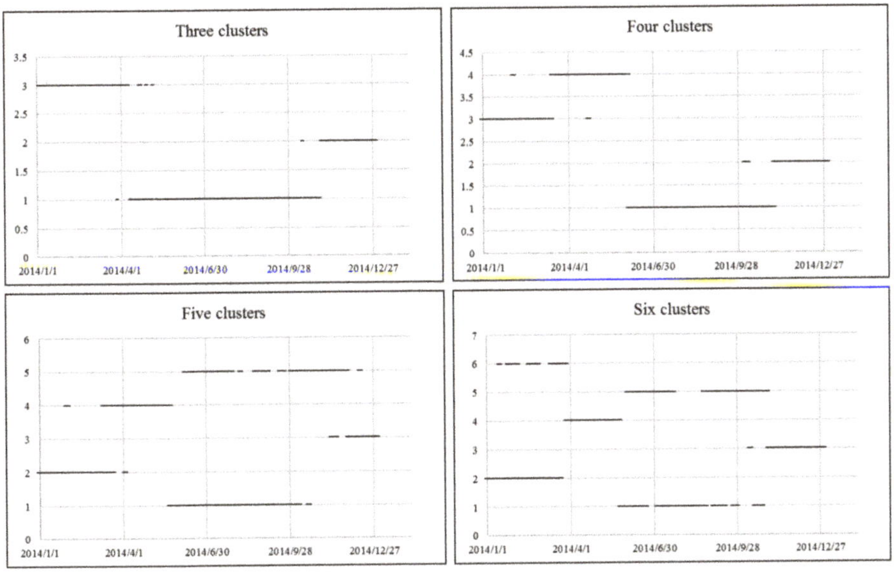

Figure 4. Results of seasonal clustering for the data pertaining to 2014.

In the figure, we use the number 1 to represent cluster 0, number 2 for cluster 1, number 3 for cluster 2 and so on. As is evident, when the year is divided into three seasons, some sample points are clustered into very few clusters. When it is divided into five classes, some objects belong to more than one category. However, when it is divided into six classes, only a few objects belong to each class, which is insufficient to form a category. Tables 5 and 6 present the specific clustering results for the cases of three and four classes.

Table 5. Clustering results for three classes.

	One	Two	Three
Month	January, February, March	April, May, June, July, August, September, October	November, December

Table 6. Clustering results for four classes.

	One	Two	Three	Four
Month	January, February 1 March–14 March	15 March–31 March, April, May	June, July, August, September, October	November, December

To facilitate the presentation of the clustering results, Figure 5 is designed, from which it can be concluded that when the year is divided into three categories, April, May, June, July, August, September, and October are clustered into a single category. However, during April to October, the temperature initially increases and then decreases, affecting the daily tourist flow accordingly. After repeated trials, the results confirm that a stable state is reached when the year is divided into four seasonal classes. The final result is also presented in Figure 5, in which January, February, and 1–14 March is taken to constitute one class. During this time, the temperature is relatively low, and the daily tourist flow remains almost identical throughout the period. However, in late March, the temperature begins to gradually increase, and the climate becomes more comfortable. Thus, the daily tourist flow at mountainous destinations during this time is similar to that of April and May. Therefore, early March is classified in the same category as January and February, whereas late March is now classified in the same category as April and May. Similarly, in June, July, and August, although the surface temperature is relatively high, the temperature in mountainous spots remains relatively low; and so, they are grouped together with September and October into a single class. Meanwhile, November and December define their own category.

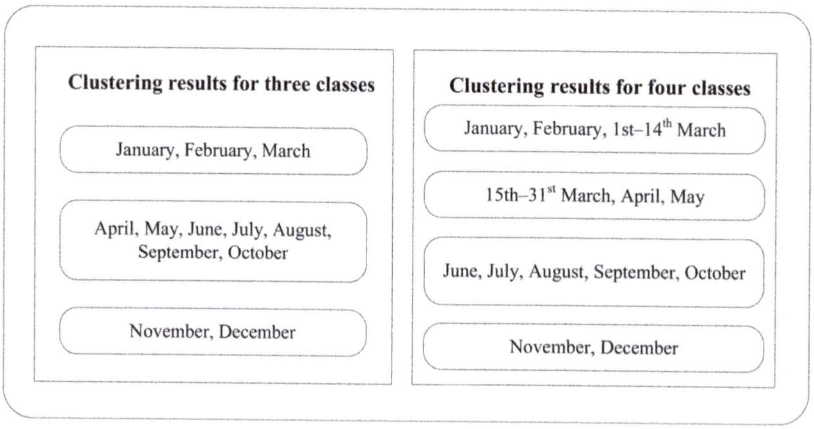

Figure 5. Clustering results of different quantity categories.

3.4. Predictions by Various Models and Their Comparison after Seasonal Clustering

To verify the effectiveness and feasibility of seasonal clustering, we use the vectors $(X_1, X_4, X_5, X_6, S_i)$ as input variables in the models, where $s_{i'} = \begin{cases} 1 \\ 0 \end{cases}$ (1 represents the ith natural season $i = 1, 2, 3, 4$) denotes the new natural seasons. In a separate experiment, we use the vectors representing the originally defined natural seasons for comparison purposes. As before, both PSO-LSSVM and LSSVM are tested with respect to both sets of vectors. Tables 7 and 8 present the results of the predictions.

Table 7. Results of the predictions by PSO-LSSVM and LSSVM using three and four seasonal classes.

	$(X_1, X_4, X_5, X_6, S_1, S_2, S_3, S_4)$			
	Three		Four	
	LSSVM	PSO-LSSVM	LSSVM	PSO-LSSVM
January	39.59%	37.92%	36.31%	35.39%
February	33.07%	31.37%	34.17%	33.38%
March	31.26%	26.50%	30.90%	26.35%
April	35.31%	32.20%	31.47%	26.52%
May	48.07%	42.57%	44.44%	42.32%
June	28.78%	29.24%	30.12%	30.95%
July	8.80%	9.26%	9.15%	8.98%
August	15.20%	13.17%	14.67%	12.98%
September	29.09%	27.33%	29.23%	27.22%
October	21.72%	21.03%	22.11%	20.03%
November	25.01%	22.24%	24.89%	22.51%
December	26.70%	31.93%	27.29%	25.29%
Average MAPE	28.49%	27.00%	27.82%	25.91%
Average RMSE	4051	3977	3967	3798

Table 8. Results of the predictions by PSO-LSSVM and LSSVM under different definitions of seasons.

	Original		New	
	LSSVM	PSO-LSSVM	LSSVM	PSO-LSSVM
January	38.65%	39.01%	36.31%	35.39%
February	34.82%	34.04%	34.17%	33.38%
March	36.81%	32.17%	30.90%	26.35%
April	31.38%	25.76%	31.47%	26.52%
May	43.06%	40.09%	44.44%	42.32%
June	30.14%	31.58%	30.12%	30.95%
July	9.10%	10.28%	9.15%	8.98%
August	13.93%	12.41%	14.67%	12.98%
September	27.05%	25.50%	29.23%	27.22%
October	22.07%	20.19%	22.11%	20.03%
November	28.18%	28.98%	24.89%	22.51%
December	24.47%	24.90%	27.29%	25.29%
Average MAPE	28.23%	27.08%	27.82%	25.91%
Average RMSE	4091	4030	3967	3798

Table 7 illustrates that when the year is divided into four seasonal classes, the MAPE/RMSE scores of both models are better corresponding to each month than those when the year is divided into three seasonal classes. Further, the reasoning behind dividing the year into four seasonal categories has already been provided. Moreover, when the year is divided into four seasonal classes, the prediction accuracy of PSO-LSSVM is observed to be better than that of LSSVM, which establishes the feasibility of the proposed model.

Table 8 shows the results of the predictions by PSO-LSSVM and LSSVM under different definitions of seasons.

(1) As is evident from Table 8, although the adoption of seasonal clustering does not reduce the monthly mean absolute percentage error, it does reduce the annual mean absolute percentage error by nearly 1.5%. Additionally, the RMSE indicator also corroborates our conclusion. This establishes that seasonal clustering is effective in enhancing the prediction accuracy.

(2) The annual MAPE/RMSE score of the PSO-LSSVM model is observed to be better than that of LSSVM overall, as can be seen from Table 8, PSO-LSSVM model has a better performance than LSSVM in most of the months, the error was reduced by an average of nearly 1.5%. This corroborates our conclusion that the PSO-LSSVM model is an effective method to forecast daily tourist flow at scenic tourist destinations.

(3) The seasonal clustering that produces the best results classifies January, February, and 1–14 March into one group, November and December into another group, and April and May into yet another group.

By comparing the predictions by PSO-LSSVM, we corroborate that the mean absolute percentage error corresponding to March decreases significantly after the seasonal adjustment. Although the MAPE scores corresponding to April and May are a little higher than those before clustering, the MAPE scores of November, December, and March are lower than those before clustering, and the value of MAPE is observed to decrease throughout the year. Therefore, the method proposed in this research is effective, moreover, the RMSE indicator also corroborates the validity of the proposed method.

4. Discussion

The prediction of daily tourist flow at scenic destinations is essential to the tourism industry, and the accuracy of forecasting is highly significant for the optimal distribution of tourism resources [8,37,43]. Mountain Huangshan is a famous scenic spot in China, and its daily tourist volume is known to exhibit complex nonlinear characteristics and the historical tourist data exhibits various trends of fluctuation during different seasons [44]. This research considers the tourist flow data at Mountain Huangshan between 2014 and 2017 as a dataset and analyzes the variation of daily tourist volumes with respect to different seasons. On the one hand, particle swarm optimization is used to optimize the least squares support vector machine; on the other hand, we focus on rearranging the seasons by clustering algorithm. In response to results in our research, it can be pointed out that the prediction performance can be improved from two aspects: the predictor itself and the input of the algorithm. The experimental results above verify the correctness of our research that the effect of classical forecasting model can be optimized by seasonal adjustment and it has an inspiration and practical value for short-term daily tourist flow forecasting.

In summary, compared with the previous research, the differences and advantages of this research are as follows:

(1) Instead of forecasting tourists flow at monthly or yearly intervals, this research is conducted at a daily time interval, and this improvement can significantly increase the efficiency of prediction.

(2) The prediction performance of the hybrid model in this research is significantly improved via the proposed optimization algorithm, which can be seen from the Section 4.

(3) Seasonal adjustment and division were included into the forecasting model as factors in our research, and it proves to be an effective method to improve the predictive performance of the model. Meanwhile, previous research works rarely considered the question, as mentioned in Section 2.

The results of this research are helpful to tourism management, and the following practical implications can be provided in management:

(1) According to the results of seasonal clustering, managers can always adopt a different hybrid model instead of using the same model. Namely, it can improve the specificity of actual management.

(2) The accurate short-term daily tourist flow forecasting can help reduce the number of crowding incidents to improve the quality of tourists' experience.

(3) In terms of resource allocation management of scenic spots, the accurate tourist flow forecasting method presented in this research can reduce the waste of resources.

In general, this research has an inspiration for tourist flow forecasting. It fills the gap of tourist flow forecasting by introducing the idea of seasonal clustering, which proved to be effective. The results of this research can also provide some practical implications.

5. Conclusions

In this research, the ambient natural season is taken to be an essential factor in the prediction of the daily tourist flow on the subsequent day, and a hybrid optimized model is proposed. The experimental results corroborate that: (1) season is a factor that profoundly affects the accuracy of prediction of the daily tourist flow, which can be supported by evidence from Table 4; (2) seasonal adjustments improve the prediction accuracy effectively by nearly 3%. In particular, it is suitable for months that exhibit significant temperature variations, e.g., March. Evidence from Tables 7 and 8 can support it; (3) the superiority of PSO-LSSVM over LSSVM is also verified and it can be supported by evidence from Tables 4, 7, and 8. This is attributed to the role of the PSO method in the determination of optimal values of LSSVM parameters based on its excellent coherence coordination. Further, the effective adjustment of natural seasons based on the K-means algorithm is another important reason behind the superiority of PSO-LSSVM. Thus, based on the idea of seasonal adjustment, PSO-LSSVM combined with the K-means algorithm was established to be a convenient and feasible method for daily tourist volume forecasting. The experimental results in this research support this conclusion.

However, the proposed method still suffers from certain limitations which could be improved in future works. First, this research was conducted with a focus on the practical utility of the method, and the underlying theory merits further research. Second, certain factors such as weather could be considered in greater complexity than was considered in this research to further improve the prediction accuracy. In addition, the method of seasonal adjustment deserves further research.

In general, this research proves the reliability of improving the prediction effect based on seasonal adjustment, and the accuracy of short-term prediction of the daily tourist flow achieved by the proposed hybrid model is beneficial to professionals in the tourism industry, enabling them to reasonably allocate appropriate resources in advance. This research also contributes to the research on short-term forecasting, which is significant as most existing studies have focused on monthly or annual prediction.

Author Contributions: Conceptualization, Keqing Li and Chu Li; Funding acquisition, Changyong Liang; Investigation, Keqing Li; Methodology, Keqing Li and Chu Li; Project administration, Changyong Liang; Resources, Changyong Liang, Wenxing Lu and Shuping Zhao; Software, Binyou Wang; Supervision, Changyong Liang, Wenxing Lu and Shuping Zhao; Validation, Keqing Li; Visualization, Keqing Li; Writing–original draft, Keqing Li; Writing–review & editing, Keqing Li. All authors have read and agreed to the published version of the manuscript.

Funding: This research was funded by National Natural Science Foundation, grant number 71771075, 71331002.

Acknowledgments: This study is part of a research work of the National Natural Science Foundation (71771075, 71331002).

Conflicts of Interest: The authors declare that there are no competing interests regarding the publication of this paper.

References

1. World Travel & Tourism Council. *Global Economic Impact & Trends 2020*; World Travel & Tourism Council: London, UK, 2020.
2. Yin, J.; Bi, Y.; Zheng, X.M.; Tsaur, R.C. Safety Forecasting and Early Warning of Highly Aggregated Tourist Crowds in China. *IEEE Access* **2019**, *7*, 119026–119040. [CrossRef]
3. China Tourism Academy. China Domestic Tourism Development Report 2020. Available online: http://www.ctaweb.org/html/2020-9/2020-9-14-13-2-83232.html (accessed on 1 October 2020).
4. Lu, W.; Wei, X. Spatio-temporal Distribution Pattern of Cable Car Passenger Flow in Panholidays: A Case Study of Huangshan Scenic Area. In Proceedings of the 2017 IEEE Second International Conference on Data Science in Cyberspace (DSC), Shenzhen, China, 26–29 June 2017; pp. 35–42.
5. Wang, Q.; Lu, L.; Yang, X.Z. Influencing factors of water resources security in water shortage mountain resorts. *J. Arid Land Resour. Environ.* **2014**, *28*, 48–53.
6. Yang, X.Z.; Wang, X. Tourism crowding characteristics and adjusting patterns of mountain scenic spots during special periods: A case study of Huangshan Mountain. *Geogr. Res.* **2019**, *38*, 961–970.

7. Feng, L. Research on Tourism Public Crisis Countermeasures Based on Big Data. In Proceedings of the 2019 IEEE 8th Joint International Information Technology and Artificial Intelligence Conference (ITAIC), Chongqing, China, 24–26 May 2019; pp. 1273–1279.
8. Song, H.; Li, G. Tourism demand modelling and forecasting—A review of recent research. *Tour. Manag.* **2008**, *29*, 203–220. [CrossRef]
9. Lim, C.; McAleer, M. Forecasting tourist arrivals. *Ann. Tour. Res.* **2001**, *28*, S0160–S7383. [CrossRef]
10. Prosper, F.; Bangwayo-Skeete, R.W.S. Can Google data improve the forecasting performance of tourist arrivals? Mixed-data sampling approach. *Tour. Manag.* **2015**, *46*, 454–464. [CrossRef]
11. Denstadli, J.M.; Jacobsen, J.K.S.; Lohmann, M. Tourist perceptions of summer weather in Scandinavia. *Ann. Tour. Res.* **2011**, *38*, 920–940. [CrossRef]
12. Gössling, S.; Scott, D.; Hall, C.M.; Ceron, J.P.; Dubois, G. Consumer behavior and demand response of tourists to climate change. *Ann. Tour. Res.* **2012**, *39*, 36–58. [CrossRef]
13. Pu, W.; Quan-sheng, G. An analysis of annual variation of tourist flows and climate change in Hainan Province. *Geogr. Res.* **2009**, *28*, 1078–1084. [CrossRef]
14. Lu, L.; Xuan, G.; Zhang, J. An approach to seasonality of tourist flows between coastland resorts and mountain resorts: Examples of Sanya, Beihai, Mt. Putuo, Mt. Huangshan and Mt. Jiuhua. *Acta Geogr. Sin.* **2002**, *57*, 731–740. [CrossRef]
15. Zheng, Q.; Chen, R.; Sun, J.S. Study on the Influencing Factors of Tourism Scenic Spots for Traffic—Taking Huangshan Scenic Area as an Example. *J. Bengbu Coll.* **2014**, *3*, 98–102. [CrossRef]
16. Chen, R.; Li, G. A Study on the Forecasting Method of AGA-SVR Modeled Holiday Tourist Flows Based on SEA. *Tour. Sci.* **2016**, *30*, 12–23. [CrossRef]
17. Huang, J.H.; Min, J.C.H. Earthquake devastation and recovery in tourism: The Taiwan case. *Tour. Manag.* **2002**, *23*, 145–154. [CrossRef]
18. Teixeira, J.P.; Fernandes, P.O. Tourism time series forecast with artificial neural networks. *Tékhne* **2014**, *12*, 26–36. [CrossRef]
19. Chen, K.Y.; Wang, C.H. Support vector regression with genetic algorithms in forecasting tourism demand. *Tour. Manag.* **2007**, *28*, 215–226. [CrossRef]
20. Palmer, A.; Montano, J.J.; Sesé, A. Designing an artificial neural network for forecasting tourism time series. *Tour. Manag.* **2006**, *27*, 781–790. [CrossRef]
21. Yan, X.; Chowdhury, N.A. Mid-term electricity market clearing price forecasting: A hybrid LSSVM and ARMAX approach. *Int. J. Electr. Power Energy Syst.* **2013**, *53*, 20–26. [CrossRef]
22. Sun, W.; Zhang, J. Forecasting Day Ahead Spot Electricity Prices Based on GASVM. Proceedings of 2008 International Conference on Internet Computing in Science and Engineering, Harbin, China, 28–29 January 2008; pp. 73–78.
23. Chen, Y.; Yang, Y.; Liu, C.; Li, C.; Li, L. A hybrid application algorithm based on the support vector machine and artificial intelligence: An example of electric load forecasting. *Appl. Math. Model.* **2015**, *39*, 2617–2632. [CrossRef]
24. Yuan, X.; Chen, C.; Yuan, Y.; Huang, Y.; Tan, Q. Short-term wind power prediction based on LSSVM–GSA model. *Energy Convers. Manag.* **2015**, *101*, 393–401. [CrossRef]
25. Zhu, X.; Xu, Q.; Tang, M. Comparison of two optimized machine learning models for predicting displacement of rainfall-induced landslide: A case study in Sichuan Province, China. *Eng. Geol.* **2017**, *218*, 213–222. [CrossRef]
26. Cong, Y.; Wang, J.; Li, X. Traffic Flow Forecasting by a Least Squares Support Vector Machine with a Fruit Fly Optimization Algorithm. *Procedia Eng.* **2016**, *137*. [CrossRef]
27. Eberhart, J.K.R. Particle swarm optimization. Proceedings of 1995 IEEE International Conference on Neural Networks, Perth, Australia, 27 November–1 December 1995.
28. Mohana, S.J.; Saroja, M.; Venkatachalam, M. Comparative analysis of swarm intelligence optimization techniques for cloud scheduling. *Int. J. Innov. Sci. Eng. Technol.* **2015**, *1*, 15–19.
29. Zeng, N.; Zhang, H.; Liu, W.; Liang, J.; Alsaadi, F.E. A switching delayed PSO optimized extreme learning machine for short-term load forecasting. *Neurocomputing* **2017**, *240*, 175–182. [CrossRef]
30. Yang, Y.; Pan, B.; Song, H. Predicting hotel demand using destination marketing organization's web traffic data. *J. Travel Res.* **2014**, *53*, 433–447. [CrossRef]

31. Pan, B.; Wu, D.C.; Song, H. Forecasting hotel room demand using search engine data. *J. Hosp. Tour. Technol.* **2012**, *3*, 196–210. [CrossRef]
32. Li, X.; Pan, B.; Law, R.; Huang, X. Forecasting tourism demand with composite search index. *Tour. Manag.* **2017**, *59*, 57–66. [CrossRef]
33. Artola, C.; Pinto, F.; de Pedraza García, P. Can internet searches forecast tourism inflows? *Int. J. Manpow.* **2015**, *36*, 103–116. [CrossRef]
34. Choi, H.; Varian, H. Predicting the Present with Google Trends. *Econ. Rec.* **2012**, *88*, 2–9. [CrossRef]
35. Shen, S.; Li, G.; Song, H. An Assessment of Combining Tourism Demand Forecasts over Different Time Horizons. *J. Travel Res.* **2008**, *47*, 197–207. [CrossRef]
36. Höpken, W.; Ernesti, D.; Fuchs, M.; Kronenberg, K.; Lexhagen, M. Big data as input for predicting tourist arrivals. *Ann. Tour. Res.* **2017**, *28*, 1070–1072. [CrossRef]
37. Colladon, A.F.; Guardabascio, B.; Innarella, R. Using social network and semantic analysis to analyze online travel forums and forecast tourism demand. *Decis. Support Syst.* **2019**, *123*, 113075. [CrossRef]
38. Oktar, Y.; Turkan, M. A Review of Sparsity-based Clustering Methods. *Signal Process.* **2018**, *148*, 20–30. [CrossRef]
39. Gordon, A.D. A Review of Hierarchical Classification. *J. R. Stat. Soc. Ser. A* **1987**, *150*, 119–137. [CrossRef]
40. Davé, A.B.R.N. Robust clustering. *Wiley Interdiscip. Rev. Data Min. Knowl. Discov.* **2012**, *2*, 29–59. [CrossRef]
41. Peña, M. Robust clustering methodology for multi-frequency acoustic data: A review of standardization, initialization and cluster geometry. *Fish. Res.* **2018**, *200*, 49–60. [CrossRef]
42. Lin, L. A Study on the seasonal changes in the tourism in mountain resorts—A case study of the Huangshan Mountain. *Geogr. Res.* **1994**, *4*, 50–58. [CrossRef]
43. Shao-Jiang, L.; Jia-Ying, C.; Zhi-Xue, L. A EMD-BP integrated model to forecast tourist number and applied to Jiuzhaigou. *J. Intell. Fuzzy Syst.* **2018**, *34*, 1045–1052. [CrossRef]
44. Li, K.; Lu, W.; Liang, C.; Wang, B. Intelligence in Tourism Management: A Hybrid FOA-BP Method on Daily Tourism Demand Forecasting with Web Search Data. *Mathematics* **2019**, *7*, 531. [CrossRef]

MDPI
St. Alban-Anlage 66
4052 Basel
Switzerland
Tel. +41 61 683 77 34
Fax +41 61 302 89 18
www.mdpi.com

ISPRS International Journal of Geo-Information Editorial Office
E-mail: ijgi@mdpi.com
www.mdpi.com/journal/ijgi

www.ingramcontent.com/pod-product-compliance
Lightning Source LLC
Chambersburg PA
CBHW061759210926
51850CB00006B/598